AF352782

Ten Years of Remote Sensing at Barcelona Expert Center

Ten Years of Remote Sensing at Barcelona Expert Center

Editors

Justino Martínez
Veronica González-Gambau
Carolina Gabarro
Estrella Olmedo

MDPI • Basel • Beijing • Wuhan • Barcelona • Belgrade • Manchester • Tokyo • Cluj • Tianjin

Editors
Justino Martínez
BEC & Institute of Marine
Science/CSIC
Spain

Veronica González-Gambau
BEC & Institute of Marine
Science/CSIC
Spain

Carolina Gabarro
BEC & Institute of Marine
Science/CSIC
Spain

Estrella Olmedo
BEC & Institute of Marine
Science/CSIC
Spain

Editorial Office
MDPI
St. Alban-Anlage 66
4052 Basel, Switzerland

This is a reprint of articles from the Special Issue published online in the open access journal *Remote Sensing* (ISSN 2072-4292) (available at: https://www.mdpi.com/journal/remotesensing/special_issues/BEC_rs).

For citation purposes, cite each article independently as indicated on the article page online and as indicated below:

LastName, A.A.; LastName, B.B.; LastName, C.C. Article Title. *Journal Name* **Year**, *Article Number*, Page Range.

ISBN 978-3-03936-982-9 (Hbk)
ISBN 978-3-03936-983-6 (PDF)

Contents

About the Editors

Justino Martínez was born in Girona, Spain, in 1966. He received his B.Sc. and Ph.D. degrees in physics from the Autonomous University of Barcelona, Barcelona, Spain, in 1991 and 1995, respectively. His research started in the field of relativistic thermodynamics in the context of gravitational collapse. Currently, he develops his task as a Scientific Researcher with the Physical Oceanography Department, Institut de Ciencies del Mar, Consejo Superior de Investigaciones Cientificas, Barcelona, Spain. He has been involved in the SMOS mission since 2009. During this period, his research has been developed in SMOS L1 corrections, the forward model implementation, and he has been involved in the generation of L3 and L4 SMOS products, as well as in their validation tasks. His recent work has centered on retrieving the Arctic sea surface salinity from SMOS data.

Verónica González-Gambau was born in Huesca, Spain, in 1981. She received her M.S. and Ph.D. degrees in telecommunication engineering from the School of Telecommunication Engineering, Universitat Politcnica de Catalunya (UPC), Barcelona, Spain, in 2006 and 2012, respectively. In 2006, she joined the Passive Remote Sensing Group of the Signal Theory and Communications Department, UPC, where she was involved in the Soil Moisture and Ocean Salinity (SMOS) payload on- ground characterization, in the framework of SMOS pre-commissioning activities. She also collaborated with the SMOS Level 1 software development and data analysis of the Microwave Imaging Radiometer by Aperture Synthesis (MIRAS) instrument validation campaigns. In 2007, she was a UPC Researcher, in collaboration with the Barcelona Expert Center (BEC), Barcelona, Spain. She is currently a Research Scientist with the Institut de Ciéncies del Mar-Consejo Superior de Investigaciones Científicas (ICM- CSIC), the SMOS mission co-leading institution, Barcelona, Spain, where she is responsible for the advance Level 1 error correction techniques and in-orbit validation procedures. She is working on the development of algorithms for the SMOS brightness temperatures improvement, and assessing the impact of these techniques on the quality of the geophysical retrievals.

Carolina Gabarro received her B. Eng. degree in Telecommunications Engineering (1998) and her Ph.D. degree in Ocean Science (2004) from the Universitat Politècnica de Catalunya, Spain. Stagiaire, Young Graduate Trainee and contracted at ESTEC, ESA from 1997 to 1999 in the Netherlands, and later at ACRI-SA, France (1999), working on ocean color remote sensing. Since 2000, she has been working at the Institute of Marine Sciences ICM-CSIC and Barcelona Expert Centre on Remote Sensing, and has been on its permanent staff since 2013. Her main work has focused on the ocean and ice products from SMOS mission and other radiometers. From 2005 to 2010, she was part of the Expert Support Laboratories team of SMOS selected by ESA. She is now studying the capability of low frequency and L-band radiometers to measure several characteristics of the Arctic ice (ice thickness, ice concentration, salinity). She is currently a Spanish delegate for the International Arctic Science Committee (IASC). Her research interest includes microwave remote sensing of the ocean and the Cryosphere. She has 35 publications in CSI journals and has made more than 50 contributions to international conferences.

Estrella Olmedo was born in La Línea de la Concepción, Spain, in 1979. She received her B.Sc. and Ph.D. degrees in mathematics from the University of Barcelona, Barcelona, Spain, in

2001 and 2007, respectively. Her research started in the field of dynamical systems and applied mathematics in the context of computation of quasi-invariant tori. Then, she was with Deimos Space S.L.U in European Space Agency's (ESA) projects related to the design of the future European Space Surveillance System. Since 2012, she has been a Researcher with the Physical Oceanography Department, Institut de Ciencies del Mar, Consejo Superior de Investigaciones Científicas, Barcelona, Spain. Her current research includes signal and data processing of the Soil Moisture and Ocean Salinity (SMOS) data, studies of oceanographic applications of the remotely sensed salinity, and analysis of the marine turbulence.

Editorial

Editorial for the Special Issue: "Ten Years of Remote Sensing at Barcelona Expert Center"

Justino Martínez *, Verónica González-Gambau, Carolina Gabarró and Estrella Olmedo

Department of Physical Oceanography, Institute of Marine Sciences, CSIC and Barcelona Expert Center, Passeig Marítim de la Barceloneta, 08003 Barcelona, Spain; vgonzalez@icm.csic.es (V.G.-G.); cgabarro@icm.csic.es (C.G.); olmedo@icm.csic.es (E.O.)
* Correspondence: justino@icm.csic.es

Received: 23 July 2020; Accepted: 28 July 2020; Published: 29 July 2020

Abstract: This book celebrates the ten year anniversary of the Barcelona Expert Center by presenting recent contributions related to the topics on which the team has been working during those years. The Barcelona Expert Center's expertise covers a wide variety of remote sensing fields, but the main focus of the research is on the SMOS data processing and its ocean, land, and ice applications. This book contains 14 scientific papers addressing topics that go from the description of the new data processing algorithms that are implemented in the last version of the operational SMOS level 1 processor to scientific applications derived from SMOS: results on the sea-surface salinity assimilation in coastal models, synergies of the sea-surface salinity with temperature and chlorophyll and their impact on the better retrieval of ocean surface currents, quality assessment of SMOS-derived sea ice thickness, sea-surface salinity, and soil moisture products, among others. Moreover, one of the papers verifies the potential of the future Copernicus Imaging Microwave Radiometer (CIMR) mission within the CMEMS sea-surface salinity (SSS) operational production after the SMOS era.

Keywords: BEC; SMOS; radiometry; remote sensing; oceanography; soil moisture; cryosphere; processing; sensor calibration; image reconstruction

---

1. Introduction

The Barcelona Expert Centre (BEC) is a joint initiative between the Spanish Research Council (CSIC) and the Universitat Politècnica de Catalunya (UPC) that was created in 2007. Since its foundation until January 2016, the head of the BEC was professor Jordi Font. The original purpose of the BEC was to provide support to the Spanish activities on the Soil Moisture and Ocean Salinity (SMOS) European mission, professor Font being the co-principal investigator of SMOS and being in charge of the sea salinity part of the mission. SMOS is a European Space Agency (ESA) mission and was the first 2D synthetic aperture interferometric radiometer operating in the microwave L-band ever put in orbit. SMOS was put in orbit on 2 November 2009 and was designed to fulfill two specific goals: to measure sea-surface salinity (SSS) with 1° spatial resolution and monthly temporal resolution and an accuracy of 0.1 PSU over the sea [1], and to measure soil moisture (SM) with 25 km spatial resolution and daily temporal resolution within a maximum uncertainty of 4% [2]. Initially, BEC missions were to provide assessments to the ESA as a level 2 ocean salinity expert support laboratory, to contribute to SMOS radiometric calibration and validation activities, and to develop and validate new algorithms for the generation of added-value products at levels 3 and 4.

Since 2016, Antonio Turiel has been the head of BEC that nowadays consists of four departments, whose areas are SMOS (salinity and land), ocean winds, cryosphere, and ocean currents. The activities developed at BEC has been diversified, as hav its main objectives, which currently are:

- Research and development in Earth observation, with a special focus on microwave remote sensing and more specifically on SMOS and follow-on missions.
- Support to the European Space Agency through expert support laboratory contracts.
- Continuous generation and distribution of high-level remote sensing products.
- Geophysical exploitation of dedicated remote sensing data, with a special interest in scientific applications.
- Fostering the use of BEC remote sensing data among academia, enterprises, and stakeholders.

BEC is now a large team of highly motivated and remote sensing specialist people carrying out the scientific and operational activities in all processing levels of different missions. At the moment we are around 20 people; among us are permanent staff, post-doctoral contracts, visiting scientists, and Ph.D. students working on a wide variety of topics:

- Improvements in calibration, image reconstruction, and stability of radiometric data.
- Synergy of observations from different sensors and data sources.
- Retrieval of geophysical variables: forward modeling and non-linear inversion.
- Validation and quality control.
- Assimilation into atmospheric and ocean models.
- Generation of added-value products at levels 3 and 4.

This book aims to be a tribute to the work done over the last 10 years at the BEC and to the entire team that has made the BEC one of the main remote sensing centers in Spain.

2. Overview of Contributions

The contributions reported in this book are structured in three different blocks: contributions to remote sensing data processing; ocean remote sensing applications; and land remote sensing applications,

2.1. Remote Sensing Data Processing Algorithms

The book starts with an analysis of the different methods for reducing the reconstruction error in microwave interferometric radiometers with a large field of view in Corbella et al. [3]. First, the authors propose estimating the reconstruction error contribution through the application of a brightness temperature model outside the fundamental period. Second, they present image reconstruction algorithms implemented on a minimum grid size that allows maximizing the efficiency of numerical processing. Last, they describe a method to reduce Gibbs oscillations based on an improved apodization window over the reconstructed image. The proposed algorithm shows similar performance with respect to the nominal one.

The second chapter, from Oliva et al. [4], is dedicated to describing the improvements gained by the SMOS level 1 operational processor. The authors performed a quality analysis of the enhanced algorithms that included an end-to-end processing of three years. The results confirmed that the new version of the SMOS level 1 operational processor deals with improvements in the SMOS measurements. The new version of the processor is foreseen to be used in the third reprocessing campaign for the SMOS measurements.

In the third chapter, Rubino et al. [5] present a new methodology to derive vertical total electron content (VTEC) maps from the radiometric measurements. The proposed methodology is an alternative approach to the one currently implemented in the SMOS data processor, which has the advantage of being independent of external databases and models. This new approach uses spatiotemporal filtering techniques with optimized filters to be robust against the thermal noise and image reconstruction artifacts present in SMOS images.

The block finishes with a chapter dedicated to analyzing the potential of the Copernicus Imaging Microwave Radiometer (CIMR) mission for the global monitoring of sea-surface salinity (SSS) using level 4 (gap-free) analysis processing, lead by Ciani [6]. Since there are no planned missions to guarantee continuity in the remote SSS measurements after SMOS and SMAP, CIMR could cover that gap, since it will carry an L-band radiometer. The CIMR mission is in a preparatory phase with an expected launch in 2026. In this paper, they study the potential of CIMR within the CMEMS SSS operational production after the SMOS era. They demonstrate that the combined use of in situ and CIMR observations improves the global SSS retrieval compared to a processing wherein only in situ observations are ingested. Therefore, they conclude that CIMR can guarantee continuity for accurate monitoring of the ocean surface salinity from space.

2.2. Ocean Remote Sensing Applications

This section starts with a chapter in which an impact assessment of assimilating SMOS sea-surface salinity data into a coastal ocean model is presented in Phillipson and Toumi [7]. The results of this study show that the assimilation of SSS and SSS combined with SSH consistently provides the best results in the Congo river plume analysis.

The next chapter from Alucino et al. [8] is dedicated to a comparison of the performance of the SMOS SSS maps within the performance of in situ high-resolution glider measurements collected in the framework of the Algerian Basin Circulation Unmanned Survey (ABACUS). The Algerian Basin (located in the Mediterranean Sea) presents complex ocean circulation, wherein the fresh Atlantic water is mixed with the more saline Mediterranean water. The study shows some limitations of the satellite data to describe small spatial structures that are captured by in situ. However, at the spatial scales resolved by the satellite, the SSS in situ and satellite measurements are in a good agreement, providing an averaged difference of −0.11 PSU with a standard deviation of 0.26 PSU.

The following two chapters deal with sea-surface currents. In the first, from Isern-Fontanet et al. [9], the relationship between satellite salinity and temperature data to correct the buoyancy and to retrieve ocean currents by using the SQG approach is explored. The study is focused in the Alboran Sea where the altimetric maps have some limitations for capturing the ocean currents. The results of this study show that the good sampling of infrared radiometers allows at least retrieving the direction of ocean currents in this area. The second paper, from Ciani et al. [10], presents a method for the retrieval of the sea-surface currents in the Mediterranean Sea. They combine the altimeter-derived currents with sea-surface temperature information, to create daily, gap-free, high-resolution maps of sea-surface currents for the period 2012–2016. The quality of the new multi-sensor current maps has been assessed through comparisons to other surface-currents estimates, as drifting buoys trajectories, HF-Radar platforms, and ocean numerical model outputs. The study yielded that the synergetic approach can improve the present-day derivation of the surface currents in the Mediterranean area.

Chapter nine, from Umbert et al. [11], aims at analyzing the similarity of mesoscale and submesoscale features observed in different ocean scalars. This study indicates that they undergo some common non-linear processes. The results show that it is possible to assume a local correspondence of SST and Chl-a multifractal singularities, due to the existence of a common cascade process which makes it possible to use SST data to infer Chl-a concentration where data are lacking. Therefore, the data fusion method was used to improve the coverage of daily Aqua MODIS level 3 chlorophyll maps by using MODIS SST maps as a template. An assessment of the quality of the inference of level 4 Chl-a maps is also performed.

The next chapter deals with tidal fluctuations observed from remote sensing infrared SST, from Gonzalez-Haro et al. [12]. The expected amplitude of fixed-point sea-surface temperature (SST) fluctuations induced by barotropic and baroclinic tidal flows is estimated from tidal current atlases and SST observations. The fluctuations considered are the result of the advection of pre-existing SST fronts by tidal currents. In this study, regional and global estimations of these expected amplitudes are presented. The results show that barotropic tidal motions produce SST fluctuations that may reach amplitudes of 0.3 K, while baroclinic (internal) tides produce SST fluctuations weaker than 0.1 K.

The amplitudes and the detectability of tidally induced fluctuations of SST are discussed in light of the expected SST fluctuations due to other geophysical processes and instrumental noise.

The remote sensing ocean application block ends with a validation of sea ice thickness (SIT) products from Sánchez-Gámez et al. [13]. In this study, in situ SIT data acquired with upward looking sonar (ULS) instruments on buoys from the Woods Hole Oceanographic Institution (WHOI) were used to validate the thin SIT maps from SMOS and SMAP missions. These buoys acquired data all year round, permitting them to overcome several limitations, thereby improving the characterization of the L-band brightness temperature response to changes in thin SIT. State-of-the-art satellite SIT products and the cumulative freezing degree days (CFDD) model were verified against the ULS ground truth.

2.3. Land Remote Sensing Applications

This section starts with a chapter dedicated to a comparison of six operational surface soil moisture (SSM) products derived from SMOS and SMAP in order to diagnose their distinct features, by analyzing their temporal and spatial characteristics, from Partal et al. [14]. The study was focused on the Iberian Peninsula and covers the period from April 2015 to December 2017. A temporal inter-comparison analysis was carried out using in situ SSM data from the Soil Moisture Measurements Station Network of the University of Salamanca (REMEDHUS). Spatial analysis was conducted for the whole Iberian Peninsula with an emphasis on the added-value that the enhanced resolution products provide. They show an overall agreement among time series of the products regardless of their spatial scale when compared to in situ measurements. The largest disparities between these products occur in forested areas, which may be related to the reduced sensitivity of high-resolution active microwave and optical data to soil properties under dense vegetation. Still, higher spatial resolutions would be needed to capture local features, such as small irrigated areas that are not dominant at the 1 km pixel scale.

In the next paper, Pablos et al. [15] presents an evaluation, both temporally and spatially, of six satellite-based root zone soil moisture (RZSM) estimates obtained from SMAP, SMOS, and the Moderate Resolution Imaging Spectroradiometer (MODIS) from March 2015 to December 2016. The RZSM estimates are compared to in situ data from 14 stations of the soil moisture measurements from the University of Salamanca (REMEDHUS), to assess the temporal analysis. Regarding the spatial assessment, the resulting RZSM maps of the Iberian Peninsula were compared between them. All RZSM values followed the temporal evolution of the ground-based measurements well, although SMOS and MODIS showed underestimation while SMAP displayed overestimation. The good results obtained from MODIS are notable, but it should be remarked that it uses optical bands, which are affected by clouds. A very high agreement was found in terms of spatial patterns for the whole Iberian Peninsula except for the extreme north area, which is dominated by high mountains and dense forests.

The last chapter of this book, from Piles et al. [16], aims at investigating the temporal variability of global surface soil moisture acquired with SMOS and two SM products derived from the models: LDAS-Noah and ERA5. The soil moisture time series are decomposed into a linear trend, interannual, seasonal, and high-frequency residual components. The relative distribution of soil moisture variance among its temporal components is illustrated at selected target sites with distinct vegetation type and seasonality. A global assessment of the dominant features and the spatial distribution of soil moisture variability are also provided. Results show that SMOS data provide coherent and reliable variability patterns at both seasonal and interannual scales. The observed linear trends, based upon one strong El Niño event in 2016, are consistent with the known El Niño Southern Oscillation (ENSO) teleconnections. This work can help further our understanding of the terrestrial branch of the water cycle and of global patterns of climate anomalies.

3. Conclusions

The contributions reported in this special issue are based on research on remote sensing techniques and its applications to better monitor and understand the changes the oceans, soils, and cryosphere are undergoing on a warming planet. These are the topics on which the BEC team has been working for the last ten years, and they also drive the future lines of research on which the BEC's strategic plan is based. This book is a tribute to the efforts made by all the people that have contributed to building up the BEC and also aims to be an inspiration for the work to be done in the future at the BEC.

Author Contributions: The four authors contributed equally to all aspects of this editorial. All authors have read and agreed to the published version of the manuscript.

Acknowledgments: The guest editors would like to thank the authors who contributed to this Special Issue and the reviewers who dedicated their time and provided the authors with valuable and constructive recommendations. They would also like to thank the editorial team of Remote Sensing for their support.

Conflicts of Interest: The authors declare no conflict of interest.

References

1. Font, J.; Camps, A.; Borges, A.; Martin-Neira, M.; Boutin, J.; Reul, N.; Kerr, Y.; Hahne, A.; Mecklenburg, S. SMOS: The challenging sea surface salinity measurement from space. *Proc. IEEE* **2010**, *98*, 649–665. [CrossRef]
2. Kerr, Y.; Waldteufel, P.; Wigneron, J.P.; Delwart, S.; Cabot, F.; Boutin, J.; Escorihuela, M.J.; Font, J.; Reul, N.; Gruhier, C.; et al. The SMOS mission: new tool for monitoring key elements of the global water cycle. *Proc. IEEE* **2010**, *98*, 666–687. [CrossRef]
3. Corbella, I.; Torres, F.; Duffo, N.; Duran, I.; González-Gambau, V.; Martín-Neira, M. Wide Field of View Microwave Interferometric Radiometer Imaging. *Remote Sens.* **2019**, *11*, 682. [CrossRef]
4. Oliva, R.; Martín-Neira, M.; Corbella, I.; Closa, J.; Zurita, A.; Cabot, F.; Khazaal, A.; Richaume, P.; Kainulainen, J.; Barbosa, J.; et al. SMOS Third Mission Reprocessing after 10 Years in Orbit. *Remote. Sens.* **2020**, *12*, 1645. [CrossRef]
5. Rubino, R.; Duffo, N.; González-Gambau, V.; Corbella, I.; Torres, F.; Durán, I.; Martín-Neira, M. Deriving VTEC Maps from SMOS Radiometric Data. *Remote Sens.* **2020**, *12*, 1604. [CrossRef]
6. Ciani, D.; Santoleri, R.; Liberti, G.L.; Prigent, C.; Donlon, C.; Buongiorno Nardelli, B. Copernicus Imaging Microwave Radiometer (CIMR) Benefits for the Copernicus Level 4 Sea-Surface Salinity Processing Chain. *Remote Sens.* **2019**, *11*, 1818. [CrossRef]
7. Phillipson, L.; Toumi, R. Assimilation of satellite salinity for modelling the congo river plume. *Remote Sens.* **2020**, *12*, 11. [CrossRef]
8. Aulicino, G.; Cotroneo, Y.; Olmedo, E.; Cesarano, C.; Fusco, G.; Budillon, G. In Situ and Satellite Sea Surface Salinity in the Algerian Basin Observed through ABACUS Glider Measurements and BEC SMOS Regional Products. *Remote Sens.* **2019**, *11*, 1361. [CrossRef]
9. Isern-Fontanet, J.; García-Ladona, E.; Jiménez-Madrid, J.A.; Olmedo, E.; García-Sotillo, M.; Orfila, A.; Turiel, A. Real-time Reconstruction of Surface Velocities from Satellite Observations in the Alboran Sea. *Remote Sens.* **2020**, *12*, 724. [CrossRef]
10. Ciani, D.; Rio, M.; Menna, M.; Santoleri, R. A Synergetic Approach for the Space-Based Sea Surface Currents Retrieval in the Mediterranean Sea. *Remote Sens.* **2019**, *11*, 1285. [CrossRef]
11. Umbert, M.; Guimbard, S.; Ballabrera-Poy, J.; Turiel, A. Synergy between Ocean Variables: Remotely Sensed Surface Temperature and Chlorophyll Concentration Coherence. *Remote Sens.* **2019**, *11*, 1153. [CrossRef]
12. González-Haro, C.; Ponte, A.; Autret, E. Quantifying Tidal Fluctuations in Remote Sensing Infrared SST Observations. *Remote Sens.* **2019**, *11*, 2313. [CrossRef]
13. Sánchez-Gámez, P.; Gabarró, C.; Turiel, A.; Portabella, M. Assessment with Controlled In-Situ Data of the Dependence of L-Band Radiometry on Sea-Ice Thickness. *Remote Sens.* **2019**, *11*, 650. [CrossRef]
14. Portal, G.; Jagdhuber, T.; Vall-llossera, M.; Camps, A. Pablos, M.; Entekhabi, D.; Piles, M. Assessment of Multi-Scale SMOS and SMAP Soil Moisture Products across the Iberian Peninsula. *Remote Sens.* **2020**, *12*, 570. [CrossRef]

15. Pablos, M.; González-Zamora, A.; Sánchez, N.; Martínez-Fernández, J. Assessment of Root Zone Soil Moisture Estimations from SMAP, SMOS and MODIS Observations. *Remote Sens.* **2018**, *10*, 981. [CrossRef]
16. Piles, M.; Ballabrera-Poy, J.; Muñoz-Sabater, J. Dominant Features of Global Surface Soil Moisture Variability Observed by the SMOS Satellite. *Remote Sens.* **2019**, *11*, 95. [CrossRef]

 *remote sensing*

Article

Wide Field of View Microwave Interferometric Radiometer Imaging

Ignasi Corbella [1,*], Francesc Torres [1], Nuria Duffo [1], Israel Duran [1], Veronica Gonzalez-Gambau [2] and Manuel Martin-Neira [3]

1   Remote Sensing Laboratory, Universitat Politecnica de Catalunya, c/ Jordi Girona 1–3,
    08034 Barcelona, Spain; xtorres@tsc.upc.edu (F.T.); duffo@tsc.upc.edu (N.D.); israel.duran@tsc.upc.edu (I.D.)
2   Department of Physical Oceanography, Institute of Marine Sciences (ICM), CSIC and Barcelona Expert
    Center, Passeig Maritim de la Barceloneta, 37–49, 08003 Barcelona, Spain; vgonzalez@icm.csic.es
3   European Space Research and Technology Centre, European Space Agency,
    2200 AG Noordwijk, The Netherlands; Manuel.Martin-Neira@esa.int
*   Correspondence: corbella@tsc.upc.edu; Tel.: +34-934-017-228

Received: 4 February 2019; Accepted: 18 March 2019; Published: 21 March 2019

Abstract: In microwave interferometric radiometers with a large field of view, as for example the Microwave Imaging Radiometer with Aperture Synthesis (MIRAS) onboard the Soil Moisture and Ocean Salinity (SMOS) satellite, one of the major causes of reconstruction error is the contribution to the visibility of the brightness temperature outside the fundamental period, defined on the basis of reciprocal grids. A mitigation method consisting of estimating this contribution through the application of a brightness temperature model outside the fundamental period is proposed. The main advantage is that it does not require any a posteriori addition of artificial scenes to the reconstructed image. Additionally, a method to avoid the sophisticated matrix regularization and inversion techniques usually applied in microwave interferometry is presented. Image reconstruction algorithms are implemented on a minimum grid size in order to maximize their numerical efficiency. An improved method to apply an apodization window to the reconstructed image for reducing Gibbs oscillations is also proposed. All procedures are generally described considering the single polarization case and successively implemented applying the MIRAS layout in both its single polarization and full polarimetric modes. Results show similar performance of the proposed algorithm with respect to the nominal one applied by SMOS. All algorithms are implemented in the MIRAS Testing Software and have been successfully used for scientific studies by other teams.

Keywords: interferometric radiometry; image reconstruction; error correction

1. Introduction

Interferometric radiometers are passive imaging instruments whose operation is based on the Van Cittert–Zernike theorem. Their main advantage with respect to other kinds of radiometers is that they do not need moving elements to produce images, as these are entirely formed through data processing of the raw measurements. This technique was firstly proposed for earth observation in Le Vine et al. [1] and Ruf et al. [2]. One of the most representative examples is the Microwave Imaging Radiometer with Aperture Synthesis (MIRAS) [3,4] embarked on board the SMOS (Soil Moisture and Ocean Salinity) satellite [5], launched by the European Space Agency in 2009 and still providing useful geophysical data to the scientific community.

As originally derived for optical signals, the Van Cittert–Zernike theorem states that the mutual intensity of a radiation is the two-dimensional Fourier Transform of the intensity distribution across the source. In microwave radiometry terms, the visibility function is the two-dimensional Fourier transform of the brightness temperature image. An inverse Fourier transform should then allow recovering of the brightness temperature from the calibrated visibility measurements. Nevertheless, in wide field of view instruments, those imaging an extended source covering most of the space in front of the antenna, there are non-negligible effects such as antenna patterns differences, obliquity factor, decorrelation, crosstalk and others that alter substantially this basic relation [6,7].

The visibility function is measured by cross-correlating all pairs of analytic signals collected by individual antennas. Assuming these ones evenly distributed on a fixed structure (as in the MIRAS case, shown in Figure 1) the visibility function becomes sampled at discrete points (u, v) on a space-limited regular grid [8]. Since the visibility equation is ultimately a Fourier transform, the recovered brightness temperature is affected by aliasing in case the antenna separation fails to meet the Nyquist rate, which is usually the case. In addition, the fact that the visibility is limited in space (due to the instrument's finite size) is equivalent to having a spatial filter that sets the spatial resolution of the recovered map and produces ripples in sharp transitions (Gibbs effect). Moreover, the combined effects of antenna pattern differences and spatial decorrelation make the visibility equation depart from a simple Fourier transform, inducing errors even in the alias-free field of view [9].

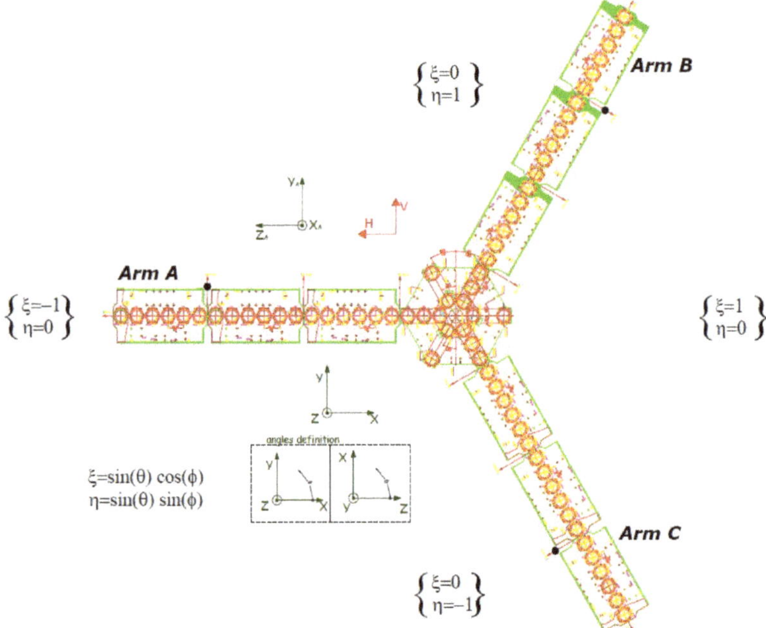

Figure 1. Microwave Imaging Radiometer with Aperture Synthesis (MIRAS) instrument layout and coordinate definition. Courtesy of AIRBUS Defence and Space [formerly EADS CASA Espacio].

The objective of this paper is to present an alternative image reconstruction algorithm for 2D interferometric radiometers. The algorithm is tested on a set of real measurements acquired by the MIRAS sensor, from which good results consistent with those obtained through the SMOS nominal processing chain are obtained. The paper is organized as follows: The main steps of the algorithm as well as MIRAS characteristics relevant to its application are described in Section 2; the results of the image reconstruction relying on the proposed algorithm are presented in Section 3; the discussion about these latter and those used by the SMOS nominal processing is given in Section 4; and finally the main conclusions are summarized in Section 5.

2. Methods

The main steps of the proposed image reconstruction algorithm are here presented by referring to the case of a single polarization measurement, and later extended to the case of a full polarimetric one. The algorithm is applied to the specific layout of the MIRAS radiometer (Figure 1). After the introduction of the visibility Equation (Section 2.1) and its conversion into a linear system of equations (Section 2.2), the algorithm develops through regularization of the matrix associated to the linear system of equations (Section 2.3), its inversion (Sections 2.4 and 2.5) and image reconstruction (Section 2.6). Lastly, the case of including the apodization into the processing chain is evaluated in Section 2.7 and the case of full polarimetric measurements is assessed in Section 2.8. Application to the MIRAS radiometer is illustrated throughout all the sections when needed.

2.1. Visibility Equation

The visibility equation to be used in aperture synthesis radiometry, derived in Corbella et al. [6], is a modified version of the Van Cittert–Zernike theorem to include the effect of the coupling between receivers and to fulfill the principle of energy conservation. In the single polarization case, after canceling the contribution of the receivers' physical temperature (i.e., approach 2 of Corbella et al. [8]) it is given by:

$$V(u,v) = \iint\limits_{\xi^2+\eta^2<1} T'(\xi,\eta)e^{-j2\pi(u\xi+v\eta)}\,d\xi d\eta, \tag{1}$$

where $T'(\xi,\eta)$ is the so-called "modified Brightness Temperature", expressed as:

$$T'(\xi,\eta) = T(\xi,\eta)\frac{F_k(\xi,\eta)F_j^*(\xi,\eta)}{\sqrt{1-\xi^2-\eta^2}\sqrt{\Omega_k\Omega_j}}\bar{\tilde{r}}_{kj}\left(-\frac{u\xi+v\eta}{f_0}\right), \tag{2}$$

in which $T(\xi,\eta)$ is the scene brightness temperature, $F_{k,j}$ are the complex field antenna patterns for the two elements k and j, $\Omega_{k,j}$ is their corresponding antenna solid angles and $\bar{\tilde{r}}_{kj}(\ )$ is the normalized fringe washing function [6], which depends on the receivers' frequency responses. Only in the case of having identical antennas and neglecting the fringe washing function, the modified brightness temperature (Equation (2)) becomes independent of the specific antenna pair and Equation (1) reduces to a two-dimensional Fourier transform $V(u,v) = \mathscr{F}[T'(\xi,\eta)]$.

The domain variables for visibility (u,v) and brightness temperature (ξ,η) are defined as

$$\begin{aligned} u &= (x_j - x_k)/\lambda_0 & v &= (y_j - y_k)/\lambda_0 \\ \xi &= x/r & \eta &= y/r, \end{aligned} \tag{3}$$

where (x,y) are the Cartesian coordinates of the observation point located at a distance r from the instrument. This latter is assumed to be centered on the origin of coordinates and aligned with the $z=0$ plane, with the antennas at coordinates $(x_{k,j}, y_{k,j})$ (see discussion below). Finally, λ_0 is the wavelength at the center frequency f_0, and ξ and η are the director cosines of the observation point with respect to axes x and y respectively, often expressed as a function of the elevation and azimuth angles (θ,ϕ) as $\xi = \sin\theta\cos\phi$ and $\eta = \sin\theta\sin\phi$.

The antenna pattern of a given element $F_k(\xi,\eta)$ characterizes the electromagnetic field radiated by the whole structure when this particular element is active and no signal is applied to the rest. The antenna position (x_k, y_k) in Equation (3) is the point at which its phase pattern is referenced to. To measure the embedded antenna pattern, the whole structure must rotate around a mechanical center of coordinates, so the radiated field becomes proportional to $F_{k0}(\xi,\eta)e^{-jkr}/r$ where $k = 2\pi/\lambda$ is the wave number and $F_{k0}(\xi,\eta)$ the pattern referenced to the coordinate center. The antenna phase pattern can then be referenced to any arbitrary position (x_k, y_k, z_k) by expressing r as $r = r_k + (r - r_k)$, with r_k the distance from this position to the observation point. For large distances, the differential length can be approximated by $r - r_k \approx \xi x_k + \eta y_k + \gamma z_k$ where $\gamma = z/r = \cos\theta$ is the third director cosine. The antenna pattern with phase referenced to coordinates (x_k, y_k, z_k) is then

$$F_k(\xi, \eta) = F_{k0}(\xi, \eta) e^{-jk(x_k\xi + y_k\eta + z_k\gamma)}. \tag{4}$$

Using this equation, the "antenna position" (x_k, y_k, z_k) can be chosen arbitrarily as long as the pattern phase is referenced to it. The position of the center of a sphere on which the phase variation of $F_k(\xi, \eta)$ is minimum is the antenna phase center, but this is not necessarily the best choice. In what follows, without loss of generality, the antenna positions are assumed to be equal to the nominal values and patterns are referenced to them. If antennas are properly designed, these positions should not be far away from their respective phase centers. And this is the case for SMOS.

In consequence the visibility (Equation (1)) is sampled at the the (u, v) coordinates corresponding to the nominal antenna positions (Equation (3)) using the regular distribution of antennas within the instrument in the $z = 0$ plane, as shown in Figure 1 for the MIRAS case.

2.2. Discretization and G-Matrix

To solve Equation (1) the director cosine domain (ξ, η) must also be discretized. The visibility equation becomes then a linear system of equations $V = GT$, where G is a complex matrix whose elements are function of the individual antenna patterns [8], V is the vector of visibilities in the (u, v) space and T the vector of brightness temperatures in the director cosine space (ξ, η). The G-matrix, defined as linear operator relating visibility to brightness temperature, was originally proposed in Tanner et al. [10]. In principle, recovering the brightness temperature requires only inverting the system of equations: $T = G^{-1}V$. However, the G-matrix just defined happens to be ill-conditioned [11], so the solution is not straightforward.

The G-matrix has as many rows as visibility samples, including those corresponding to zero spacing (single antenna). For an instrument having N antennas there are $N(N-1)/2$ complex rows (MIRAS, with $N = 69$, has 2346) and as many real rows as number of antennas used to measure the antenna temperature (visibility at zero spacing). The current nominal SMOS processing uses only one, but there is a backup mode that uses all or a selected set of antennas for the visibility at the origin [12]. The number of columns of the G-matrix is the total number of grid points (ξ, η) that fall inside the unit circle defined as $\xi^2 + \eta^2 < 1$.

Using Equations (1) and (2), the elements of the G-matrix are written as

$$G_{lm} = \Delta\xi\Delta\eta \; \frac{F_k(\xi, \eta)F_j^*(\xi, \eta)}{\sqrt{1 - \xi^2 - \eta^2}\sqrt{\Omega_k\Omega_j}} \tilde{r}_{kj}\left(-\frac{u\xi + v\eta}{f_0}\right) e^{-j2\pi(u\xi + v\eta)} \tag{5}$$

where the values of (u, v) and (ξ, η) are those of their respective grids and $\Delta\xi\Delta\eta$ is the elementary area (see Section 2.4).

2.3. Hermiticity and Redundant Baselines

Given the hermiticity property of the visibility $V(-u, -v) = V^*(u, v)$, for each complex row of the G-matrix an additional one can be added by changing the signs of u and v, provided the corresponding row of the visibility vector is conjugated. This operation has to be performed before dealing with the redundant baselines.

Redundant baselines are those having identical (u, v) values. Even though they correspond to the same visibility sample, they provide slightly different measurements with respect to each other because of the diverse antenna patterns of the involved elements. Considering two redundant baselines, the corresponding two distinct rows of the discretized visibility equation are:

$$V_l(u, v) = \sum_m G_{lm}T_m \quad ; \quad V_n(u, v) = \sum_m G_{nm}T_m, \tag{6}$$

where the subscripts l and n refer to two (k, j) pairs of redundant baselines and the subscript m ranges all (ξ, η) pairs in the unit circle. Note that (u, v) is the same in both by definition of redundant baselines. These two equations can be averaged to form a third one relating to the visibility of the same (u, v) point to the scene brightness temperature

$$\bar{V}_l(u, v) = \sum_m \bar{G}_{lm} T_m,$$

(7)

where $\bar{V}_l(u, v) = (V_l(u, v) + V_n(u, v))/2$ and $\bar{G}_{lm} = (G_{lm} + G_{nm})/2$. This last equation can be used for inversion without any loss of information. As a matter of fact, different complete sets of visibility samples $V(u, v)$ are obtained by randomly choosing unique sets of non-redundant baselines. For each one, the corresponding visibility function becomes related to the same brightness temperature image, so the image reconstruction algorithm for each of them should yield the same result in the absence of noise and errors. So only one set of non-redundant visibilities is enough to fully recover the brightness temperature. Averaging all measurements of the same (u, v) is not needed in the ideal case but has the effect of thermal noise reduction in practice.

The averaging operation must also be performed for the zero spacing visibility, which has a redundancy order equal to the number of antennas used to measure the antenna temperature.

After hermiticity extension and averaging of redundant visibilities, the number of rows of the G-matrix becomes equal to the total unique points in the (u, v) domain. In MIRAS it is equal to 2791, of which one is real and the rest are complex. These two operations notably improve the G-matrix condition number, acting as a regularization method to make image reconstruction feasible. This method was used in both references [8] and [13], although these references do not mention it explicitly.

2.4. Aliasing and Floor Error

Since the visibility equation is fundamentally a Fourier transform, discretization grids for regular sampling in both domains must be reciprocal to each other. The lattice depends on the overall geometry: Rectangular for U-shaped instruments [14], or hexagonal [15] for Y-shaped (MIRAS), hexagonal [16] or triangular ones. In any case, the visibility sampling coordinates (u, v) are a subset of the grid points in the fundamental period (a square for rectangular grids and a star or an hexagon for hexagonal grids). The minimum number of grid points in the fundamental period needed to include all measured samples is N_T^2 where $N_T = 4N_{EL} + 1$ for a hexagonal or triangular instrument, $N_T = 3N_{EL} + 1$ for a Y-shaped instrument or $N_T = 2N_{EL} + 1$ for a rectangular instrument [8]. In all cases, N_{EL} is the number of elements in one arm. Since MIRAS has $N_{EL} = 21$, it follows that $N_T = 64$ in this case (The SMOS Level 1 Operational Processor uses $N_T = 196$). The number of points in the fundamental period of the corresponding (ξ, η) reciprocal grid is also N_T^2. If reciprocal grids are used, the elementary area $\Delta\xi\Delta\eta$ in Equation (5) becomes then equal to $1/(N_T^2 d^2)$ for rectangular grids or $1/(N_T^2 d^2 \sin 60°)$ for hexagonal grids [8] where d is the minimum antenna spacing normalized to the center wavelength.

Figure 2 shows the MIRAS reciprocal grids for $N_T = 64$. The fundamental hexagon is depicted in blue in both domains, the green star in (u, v) includes all measured visibility points and their conjugate ones; and the extension of the (ξ, η) grid to the unit circle is drawn in gray. Both fundamental hexagons have the same number of points, in this case equal to $64^2 = 4096$. A larger value just enlarges the (u, v) hexagon while keeping the star shape identical, and provides a thicker grid in (ξ, η) [8]. The fundamental hexagon in this domain has a fixed side, independent of the number of points, equal to $2/(3d)$ where d is defined in the previous paragraph. This hexagon always falls inside the unit circle if $d > 1/\sqrt{3}$.

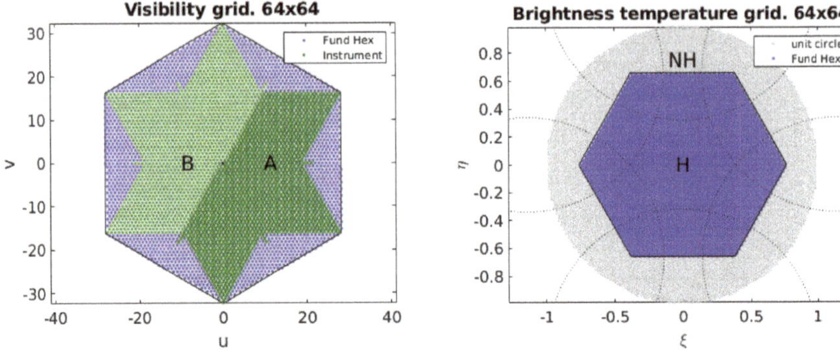

Figure 2. MIRAS reciprocal grids for visibility (**left**) and brightness temperature (**right**) using $N_T = 64$. Fundamental hexagons in both domains are drawn in blue. Two areas with hermitic (u, v) points are highlighted in the left plot. Unit circles aliases are added in the right plot.

A Fourier transform needs zero padding to complete the (u, v) fundamental period. The resultant modified brightness temperature is then obtained in the (ξ, η) fundamental period. The periodic repetition of the phase produces the well known phenomenon of aliasing, resulting in that the same image is repeated at all adjacent periods. All unit circle aliases for the MIRAS case are depicted in Figure 2. The zone in which these circles do not overlap is the alias-free field of view. Out of it, the image is always contaminated with replicas of other areas of the same image. In principle, error-free imaging is only possible in the alias-free field of view unless the overlapped image content is null, as for example in the case of having a small target in the center of the field of view surrounded by a very low background. Contrarily, for wide field of view imagers, aliasing is a strong source of errors. Changing the normalized antenna spacing d to a value lower than the Nyquist sampling rate ($1/2$ for rectangular grids and $1/\sqrt{3}$ for hexagonal grids) would make the circle be inscribed within the fundamental period, so eliminating the aliases.

For the minimum MIRAS reciprocal grid of Figure 2, the total number of (ξ, η) points within the unit circle is $N_p = 8491$, so this is the number of columns of the G-matrix in this case. Splitting the columns into the fundamental hexagon G_H and the rest of the unit circle G_{NH} (see Figure 2), the discretized visibility equation can be written as

$$V = \begin{bmatrix} G_H & G_{NH} \end{bmatrix} \begin{bmatrix} T_H \\ T_{NH} \end{bmatrix} = G_H T_H + G_{NH} T_{NH}, \tag{8}$$

where the same nomenclature applies to T_H and T_{NH}. Clearly, inverting only the G-matrix in the fundamental hexagon $T = G_H^{-1} V$ neglects the term $G_{NH} T_{NH}$ and produces what is sometimes called "floor error" [17,18]. Contrarily to the case of aliases in Fourier inversion, this one also spreads into the alias-free field of view unless considering identical antenna patterns and no fringe washing function. In this limiting case, imaging with G-matrix is equivalent to an inverse Fourier transform and the floor error is reduced to the aforementioned aliasing error.

In any case, the floor error can be mitigated by subtracting from the visibilities an estimation based on a forward a priori brightness temperature model outside the hexagon M_{NH}. Using Equation (8), the equation to invert becomes

$$V - G_{NH} M_{NH} \approx G_H T_H, \tag{9}$$

which leads to

$$T_H \approx G_H^{-1}(V - G_{NH} M_{NH}) = G_H^{-1} V - FE\, M_{NH}, \tag{10}$$

where $FE = G_H^{-1} G_{NH}$ is the floor error matrix. Even though computationally expensive, the floor error matrix is specific of the sensor and thus needs to be computed only once.

In conclusion: Image reconstruction is carried out by multiplying the inverse of the regularized G-matrix in the fundamental hexagon by the measured visibilities and substracting from the result an estimation of the floor error, which is equal to the product of the floor error matrix times a scene model outside the fundamental hexagon. Needless to say, the closer the model to the actual image, the lower the reconstruction error.

2.5. Matrix Extension and Inversion

Assuming that the regularization described in Section 2.3 has been applied and that the minimum reciprocal grids are used, the complex matrix G_H has, in the MIRAS case, 2791 rows and 4096 columns, corresponding respectively to the (u, v) unique grid points with measured visibilities (green star of Figure 2) and the fundamental hexagon in (ξ, η). Its condition number is about 3.2. The matrix G_H can thus be inverted using a Moore–Penrose pseudoinverse algorithm, as in Corbella et al. [8], by a conjugate-gradient method as in Camps et al. [13] or by other methods listed also in this reference.

A different approach is proposed here. First, the matrix is extended in the (u, v) domain (rows) up to the whole principal hexagon (blue dots of Figure 2) using an average antenna pattern and unit fringe washing function. Using Equation (5), the G-matrix rows corresponding to the blue dots in Figure 2 (left), that is outside the star, are computed as

$$G_{lm} = \Delta\xi\Delta\eta \, \frac{|\bar{F}_n(\xi, \eta)|^2}{\sqrt{1 - \xi^2 - \eta^2}\,\Omega} e^{-j2\pi(u\xi + v\eta)}. \tag{11}$$

The extended G-matrix becomes then square with size $N_T^2 \times N_T^2$ and keeps the condition number, so it can be straightforwardly inverted with standard algorithms. Note that, if all antenna patterns were substituted by the average pattern and the fringe washing function was neglected, this extended G-matrix would be a Fourier matrix with all columns multiplied by the average antenna pattern. The product $G_H T$ would become then equivalent to the product of a Fourier matrix and the modified brightness temperature, as expected. The proposed method can be viewed as a modification of the Fourier inversion to include different antenna patterns. Although not specifically reported elsewhere, this inversion method has been successfully implemented in the MIRAS Testing Software [19] since its first version.

2.6. Image Reconstruction

The brightness temperature map is recovered by multiplying the calibrated visibility by the inverted extended G_H matrix. For X or Y polarization the brightness temperature is real and G_H^{-1} is hermitic, so the first term of Equation (10) can be written as

$$G_H^{-1}V = \left[G_H^{-1}\big|_A \;\; G_H^{-1}\big|_0 \;\; G_H^{-1}\big|_B \right] \begin{bmatrix} V_A \\ V_0 \\ V_B \end{bmatrix} = G_H^{-1}\big|_0 V_0 + 2\Re e \left[G_H^{-1}\big|_A V_A \right], \tag{12}$$

where the subscript 0 refers to the origin ($u = v = 0$) and A and B are two grid point subsets (matrix columns) having opposite (u, v) signs. The ones used in the MIRAS Testing Software are shown in Figure 2 but the splitting is arbitrary. For points outside the star the visibility is ignored (see comment below about zero padding), so the corresponding columns of G_H^{-1} are not used. The last equality in the above equation holds because of the hermiticity property of visibility function. Since both $G_H^{-1}\big|_0$ and V_0 are real, this equation can be written in a more compact form as

$$G_H^{-1}V = \Re e \left\{ \left[G_H^{-1}\big|_0 \;\; 2G_H^{-1}\big|_A \right] \begin{bmatrix} V_0 \\ V_A \end{bmatrix} \right\}. \tag{13}$$

In MIRAS, using the minimum reciprocal grids, this operation involves the multiplication of a 4096×1395 complex matrix by a complex vector of 1395 elements.

If the complex polarimetric brightness temperature T_{xy} is being imaged, the full G_H^{-1} matrix should be used instead of its real part, although points outside the star should also be discarded.

The second term of Equation (10) does not depend on the measurement since it is computed using a model outside the hexagon. The hermiticity property of the X and Y polarizations can also be used to reduce the size of the floor error matrix in this case.

$$ FE = \Re\left\{ \left[\left. G_H^{-1}\right|_0 \quad 2G_H^{-1}\big|_A \right] \begin{bmatrix} G_{NH}\big|_0 \\ G_{NH}\big|_A \end{bmatrix} \right\}. \tag{14} $$

The size of the MIRAS single polarization floor error matrix using the minimum size grids is always 4096×4395 corresponding to the (ξ, η) points in both the fundamental hexagon and outside it respectively. This matrix is real for T_x and T_y and complex for T_{xy}.

A final comment about zero-padding is worth mentioning. The above equations detail the computation of each one of the two terms of Equation (10) separately but in a consistent way. Considering the version of this equation written at the first equal sign (that is $T_H = G_H^{-1}(V - G_{NH}M_{NH})$), it comes out that zero padding outside the star means using in these points the product $G_{NH}M_{NH}$, but not zero. In practice this needs not to be done explicitly, as it suffices just to ignore the columns of G_H^{-1} outside the star.

2.7. Apodization

Due to the limited visibility coverage in the (u, v) plane, the reconstructed brightness temperature is affected by the Gibbs effect showing ripples around abrupt changes in the original scene, as for example coastlines. As it is well known from Fourier imaging, ripples can be reduced at the expense of degrading spatial resolution by using an apodization window in the original domain. In Corbella et al. [8] the apodization window was directly applied to the measured visibilities, which is correct if a Fourier inversion is used but it is at least questionable for the G-matrix technique. A more rigorous approach is to window the Fourier components of the reconstructed brightness temperature image. In this case, the apodized brightness temperature is related to the reconstructed brightness temperature T by

$$ T_{apodized} = \mathscr{F}^{-1}\{W\mathscr{F}\{T\}\}, \tag{15} $$

where W is the window function, which in SMOS is always of Blackman type.

The reconstructed brightness temperature is defined in the (ξ, η) fundamental hexagon, so its Fourier components, to which the window function is applied, are defined in the (u, v) fundamental hexagon. The DFT operation involved in Equation (15) assumes implicitly that the reconstructed brightness temperature (T) is a periodic function in (ξ, η) replicating itself in hexagons adjacent to the fundamental one. Since in typical SMOS images the earth disk is at the bottom of the hexagon and the sky at the top, there are abrupt changes at the border of the fundamental period that may induce ripples. To mitigate them, constant temperature levels are subtracted from the sky and earth zones so as to have a zero mean image. Specifically, the constant temperature subtracted to the sky pixels is computed as the median of the recovered image in them, while the value subtracted to the Earth pixels is computed so as to cancel the Fourier component at the origin. The consequent reduction of the contrasts within the image cause the minimization of the associated ripples. The constant temperatures are added back after Fourier inversion.

This approach is strongly inspired on the incremental visibility image reconstruction method proposed in Camps et al. [13]. In that case, however, the method was applied to visibilities instead to frequency components, but the idea is the same.

2.8. Full Polarimetric Case

Considering a baseline formed by two dual-polarization antennas, the full polarimetric discretized visibility equation [20] can be written in terms of G-matrices as

$$V_{xx} = \mathbf{G_{xx}^{RR}T_x} + G_{xx}^{CC}T_y + G_{xx}^{RC}T_{xy} + G_{xx}^{CR}T_{yx} \tag{16}$$

$$V_{yy} = G_{yy}^{CC}T_x + \mathbf{G_{yy}^{RR}T_y} + G_{yy}^{CR}T_{xy} + G_{yy}^{RC}T_{yx} \tag{17}$$

$$V_{xy} = G_{xy}^{RC}T_x + G_{xy}^{CR}T_y + \mathbf{G_{xy}^{RR}T_{xy}} + G_{xy}^{CC}T_{yx} \tag{18}$$

$$V_{yx} = G_{yx}^{CR}T_x + G_{yx}^{RC}T_y + G_{yx}^{CC}T_{xy} + \mathbf{G_{yx}^{RR}T_{yx}}, \tag{19}$$

where, for example, G_{xy}^{RC} denotes the G-matrix computed according to Equation (5) using the Reference (co-polar) pattern of the X polarization antenna and the Cross-polar pattern of the Y polarization antenna. The terms marked in boldface are the dominant ones, in case of antennas with negligible cross-polar patterns, as considered in Equations (1) and (2).

Averaging redundant baselines, adding hermitic points and extending the matrix to the full hexagon is here carried out for each of the sub-matrices using the procedures detailed in Sections 2.3 and 2.5. Once this is done, the combination of the four equations can be written as $V = GT$ where now G is a matrix with dimension $4N_T^2 \times 4N_p$, where N_T^2 is the total number of points in the fundamental hexagon and N_p the number of points in the unit circle (4096 and 8491 respectively for the MIRAS minimum grid). The floor error mitigation through the application of a model outside the fundamental period can be carried out in the polarimetric case in exactly the same manner as done for a single polarization. The part of the extended G-matrix inside the fundamental period has now a size of $4N_T^2 \times 4N_T^2$ and, when inverted, provides as result a square matrix:

$$G_H^{-1} = \begin{bmatrix} IG_{11} & IG_{12} & IG_{13} & IG_{14} \\ IG_{21} & IG_{22} & IG_{23} & IG_{24} \\ IG_{31} & IG_{32} & IG_{33} & IG_{34} \\ IG_{41} & IG_{42} & IG_{43} & IG_{44} \end{bmatrix}, \tag{20}$$

where IG_{ij} are submatrices of size $N_T^2 \times N_T^2$.

Applying the hermiticity property to the matrices of the first two rows using the procedures of Section 2.6 the following equations are obtained:

$$T_x = \Re e \left\{ \left[IG_{11}\big|_0 \ 2IG_{11}\big|_A \right] \begin{bmatrix} V_{x_0} \\ V_{x_A} \end{bmatrix} + \left[IG_{12}\big|_0 \ 2IG_{12}\big|_A \right] \begin{bmatrix} V_{y_0} \\ V_{y_A} \end{bmatrix} + \right. \tag{21}$$
$$\left. + \left[IG_{13}\big|_0 \ 2IG_{13}\big|_A \right] \begin{bmatrix} V_{xy_0} \\ V_{xy_A} \end{bmatrix} + \left[IG_{14}\big|_0 \ 2IG_{14}\big|_A \right] \begin{bmatrix} V_{yx_0} \\ V_{yx_A} \end{bmatrix} \right\},$$

$$T_y = \Re e \left\{ \left[IG_{21}\big|_0 \ 2IG_{21}\big|_A \right] \begin{bmatrix} V_{x_0} \\ V_{x_A} \end{bmatrix} + \left[IG_{22}\big|_0 \ 2IG_{22}\big|_A \right] \begin{bmatrix} V_{y_0} \\ V_{y_A} \end{bmatrix} + \right. \tag{22}$$
$$\left. + \left[IG_{23}\big|_0 \ 2IG_{23}\big|_A \right] \begin{bmatrix} V_{xy_0} \\ V_{xy_A} \end{bmatrix} + \left[IG_{24}\big|_0 \ 2IG_{24}\big|_A \right] \begin{bmatrix} V_{yx_0} \\ V_{yx_A} \end{bmatrix} \right\},$$

$$T_{xy} = IG_{31}V_x + IG_{32}V_y + IG_{33}V_{xy} + IG_{34}V_{yx}, \tag{23}$$

in which the rows of the inverted G-matrix outside the star are always discarded. Note that, by definition, $T_{yx} = T_{xy}^*$, so there is no additional equation for this term. Implementation wise it is found that this identity holds to the machine precision, which is a consistency indicator.

The floor error matrix is computed analogously to the case of single polarization measurements (Section 2.6) but using the larger G-matrix defined here. The result is a matrix four times the size of that for single polarization.

3. Results

The image reconstruction methodology outlined in the previous sections is fully implemented in the MIRAS Testing Software [19]. This tool is being systematically used for scientific studies as a flexible alternative to the nominal SMOS Level 1 data products and with successful results, as reported for example in González-Gambau et al. [21].

Exploiting the processing capabilities of this software, the proposed algorithm was tested relying on a set of real MIRAS measurements. Specifically, a SMOS orbit was processed in full polarimetric mode (Section 2.8) using the all-LICEF calibration approach [12]. Figure 3 shows the retrieved geo-located images at X- and Y-polarizations corresponding to a single snapshot acquired on 24 January 2019 at 21:44 UTC, when the satellite was passing over the coast of Australia during an ascending orbit. The Blackman window is applied using the methods in Section 2.7.

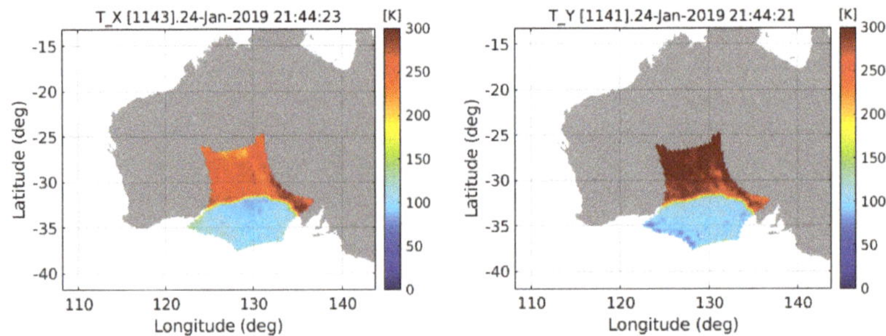

Figure 3. Geo-located images, in the extended alias-free field of view, of the two snapshots of Figure 4.

Figure 4 shows the brightness temperatures of same snapshot in the (ξ, η) plane. Clearly, the image reconstruction algorithm is able to capture all the scene features in the whole hexagon, not only in the alias-free field of view. Ocean, land and sky areas are clearly distinguished and, at this scale, aliases impact is minor. Differences between T_x and T_y are due to the stronger T_y increments associated to both the sea/land and Earth/sky transitions with respect to T_x. The model used to cancel the floor error [M_{NH} in Equation (10)] consists of a constant value in land zones (258 K for X-pol and 285 K for Y-pol), specular reflection in ocean areas using Fresnel reflection coefficients with climatology salinity and temperature [22], a constant 8.5K to account for atmospheric effects and the L-Band sky and Galaxy map available as auxiliary file in the SMOS data base.

Figure 5 shows the X-polarization case with both terms of Equation (10) drawn separately as well as the difference between them, which is the final reconstructed image. The left panel shows the result of multiplying the calibrated visibility by the inverse of the G-matrix (first term of Equation (10), the center panel is the floor error and the right panel the corrected image. As expected, the floor error is very small in regions where the earth aliases do not enter into the hexagon, which is the so-called "extended alias-free field of view", nominally used in SMOS images (as in the geo-located images of Figure 3). Most of the artifacts of the raw image at the left are effectively removed in the corrected one. Specifically, strong improvement in the image reconstruction is found in the sky area as well as in both the bottom left and right strips. Contrarily, much lower effects are exhibited in the extended alias-free field of view, where the floor error is minimum. Later it will be shown that there is indeed a small improvement in this area. Apodization of the corrected image with a Blackman window, using the procedure of Section 2.7, provides the final result seen at the left of Figure 4.

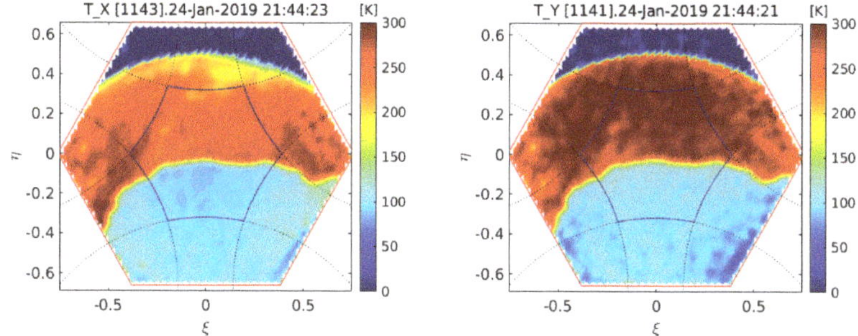

Figure 4. Image reconstruction result of two individual Soil Moisture and Ocean Salinity (SMOS) snapshots over the coast of Australia. Left is for X-pol and right is for Y-pol.

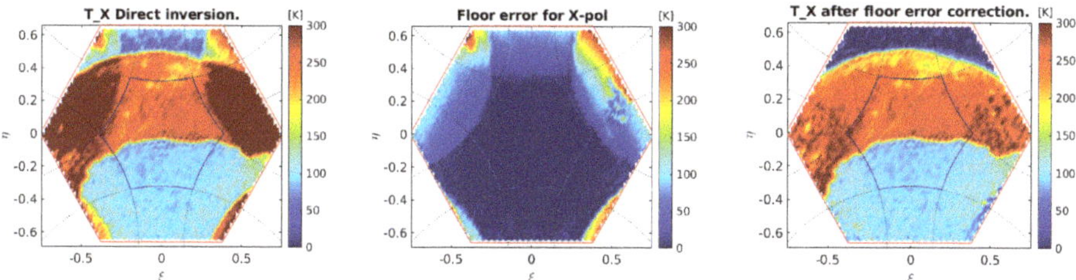

Figure 5. Floor error correction of the X-pol snapshot of Figure 4. Left and center images are first and second terms of Equation (10) respectively, and the difference is shown at right.

The scale of Figure 4 does not allow one to assess the effect of the floor error correction in the alias-free field of view. This is only possible using a differential image, plotting the brightness temperature bias with respect to its expected value. Additionally, averaging several snapshots is desirable in order to reduce thermal noise. Both requirements can be met if the scene is limited to snapshots over the ocean, for which a very comprehensive model is available from the SMOS science community (The authors would like to thank Joseph Tenerelli (OceanDataLab, France) for providing the ocean forward model).

Figure 6 shows the bias with respect to the model of all snapshots ranging in a latitudinal range of $[-40°, -5°]$ over the Pacific ocean in an ascending orbit of 28 January 2011 (chosen quite often for these kinds of analysis by the SMOS Level 1 team). In the area imaging the sky, the model is the standard SMOS Galaxy map. All four polarimetric products are included in Figure 6, brightness temperature at X and Y polarizations and real and imaginary parts of the complex brightness temperature. Basic statistics, referring to the alias-free field of view only, are shown at the bottom of each panel of Figure 6. Namely, spatial standard deviations and mean values are indicated by σ and T respectively. The relatively large negative bias in the Y-pol image may be due to poor modeling at high incident angles. In any case, these residual error images are compatible with the ones obtained with the SMOS Level-1 Operational Processor and have similar statistics.

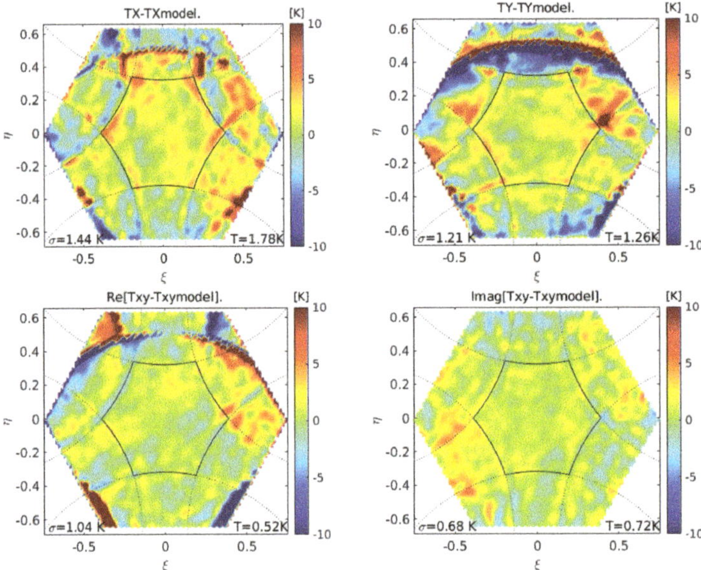

Figure 6. Spatial bias computed as the difference between the reconstructed brightness temperature and a model for the ocean. Blackman window is applied.

The same images have been produced without correcting the floor error, that is using only the first term of Equation (10). Results are shown in Figure 7. In this case, the error outside the alias-free field of view increases dramatically, especially in the areas in which the earth enters the hexagon. In the alias-free field of view there is a small impact in spatial bias, quantified in the standard deviation shown at the bottom. In all cases it increases with respect to the numbers provided in Figure 6. Note that the extended alias-free field of view is quite well recovered even if no correction is applied.

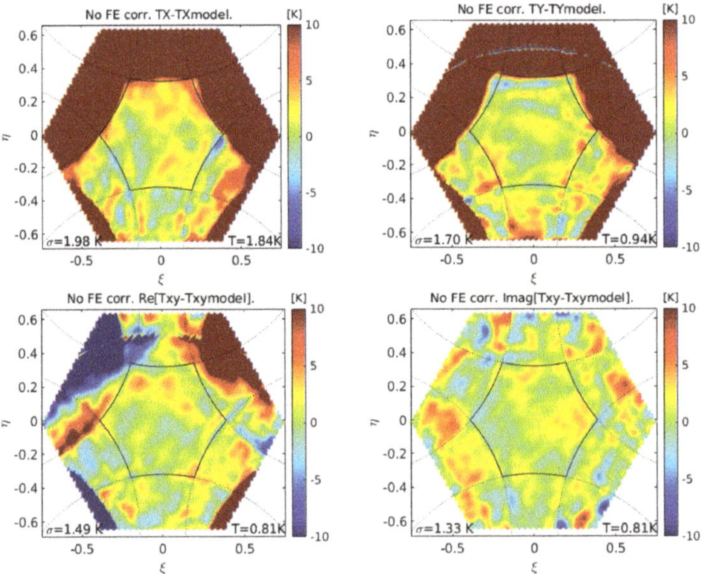

Figure 7. Spatial bias of Figure 6 without removing the floor error.

4. Discussion

Years before SMOS launch in November 2009, methods for 2D interferometric radiometer imaging were derived by different groups with the aim of having working algorithms as soon as visibility measurements were provided by MIRAS [8,11,23]. Radioastronomy heritage showed to be not directly applicable due to different instrument layouts, as for example the reduced antenna spacing, and especially because of the larger instantaneous field of view typical of earth observation. Working with simulated data it was early discovered that after a complete forward-backward simulation of a known scene, the original brightness temperature was not perfectly recovered even in the alias-free field of view. This misfit was called "scene-dependent bias" and was attributed to an underdetermination of the mathematical problem. An efficient mitigation algorithm was proposed in Corbella et al. [24], used in Anterrieu et al. [25] and improved later in Camps et al. [13]. This consists of subtracting from the visibility measurements different contributions estimated by simulation, and inverting the resultant "differential visibilities". Some of the contributions are later added back to the reconstructed image to recover the final brightness temperature map. The most recent implementation of this idea, described in Khazâal et al. [26], is included in the version 7 of the SMOS Level 1 Operational Processor. The method, even in its simplest version, is highly effective in canceling the sky aliases so expanding the alias-free field of view—limited by the unit circle aliases—to the extended alias-free field of view, limited by the earth shape. In its latest version [26], using a sophisticated model as "artificial scene" it is able to further reduce the error in the alias-free field of view.

All these methods are based on inverting only the portion of the G-matrix inside the fundamental hexagon discarding the rest. As already pointed out in Corbella et al. [17], discarding the G-matrix outside the fundamental hexagon is responsible for the appearance of the aliases. If all antenna patters were identical, this error would become limited to only the aliasing regions. In the real case, with different antenna patterns, the error is indeed larger in these regions but spreads also in the alias-free field of view (see center panel of Figure 5). The G-matrix in the whole unit circle is always used to estimate the corresponding visibility of the model or artificial scene, so imaging differential visibilities can be interpreted as removing an estimation of the aliases.

Using the concept of extended G-matrix of Section 2.5, it is easily shown that the method of imaging differential visibilities is equivalent to removing the floor error. The reconstructed brightness temperature from differential visibilities is

$$T = G_H^{-1}(V - GM) + M_H = G_H^{-1}V - (G_H^{-1}G - U)M, \tag{24}$$

where U is a matrix with the same size of G equal to the identity matrix for the columns inside the hexagon and zero outside $U = [I\ 0]$. In this equation, the G-matrix is extended, so having as many rows as number of points in the fundamental (u, v) hexagon, and thus G_H is a square and invertible matrix. The first term of this equation is the same as that of Equation (10). The second term can be expanded as

$$\left(G_H^{-1}\begin{bmatrix} G_H & G_{NH} \end{bmatrix} - \begin{bmatrix} I & 0 \end{bmatrix}\right)\begin{bmatrix} M_H \\ M_{NH} \end{bmatrix} = \begin{bmatrix} 0 & G_H^{-1}G_{NH} \end{bmatrix}\begin{bmatrix} M_H \\ M_{NH} \end{bmatrix} = G_H^{-1}G_{NH}M_{NH}, \tag{25}$$

which coincides with Equation (10). It is important to point out that this is only true if the visibility simulation of the model uses a G-matrix in the full unit circle defined in the same grid as the one used for inversion. That is, the matrix inverted is a subset of the complete one. Conceptually, a different matrix could be used since $V = GT$ is just a mathematical model of what the instrument actually measures. Also, the equivalence shown is based on the use of the extended G-matrix concept defined in Section 2.5. Different implementations of matrix inversion in Equation (24) are not equivalent. The main advantage of using the floor error matrix defined in Equation (10) is that processing does not require estimating visibilities with the full G-matrix and does not require adding back any artificial scene to the result. It is a correction applied directly to the reconstructed image.

The outcome of the proposed methodology is the brightness temperature map of the scene. This is not the case for the SMOS Level 1 Operational Processor which, based in the proposal of Anterrieu et al. [11], defines the Level 1B data as the spatial frequency components of the brightness temperature. This is done through the concept of J-matrix, the concatenation of the G-matrix with the Fourier operator as dully explained in Khazâal et al. [26]. This approach is a different method to regularize the system of equations that relates all measured visibility, including redundant baselines, to brightness temperature. Due to redundancies the G-matrix in this case is ill-posed but the corresponding J-matrix is well behaved and can be inverted. As pointed out in Section 2.7, since the brightness temperature is not a periodic function, when recovering it from the corresponding frequencies, ripples can appear in the limits of the fundamental hexagon. In practice, this does not happen due to the imaging of differential visibilities used in the processor. The Level 1B data consists actually of frequency components of the difference between the brightness temperature and an artificial scene. After reconstructing the map including apodization, the latter is added back. Versions prior to 7 of the SMOS Level 1 Operational Processor use as artificial scene a constant in the earth disk and a sky map for the sky. In the end the constant earth is added to all points, and this is why the nominal SMOS images at Level 1B in the full hexagon show high temperature in the sky (see Figure 8). Here it has been shown that averaging redundant baselines improves the condition of the matrix and makes it invertible, so there is no need to compute the J-matrix. As a post-processing, however, the frequency components are computed in order to apply the Blackman window, as explained in Section 2.7.

For completeness, equivalent images as those shown in Figures 4 and 6 are provided respectively in Figures 8 and 9 using SMOS Level 1B data read from the ESA SMOS Online Dissemination Service, produced by version 6.21 of the SMOS Level 1 Operational Processor.

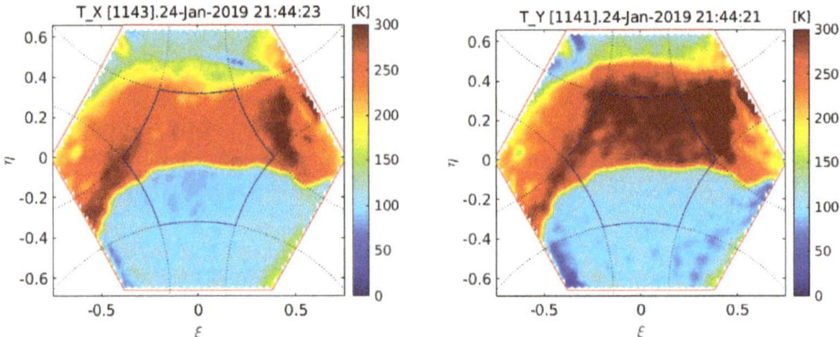

Figure 8. Same snapshots as in Figure 4 but using data from the SMOS Level 1 Operational Processor version 6.21. Deimos Engenharia, Lisbon (Portugal).

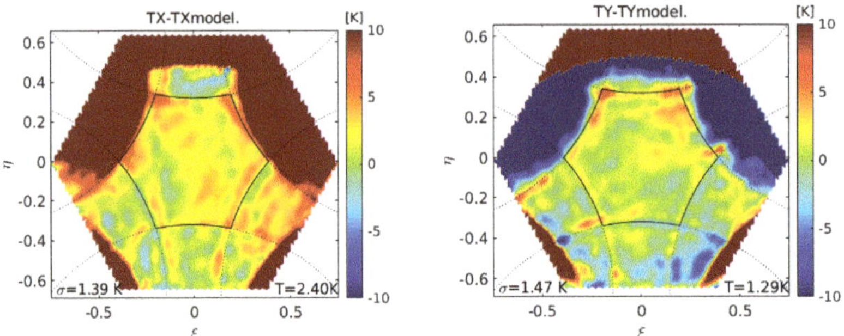

Figure 9. *Cont.*

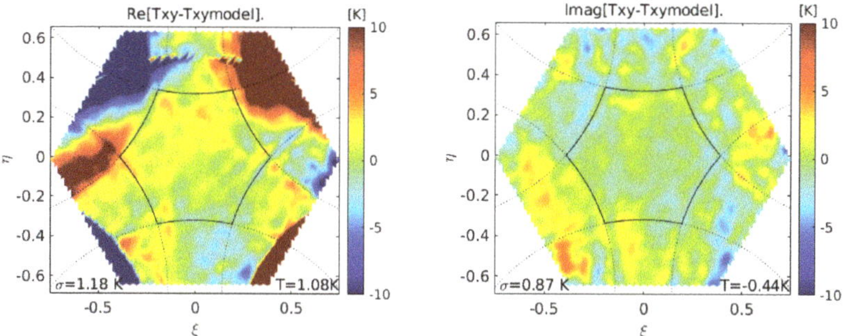

Figure 9. Same differential images as those in Figure 6 but using data from the SMOS Level 1 Operational Processor version 6.21. Deimos Engenharia, Lisbon (Portugal).

5. Conclusions

Accurate brightness temperature retrieval in 2D interferometric radiometers with a large field of view is not straightforward. The simplest approach consists of applying an inverse Fourier transform, which provides reasonable results in the alias-free field of view but large errors outside. The G-matrix approach takes into account individual differences between antenna patterns and also the fringe washing function and thus provides images of improved quality. In any case, the major error contribution is localized in the alias regions where brightness temperatures from several spatial directions overlap. For zones where the overlap is the sky the impact is low, but where aliases are produced by the earth it may become significant, affecting also the alias-free field of view. The origin of these errors is found in the contribution to the visibility of the brightness temperature from directions that fall outside the fundamental period. Subtracting an estimation of these visibilities to the measurements greatly reduces the error. Correction is carried out through the definition of the floor error matrix and relying on a brightness temperature model defined only outside the fundamental period.

Expressing the visibility equation as a linear system of equations, image reconstruction can be implemented through the inversion of the associated matrix, called G-matrix. Unfortunately, the G-matrix results are ill-conditioned and cannot be inverted without regularization. This is obtained by averaging redundant visibilities and extending the G-matrix to fill the whole hexagon in the (u, v) plane, resulting in a square and well-conditioned matrix that is easy to invert. Using this approach, combined with the floor error removal, provides high quality images in the full hexagon, and especially in the extended alias-free field of view. The method is applicable to the full polarimetric operation of SMOS, with the only difference being increasing the size of the G-matrix. Results of SMOS complex images and full polarimetric error maps over ocean demonstrate the procedure.

The inversion method presented in this paper uses the minimum number of grid points, allowing for very fast and efficient programming. It has been implemented in the MIRAS Testing Software for several years and is being successfully used by science teams for salinity and soil moisture retrievals. In any case, the procedures outlined are not intended to replace the ones used by the SMOS Level 1 Operational Processor, but they are presented to the community with the objective of reporting other means of solving the same problem with similar results.

Author Contributions: Conceptualization, I.C.; Investigation, I.C. and I.D.; Methodology, F.T. and N.D.; Software, I.C.; Supervision, M.M.-N.; Validation, V.G.-G.; Writing—original draft, I.C.; Writing—review & editing, V.G.-G. and M.M.-N.

Funding: This research was funded by the European Space Agency through SMOS P7 subcontract DME CP12 no. 2015-005 with Deimos Enginheria (Portugal) and by Ministerio de Economia, Industria y Competitividad, Gobierno de España, projects TEC2014-58582-R, TEC2017-88850-R and ESP2015-67549-C3-1-R.

Conflicts of Interest: The authors declare no conflict of interest.

Abbreviations

The following abbreviations are used in this manuscript:

SMOS   Soil Moisture and Ocean Salinity
MIRAS   Microwave Imaging Radiometer with Aperture Synthesis
DFT   Digital Fourier Transform

References

1. Le Vine, D.M.; Good, J.C. *Aperture Synthesis for Microwave Radiometers in Space*; Technical Memorandum TM-85033; NASA Goddard Space Flight Center: Greenbelt, MD, USA, 1983.
2. Ruf, C.S.; Swift, C.T.; Tanner, A.B.; Le Vine, D.M. Interferometric Synthetic Aperture Microwave Radiometry for the Remote Sensing of the Earth. *IEEE Trans. Geosci. Remote Sens.* **1988**, *26*, 597–611. [CrossRef]
3. Martín-Neira, M.; Goutoule, J.M. *MIRAS—A Two-Dimensional Aperture-Synthesis Radiometer for Soil Moisture and Ocean Salinity Observations*; ESA Bulletin Nr. 92; European Space Agency: Paris, France, 1997; pp. 95–104.
4. McMullan, K.; Brown, M.; Martín-Neira, M.; Rits, W.; Ekholm, S.; Marti, J.; Lemanzyk, J. SMOS: The payload. *IEEE Trans. Geosci. Remote Sens.* **2008**, *46*, 594–605. [CrossRef]
5. Barré, H.; Duesmann, B.; Kerr, Y. SMOS: The mission and the system. *IEEE Trans. Geosci. Remote Sens.* **2008**, *46*, 587–593. [CrossRef]
6. Corbella, I.; Duffo, N.; Vall-llossera, M.; Camps, A.; Torres, F. The visibility function in interferometric aperture synthesis radiometry. *IEEE Trans. Geosci. Remote Sens.* **2004**, *42*, 1677–1682. [CrossRef]
7. Corbella, I.; Torres, F.; Camps, A.; Duffo, N.; Vall-llossera, M.; Rautiainen, K.; Martín-Neira, M.; Colliander, A. Analysis of Correlation and Total Power Radiometer Front-Ends Using Noise Waves. *IEEE Trans. Geosci. Remote Sens.* **2005**, *43*, 2452–2459. [CrossRef]
8. Corbella, I.; Torres, F.; Camps, A.; Duffo, N.; Vall-llossera, M. Brightness Temperature Retrieval Methods in Synthetic Aperture Radiometers. *IEEE Trans. Geosci. Remote Sens.* **2009**, *47*, 285–294. [CrossRef]
9. Corbella, I.; Torres, F.; Wu, L.; Duffo, N.; Duran, I.; Martín-Neira, M. Spatial biases analysis and mitigation methods in SMOS images. In Proceedings of the International Geoscience and Remote Sensing Symposium, IGARSS 2013, Melbourne, Australia, 21–26 July 2013; pp. 3145–3418.
10. Tanner, A.; Swift, C.T. Calibration of a Synthetic Aperture Radiometer. *IEEE Trans. Geosci. Remote Sens.* **1993**, *31*, 257–267. [CrossRef]
11. Anterrieu, E. A Resolving Matrix Approach for Synthetic Aperture Imaging Radiometers. *IEEE Trans. Geosci. Remote Sens.* **2004**, *42*, 1649–1659. [CrossRef]
12. Corbella, I.; González-Gambau, V.; Torres, F.; Duffo, N.; Duran, I.; Martín-Neira, M. The MIRAS "All-LICEF" calibration mode. In Proceedings of the International Geoscience and Remote Sensing Symposium, IGARSS 2016, Beijing, China, 10–15 July 2016; pp. 2013–2016. [CrossRef]
13. Camps, A.; Vall-llossera, M.; Corbella, I.; Duffo, N.; Torres, F. Improved Image Reconstruction Algorithms for Aperture Synthesis Radiometers. *IEEE Trans. Geosci. Remote Sens.* **2008**, *46*, 146–158. [CrossRef]
14. Rautiainen, K.; Kainulainen, J.; Auer, T.; Pihlflyckt, J.; Kettunen, J.; Hallikainen, M.T. Helsinki University of Technology L-Band Airborne Synthetic Aperture Radiometer. *IEEE Trans. Geosci. Remote Sens.* **2008**, *46*, 717–726. [CrossRef]
15. Camps, A.; Bará, J.; Corbella, I.; Torres, F. The Processing of Hexagonally Sampled Signals with Standard Rectangular Techniques: Application to 2D Large Aperture Synthesis Interferometric Radiometers. *IEEE Trans. Geosci. Remote Sens.* **1997**, *GRS-35*, 183–190. [CrossRef]
16. Zurita, A.M.; Corbella, I.; Martín-Neira, M.; Plaza, M.A.; Torres, F.; Benito, F.J. Towards a SMOS Operational Mission: SMOSOps-Hexagonal. *IEEE J. Sel. Top. Appl. Earth Obs. Remote Sens.* **2013**, *6*, 1769–1780. [CrossRef]
17. Corbella, I.; Torres, F.; Wu, L.; Duffo, N.; Duran, I.; Martín-Neira, M. SMOS image reconstruction quality assessment. In Proceedings of the International Geoscience and Remote Sensing Symposium, IGARSS 2014, Québec City, QC, Canada, 13–18 July 2014; pp. 1914–1916.
18. Duran, I.; Lin, W.; Corbella, I.; Torres, F.; Duffo, N.; Martín-Neira, M. SMOS floor error impact and mitigation on ocean imaging. In Proceedings of the International Geoscience and Remote Sensing Symposium, IGARSS 2015, Milano, Italy, 26–31 July 2015; pp. 1437–1440.

Remote Sens. **2019**, *11*, 682

19. Corbella, I.; Torres, F.; Duffo, N.; González, V.; Camps, A.; Vall-llossera, M. Fast processing tool for SMOS data. In Proceedings of the International Geoscience and Remote Sensing Symposium, IGARSS 2008, Boston, MA, USA, 8–11 July 2008; Number II, pp. 1152–1155.

20. Camps, A.; Corbella, I.; Torres, F.; Vall-llossera, M.; Duffo, N. Polarimetric formulation of the visibility function equation including cross-polar antenna patterns. *IEEE Geosci. Remote Sens. Lett.* **2005**, *2*, 292–295. [CrossRef]

21. González-Gambau, V.; Olmedo, E.; Martínez, J.; Turiel, A.; Durán, I. Improvements on Calibration and Image Reconstruction of SMOS for Salinity Retrievals in Coastal Regions. *IEEE J. Sel. Top. Appl. Earth Obs. Remote Sens.* **2017**, *10*, 3064–3078. [CrossRef]

22. Antonov, J.I.; Seidov, D.; Boyer, T.P.; Locarnini, R.A.; Mishonov, A.V.; Garcia, H.E.; Baranova, O.K.; Zweng, M.M.; Johnson, D.R. *World Ocean Atlas 2009 Volume 2: Salinity*; Levitus, S., Ed.; NOAA Atlas NESDIS 69; U.S. Gov. Printing Office: Washington, DC, USA, 2010; 184p.

23. Camps, A.; Bará, J.; Torres, F.; Corbella, I. Extension of the CLEAN technique to the microwave imaging of continuous thermal sources by means of aperture synthesis radiometers. *Prog. Electromagn. Res.* **1998**, *18*, 67–83. [CrossRef]

24. Corbella, I.; Torres, F.; Camps, A.; Colliander, A.; Martín-Neira, M.; Ribó, S.; Rautiainen, K.; Duffo, N.; Vall-llossera, M. MIRAS End-to-End Calibration. Application to SMOS L1 Processor. *IEEE Trans. Geosci. Remote Sens.* **2005**, *43*, 1126–1134. [CrossRef]

25. Anterrieu, E. On the Reduction of the Reconstruction Bias in Synthetic Aperture Imaging Radiometry. *IEEE Trans. Geosci. Remote Sens.* **2007**, *45*, 592–601. [CrossRef]

26. Khazâal, A.; Richaume, P.; Cabot, F.; Anterrieu, E.; Mialon, A.; Kerr, Y.H. Improving the Spatial Bias Correction Algorithm in SMOS Image Reconstruction Processor: Validation of Soil Moisture Retrievals With In Situ Data. *IEEE Trans. Geosci. Remote Sens.* **2019**, *57*, 277–290. [CrossRef]

 *remote sensing*

Article

SMOS Third Mission Reprocessing after 10 Years in Orbit

Roger Oliva [1,*], Manuel Martín-Neira [2], Ignasi Corbella [3], Josep Closa [4], Albert Zurita [4], François Cabot [5], Ali Khazaal [5], Philippe Richaume [5], Juha Kainulainen [6], Jose Barbosa [7], Gonçalo Lopes [8], Joseph Tenerelli [9], Raul Díez-García [10], Veronica González-Gambau [11] and Raffaele Crapolicchio [12]

[1]   Zenithal Blue Technologies S.L.U., 08023 Barcelona, Spain
[2]   European Space Research and Technology Centre, European Space Agency, 2201 AZ Noordwijk, The Netherlands; Manuel.Martin-Neira@esa.int
[3]   Remote Sensing Laboratory, Universitat Politecnica de Catalunya, 08034 Barcelona, Spain; corbella@tsc.upc.edu
[4]   Airbus Defence and Space, Microwave Instruments department, 28692 Madrid, Spain; Josep.Closa@airbus.com (J.C.); alberto.Zurita@airbus.com (A.Z.);
[5]   Centre D'Etudes Spatiales de la Boisphere, 31400 Toulouse, France; francois.cabot@cesbio.cnes.fr (F.C.); Ali.Khazaal@cesbio.cnes.fr (A.K.); philippe.richaume@cesbio.cnes.fr (P.R.)
[6]   Harp Technologies, 02150 Espoo, Finland; juha.kainulainen@harptechnologies.com
[7]   Research and Development in Aerospace, 8006 Zurich, Switzerland; jose.barbosa@rdaerospace.ch
[8]   Deimos Engenheria, 1998-023 Lisbon, Portugal; goncalo.lopes@deimos.com.pt
[9]   OceanDataLab, 29280 Locmaria Plouzané, France; joseph.tenerelli@oceandatalab.com
[10] European Space Astronomy Centre, European Space Agency, 28692 Madrid, Spain; raul.diez.garcia@esa.int
[11] Department of Physical Oceanography, Institute of Marine Sciences, CSIC and Barcelona Expertise Center, 08003 Barcelona, Spain; vgonzalez@icm.csic.es
[12] European Space Research Institute, European Space Agency, 00044 Frascati, Italy; raffaele.crapolicchio@esa.int
*   Correspondence: r.oliva@zenithalblue.com

Received: 25 March 2020; Accepted: 14 May 2020; Published: 20 May 2020

Abstract: After more than 10 years in orbit, the SMOS team has started a new reprocessing campaign for the SMOS measurements, which includes the changes in calibration and image reconstruction that have been made to the Level 1 Operational Processor (L1OP) during the past few years. The current L_1 processor, version v620, was used for the second mission reprocessing in 2014. The new version, v724, is the one run in the third mission reprocessing and will become the new operational processor. The present paper explains the major changes applied and analyses the quality of the data with different metrics. The results have been obtained with numerous individual tests that have confirmed the benefits of the evolutions and an end-to-end processing campaign involving three years of data used to assess the improvements of the SMOS measurements quantitatively.

Keywords: SMOS; calibration; radiometry; reprocessing

---

1. Introduction

The Soil Moisture and Ocean Salinity (SMOS) mission is the second Earth Explorer mission of the European Space Agency (ESA). The satellite was launched in November 2009 and has been continuously operating ever since, with an excellent health status. Data acquisition is in the order of 99.88%, and processing performance is above 99%. As such, the ESA has continuously provided nominal and near-real-time data for the past 10 years since the end of the commissioning phase.

The original objectives of soil moisture [1] and sea surface salinity [2] have been complemented with new applications, such as to thin sea-ice thickness, severe winds over ocean and freeze/thaw soil state products [3]. The satellite contains a single payload, the MIRAS (Microwave Imaging Radiometer using Aperture Synthesis), the first ever space-based L-band interferometric radiometer [4]. Even though interferometric radiometers have long been used by radio-astronomers, having such an instrument space-based for earth observation missions has presented several challenges. More than ten years after launch, the SMOS team continues to improve the calibration and the image reconstruction processes. As a result of this, new processor versions are developed, and when the changes in quality are considered important, the SMOS team prepares for a new reprocessing. Currently, SMOS is preparing the third mission reprocessing with the L1OP v724. The Methods section provides an overview of the changes involved in the new version with respect to the v620 operational version used in the second mission reprocessing. The Results section assesses the end-to-end improvements of the data.

2. Methods

In this section, we present the improvements that were applied to the v620 processing baseline to form the new v724 processing baseline. A high-level overview of the SMOS Level 1 processing baseline is presented in Figure 1.

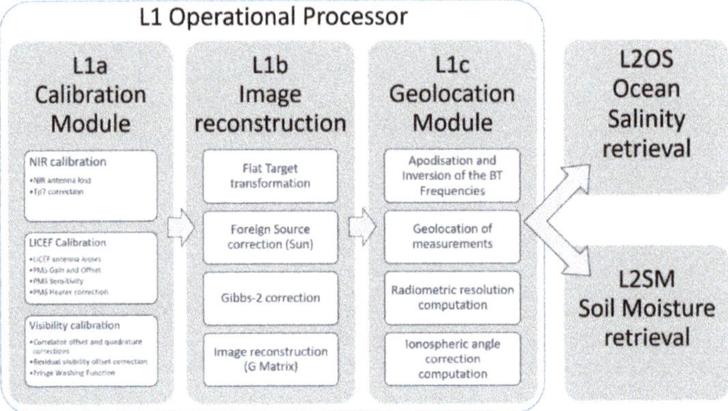

Figure 1. High-level architecture of the SMOS Operational processing baseline.

The improvements in the new L1OP were applied to the calibration process and the image reconstruction processing. The improvements for these two categories are presented below in separate sub-sections.

2.1. Changes in Calibration

Calibration is a process where raw MIRAS data, including, e.g., cross-correlations between all the pairs formed by the 72 MIRAS receivers, counts of the receivers' total power detectors, and noise injection radiometer pulse lengths, are turned into radiometric observables like antenna temperatures and power levels. The calibration of the MIRAS instrument as a whole is a complex process including several steps. An overview of the calibration process can be found, e.g., in [5].

For the v724 processing baseline, five main improvements were done in this calibration process. They are improvements related to the following:

-   Noise injection radiometer (NIR) calibration strategy.
-   NIR antenna losses.
-   Power measurement system (PMS) sensitivity factors.
-   PMS heater correction.

-    Thermal latency of the temperature sensor in NIR antennas.

In the following sub-sections, we describe these updates in detail.

2.1.1. NIR Calibration

Analysis of the second mission reprocessing led to the following conclusion: the calibration of the NIR parameters, the noise injection temperature (Tna) and the level of the noise injection (Tnr) [6] were introducing a bias in the stability of the measurements. This was evident when looking at the bias of the measurements over a large portion of the Pacific open ocean with respect to the ocean forward model [7]. The comparison of those biases showed a large negative correlation with the variation in the main NIR calibration parameter, Tna. Figure 2 shows such a comparison for X polarisation measurements in ascending orbits.

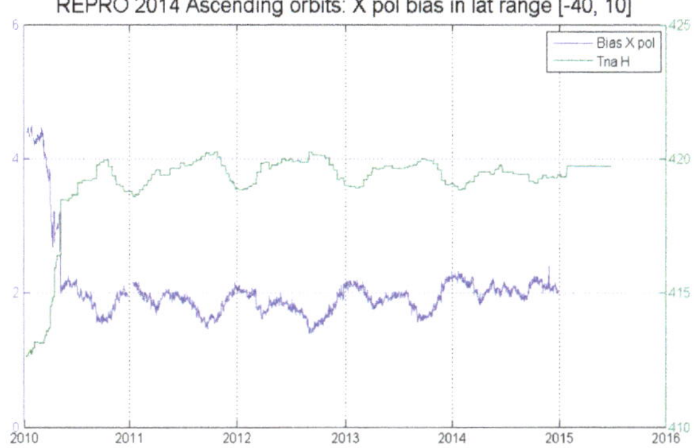

Figure 2. Bias of the Xpol SMOS measurements for ascending orbits over the Pacific Ocean (blue) for the period 2010–2015 and the NIR calibration parameter Tna (green).

A computation of the correlation factor between the two variables is provided in Table 1.

Table 1. Correlation factor between brightness temperature bias as computed over the ocean and NIR calibration parameter Tna.

Polarization	Ascending	Descending
X polarisation	−0.97	−0.88
Y polarisation	−0.96	−0.74

This high (negative) correlation factor between bias and the NIR calibration parameter suggests that the NIR variations present in NIR calibration parameter Tna are not real, but artefacts established by some non-ideality in the instrument model, and, further, that the NIR unit reference temperature Tna is extremely stable.

NIR calibration is performed during external manoeuvres, during which the instrument points upwards to the cold sky [5]. During this process, the temperature of the NIR antennas' patches gets colder and outside the nominal temperature range of the instrument. Clearly, the current NIR instrument model, and especially its thermal parametrization, is not able to account for such circumstances. This realisation introduced two main changes to the SMOS calibration. On one hand, starting in 2014, SMOS NIR calibration manoeuvre has been done keeping the Sun at approximately 10 degrees above the antenna plane to avoid getting in a thermal range different from the one during science measurements. On the other hand, the NIR parameters Tna and Tnr were set to a fixed value for the third mission reprocessing. These changes have improved the stability of the measurements.

2.1.2. Antenna Losses

In SMOS, the NIR antenna losses are divided between the antenna patch (L_1) and the feeding circuits in the innermost part of the antenna (L_2) as shown in Figure 3. They are at different physical temperatures. The innermost part of the antenna (T_{p6}) is within the thermal control, whereas the antenna patch is more exposed to the temperature fluctuations of outer space (T_{p7}).

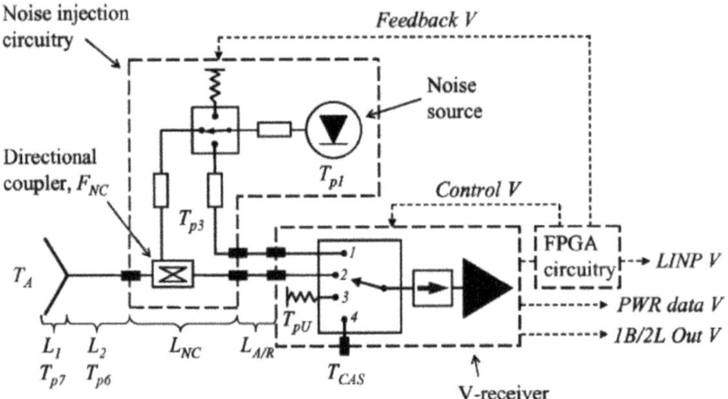

Figure 3. Schematic diagram of the structure of the V-channel of the NIR. L stands for loss, Tp for physical temperature, TA for antenna temperature and Tcas for the Calibration Subsystem (CAS) noise temperature [6].

NIR antenna patch losses have been the most challenging problem in SMOS calibration as both L_1 and L_2 are outside the radiometer's internal calibration loop. A wrong characterisation of the antenna losses introduces variations in the measurements. These variations are related to the variations in the physical temperature of the antenna.

The basic equation for the antenna temperature retrieval at NIR is

$$T_A = -L_1 L_2 T_{NA} \eta + L_1 L_2 [L_{NC} L_A L_{DA} (T_U - T_{t2}) - T_{t1}], \tag{1}$$

where T_A is the antenna temperature as measured by the NIR, L_1 is the antenna patch loss, L_2 is the intermediate layer antenna loss, η is the NIR pulse length, L_{nc}, L_A and L_{DA} are losses of different sections of the cables connecting the antenna to the receiver, T_U is the load noise temperature and T_{t1} and T_{t2} are

$$T_{t1} = \frac{L_1 - 1}{L_1 L_2} T_{p7} + \frac{L_2 - 1}{L_2} T_{p6}, \tag{2}$$

$$T_{t2} = \frac{(L_{NC} - 1)}{L_{NC} L_A L_{DA}} T_{p3} + \frac{(L_A - 1)}{L_A L_{DA}} T_{Cab} + \frac{(L_{DA} - 1)}{L_{DA}} T_{pU}, \tag{3}$$

and T_{NA} is the value measured during calibration, and corresponds to

$$T_{NA} = \frac{-T_{A,cal} + L_1 L_2 \left[L_{NC} L_A L_{DA} \left(T_{U,cal} - T_{t2,cal} \right) - T_{t1,cal} \right]}{\eta L_1 L_2}, \tag{4}$$

where "X,cal" indicates the value of parameter "X" obtained during the calibration against the cold sky.

Now, if we analyse the equation as a function of the L_1 uncertainty using error propagation, we get

$$\frac{\Delta T_A}{\Delta L_1} = \frac{\partial T_A}{\partial L_1} + \frac{\partial T_A}{\partial T_{t1}} \frac{\partial T_{t1}}{\partial L_1} + \frac{\partial T_A}{\partial T_{NA}} \frac{\partial T_{NA}}{\partial L_1}, \tag{5}$$

and finally

$$\Delta T_A = \frac{\Delta L_1}{L_1}\left[\frac{L_1 L_2[TL_D - T_{t1}] - T_A}{L_1 L_2\left[TL_{D,cal} - T_{t1,cal}\right] - T_{A,cal}}\left(T_{p7,cal} - T_{A,cal}\right) - \left(T_{p7} - T_A\right)\right], \tag{6}$$

where

$$TL_D = L_{NC}L_A L_{DA}(T_U - T_{t2}), \tag{7}$$

This equation, as given, is difficult to interpret, but by making some realistic numerical simulations, we realised that in a scenario where the calibration was obtained at a T_{p7} of 295 K, errors in the L_1 antenna patch loss would propagate to T_A at a different rate depending on the T_{p7} during measurement. Figure 4 shows how an error in the antenna losses characterisation will introduce an error in the antenna temperature that will be a function of the temperature of the antenna patch (T_{p7}).

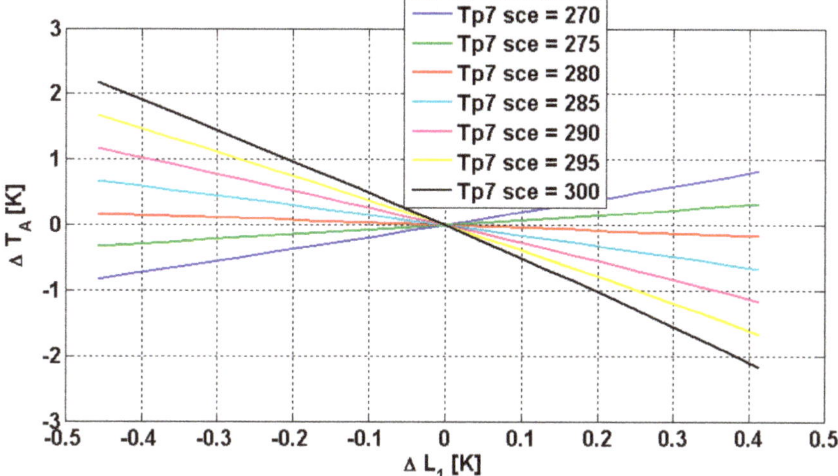

Figure 4. Expected error in the antenna temperature as a function of the error in the antenna losses (L_1), for simulated scenarios with different temperatures, when the calibration of the NIR was obtained at a temperature of $T_{p7} = 295$ K.

Therefore, errors in the antenna loss characterisation should be correlated with the variations of the antenna physical temperature, which is exactly what has been observed in SMOS.

Initially, just after launch, the on-ground characterisation values for L_1 and L_2 were used. Later, during SMOS's first mission reprocessing, an antenna thermal model was introduced to correct for variations observed during NIR external calibration manoeuvres. However, the antenna thermal model was quickly abandoned, as the instrument became more stable following the initial months in orbit. For the second mission reprocessing, the team derived a method to calibrate the antenna losses in orbit [8]. Antenna losses were measured every 15 days, and, since the values were stable, the average value was used for the entire reprocessing. This correction was key to improve the stability of the data in the second mission reprocessing. The calibration procedure could only measure the antenna losses for whole of the antenna patch and the inner part of the antenna (L_1 and L_2 losses respectively). But the antenna patch and the innermost part of the antenna in SMOS suffer different temperature excursions. Introducing the correct split in the total antenna loss between L_1 and L_2 is key for obtaining good instrument stability. This split was obtained by assessing the brightness temperature variations over the ocean against an ocean forward model for Stokes-1 measurements and applying the same antenna loss value at the H and V polarisations.

In the third mission reprocessing, it became evident that a different split was necessary for H and V polarisation, as the antenna has different patch for each polarisation. The exercise was then repeated for each of the two polarisations [9]. Figure 5 shows the variations in the brightness temperature biases over ocean as a function of the physical temperature differences between the antenna patch and the innermost part of the antenna, when the L_1 antenna loss has been set to 0 dB. The plots show a clear slope in the data, which can account for the antenna losses. NIR-CA H pol L_1 antenna loss was set to 0.27 dB, and V polarisation L_1 was set to 0.14 dB. L_2 values were set to the difference between the total loss as measured by calibration and the corresponding L_1 values (L_2 equals 0.19 for H and 0.30 for V polarisation for NIR-CA; the other two NIR units are not used to derive the antenna temperature, but their values can be found in [9]).

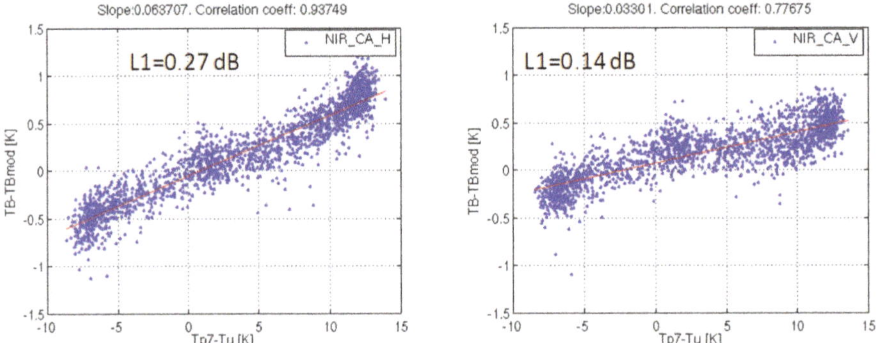

Figure 5. Variations of brightness temperature biases over the ocean as a function of temperature variations between the antenna patch and the innermost part of the antenna, for H polarisation (**left**) and V polarisation (**right**).

2.1.3. PMS Sensitivities

The sensitivity of the power measurement system (PMS) gain to physical temperature variations was first characterised by the receiver supplier and later verified on-ground at instrument level during the test in thermal vacuum conditions at the Large Space Simulator (LSS) at ESTEC [10], and then again during special calibration events in the SMOS commissioning phase [11]. The latter values have been used until now for adjusting the temperature sensitivity. However, a recent analysis of the variations of the PMS gain through the years showed that the pre-launch PMS sensitivities provide for a more natural behaviour of the PMS gain's aging with time. Figure 6 shows PMS behaviour for receiver LICEF C10 (LICEF stands for lightweight cost-effective front end). As it is seen, using the pre-launch sensitivities provides the best cancellation of the PMS gain oscillations due to physical temperature swings. Similar results are seen in other receivers. For the third mission reprocessing, PMS sensitivities' pre-launch values were used again.

2.1.4. PMS Heater Correction

A known problem in the SMOS instrument, which was detected during the thermal tests at the LSS chamber, was PMS offsets jumps following the instrument heater switching from on to off and vice versa. A correction was introduced early in the mission to mitigate this effect, which consisted of a delayed voltage offset with respect to the heater status transition, but the problems were still noticeable, particularly for a few receivers [11].

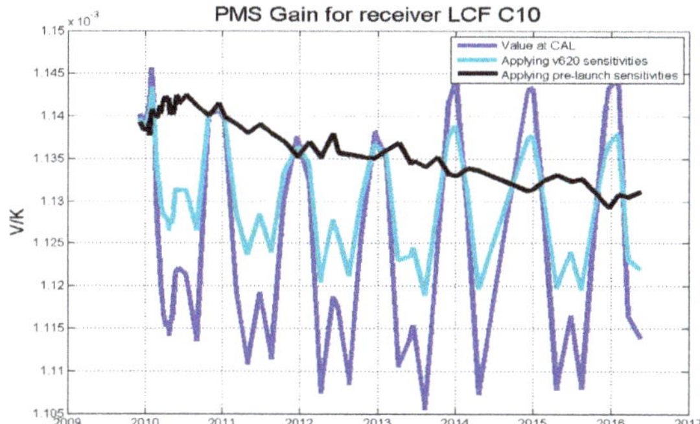

Figure 6. SMOS PMS gain for receiver LCF_C10 over the years. The blue line indicates the value as measured during the calibration event at the physical temperature during the calibration event. The cyan line and black line show the PMS gain transported to 21 degrees Celsius using the second mission reprocessing and pre-launch PMS sensitivities, respectively.

A more careful analysis showed that the jumps related to the heater status do not correspond to a simple delayed offset, but that the behaviour follows a double exponential [9]. Figure 7 shows the PMS voltages for a calibration event, where a constant noise from an internal warm load source was introduced at the receivers.

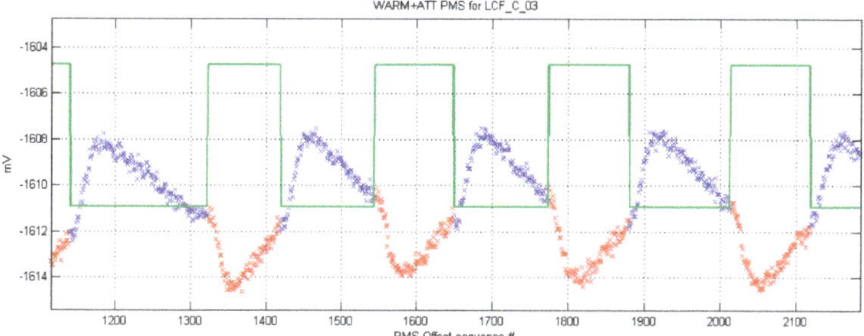

Figure 7. PMS voltages for LICEF C-03 when a constant noise source is introduced. In green, the status of the heater is depicted. Red crosses for the PMS voltage indicate that the heater is on and blue crosses show when the heater is off.

Based on this analysis, the correction applied, ΔV, was set to

$$\Delta V_{ON} = \alpha_{ONi}(V_i - V_{max})\left(1 - e^{\frac{-t}{\tau_{ON\_1i}}}\right) + \beta_{ONi}(V_i - V_{max})\left(1 - e^{\frac{-t}{\tau_{ON\_2i}}}\right), \tag{8}$$

$$\Delta V_{OFF} = \alpha_{OFFi}(V_i - V_{max})e^{\frac{-t}{\tau_{OFF\_1i}}} + \beta_{OFFi}(V_i - V_{max})e^{\frac{-t}{\tau_{OFF\_2i}}}, \tag{9}$$

where V_i is the current PMS value without correction for each of the 72 receivers, in volts. V_{max} is the maximum measurable PMS, set to a value of 2.5V. α, β, τ_{ON} and τ_{OFF} are fixed constants for each receiver that empirically determine the double exponential behaviour, and t is the number of epochs since the corresponding transition of the heater status (on to off, or vice versa).

The validation of this correction was performed by means of a relative comparison of the antenna temperature of one receiver to the average of all receivers. Figure 8 shows the behaviour for L1OP v620 (delay heater correction) and for L1OP v724 (double exponential heater correction). The new correction clearly reduces the obvious PMS offset jumps due to the heater status.

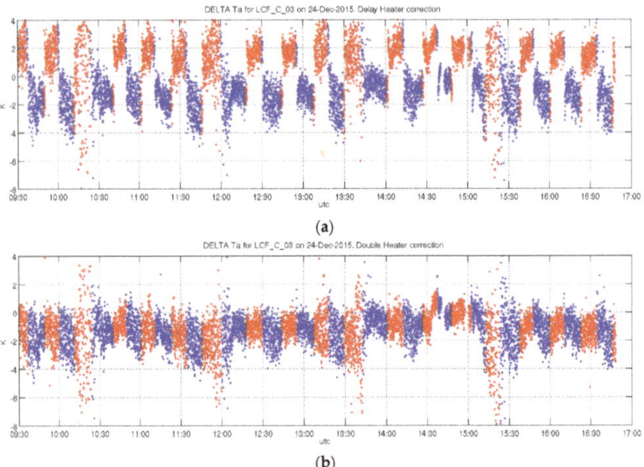

Figure 8. Difference between the antenna temperature measurement of receiver LCF_C_03, with respect to the average of all receivers, when applying the v620 delayed offset heater correction (**a**) and when applying the new v724 double exponential heater correction (**b**). The colour of the points in both plots indicates the status of the heater: red when the heater is on and blue when it is off.

2.1.5. Antenna Patch Thermistor Correction

Another aspect that was discovered during the second mission reprocessing analysis was an increased bias immediately following the Sun's eclipse by the Earth, relative to the instrument. This effect clearly pointed to another problem related to temperature variations. While the instrument backend is kept under thermal control [McMullan et al., 2008], the antenna patches suffer large thermal excursions. Those changes are monitored by three thermistors placed at the screw of each NIR antenna patch (T_{p7} in Figure 3) and are used in the NIR radiometric equation presented in Section 2.1.2.

The team considered that the reading of the thermistor did not properly describe the temperature of the antenna patch and proposed a correction [12]. The correction was introduced based on the observed thermal latency during inertial external manoeuvres. During these manoeuvres, the instrument points at the cold sky for several minutes. However, T_{p7} thermistor readings take a long time to stabilise to a constant temperature. The thermistor reading was considered to be thermally coupled to the innermost part of the antenna through the antenna screw, inside whose head the thermistor is mounted. As such, the thermistor reading is not fully representative of the antenna patch region. The team decided then to apply a correction to the thermistor reading by assuming that the temperature to which the thermistor stabilises at the end of the external manoeuvre is the actual temperature during the entire inertial manoeuvre. The following correction was derived:

$$\hat{T_{p7}} = T_{p7} - \frac{1}{LP}\frac{dT_{p7}}{dt},$$ (10)

where $\hat{T_{p7}}$ is the corrected thermistor temperature, T_{p7} the actual thermistor reading, and LP a constant that was estimated to be −0.0031.

The correction was then used to process the ocean brightness temperature, and the bias with respect to the forward model was re-assessed. The impact of this correction is a clear mitigation of the bias observed during the eclipse period. Figure 9 shows the Y polarisation brightness temperature bias observed over the ocean with and without the T_{p7} correction applied. The increased bias in the eclipse is observed around 35N to 60N degrees in latitude during the Northern Hemisphere (NH) winter months in the left plot.

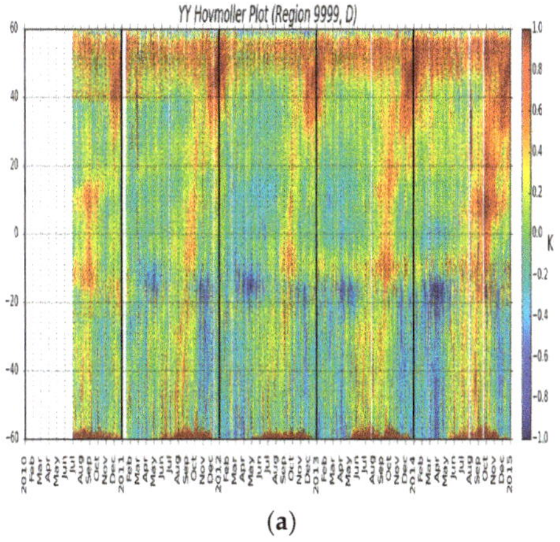

(**a**)

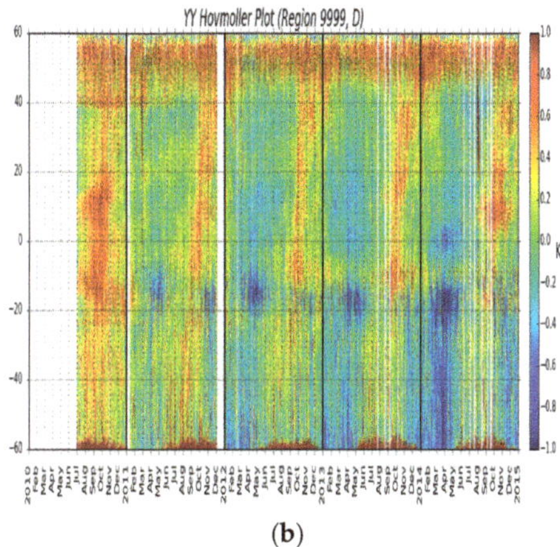

(**b**)

Figure 9. Hovmoller plot showing Y pol BT bias over the ocean as a function of time and latitude with (**a**) and without (**b**) the T_{p7} correction.

2.2. Changes in Image Reconstruction

Image reconstruction is a process where the calibrated MIRAS data are turned into radiometric maps that can be projected on the Earth's surface. For the v724 processing baseline, three main improvements were made to this process. They are improvements related to

- Gibbs phenomena correction.
- The use of a sea-ice mask.
- Correction of the Sun's influence in the images.

In the following sub-sections, we describe these updates in detail.

2.2.1. Gibbs-2 Algorithm

The so-called Gibbs-2 algorithm is an evolution of the Gibbs-1 algorithm applied to SMOS measurements since its launch. Originally, the Gibbs correction aimed to reduce the Gibbs artefacts that appeared in the image following large BT transitions between land and ocean (or sky and Earth) due to limited coverage in the visibility domain. Soon after, the team realised that the correction not only reduces the Gibbs artefacts but also a floor error induced in the retrieved images due to dissimilarities in the antenna patterns and the aliasing [13,14]. Gibbs-1 correction reduces this so-called floor error by removing a constant brightness temperature (BT) in the reconstruction process, which reduces the visibility values before inversion, and adding it back at the end of the inversion process. In Gibbs-2, the process has evolved to include the use of an artificial scene as close as possible to the observed one. This artificial scene, V_a, uses a Fresnel model over the ocean and a constant value (250 K) over land. Figure 10 shows an example of the artificial scene as used in the image reconstruction processor. The visibilities of this artificial scene are computed using an SMOS instrument model:

$$V_a = GT_a, \tag{11}$$

where T_a represents the modelled BT of the artificial scene, G is the instrument model, and V_a are the visibilities derived from the T_a scene.

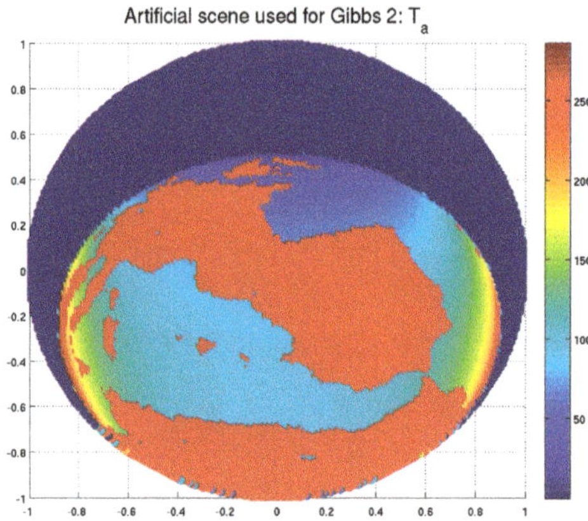

Figure 10. Artificial scene of one SMOS observation of the Iberian peninsula and northern Africa.

Then, the image reconstruction algorithm is applied over the following differential linear problem:

$$V - V_a = G(T - T_a), \tag{12}$$

yielding to the following retrieved BT:

$$T_r = U^*ZJ^+(V - V_a) + T_a, \tag{13}$$

where U is the Fourier transform operator, Z is the zero-padding operator beyond the SMOS frequency coverage, $J = GU^*Z$ is the image reconstruction operator used in the SMOS processor and J^+ is the pseudo-inverse of J [15].

2.2.2. Sea-Ice Mask

The calculation of the artificial scene used in the Gibbs-2 algorithm described above is based on the use of a fixed global land–ocean mask. In fact, we identify the land and ocean pixels within the field of view and assign a constant value over land and the Fresnel forward model over ocean. To improve the accuracy of the artificial scene in seasonal sea-ice growth, we have developed an operational strategy to measure the sea-ice extension from the actual SMOS measurements and apply this extension to the artificial scene. Measurements are collected for 10 days, then a mask of the percentage of sea ice over the ocean is derived and this mask is used in the Gibbs-2 algorithm with a constant value of 250 K (same as for land pixels), as can be seen in Figure 11. In the third mission reprocessing data, the mask computation will be aligned with the data that is applied. However, in nominal operations, the mask will be applied, typically with a 12-day delay from the moment it first started estimating the extension. Errors derived from this 12-day delay were analysed and resulted to be much lower than those present when not applying the correction at all.

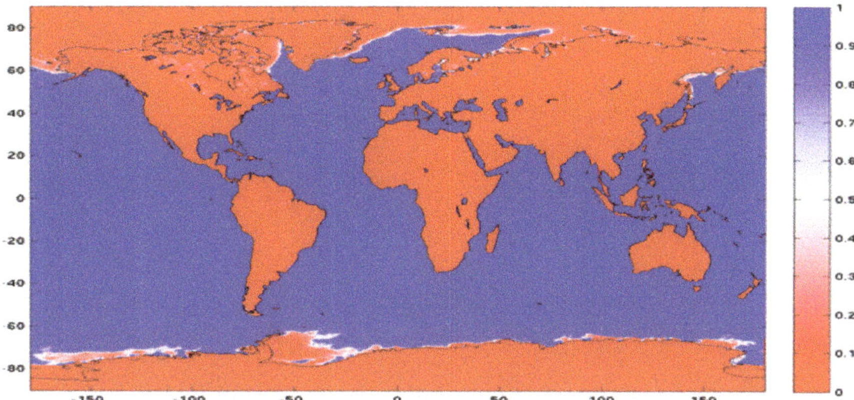

Figure 11. Mask showing the extensions of land/ice pixels versus ocean pixels, as derived from January 2010 SMOS BT measurements.

Figure 11 shows that the sea ice detected by SMOS BT measurements goes beyond the continental surfaces.

2.2.3. Super-Sampled Sun Correction

L-band observations of the Sun disk showed that its BT emissivity is not spatially homogeneous [16]. Sunspots tend to have much larger BT emissions than other Sun regions. The Sun correction applied to SMOS during the second mission reprocessing considered the Sun as a point source. The team considered this correction to be insufficient and derived a new method to correct for the Sun BT emissions to take into account those spatial inhomogeneities Figure 12.

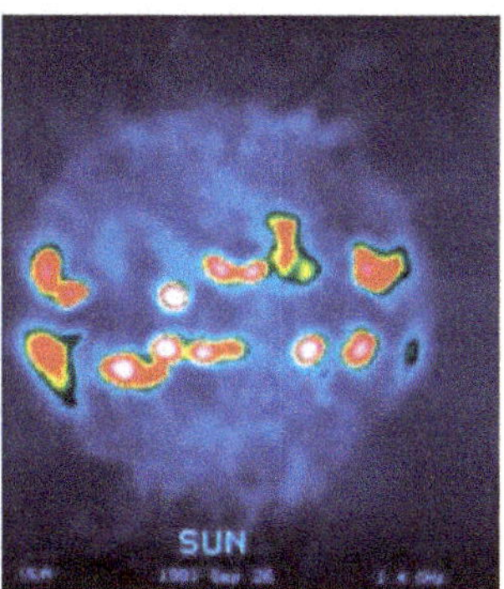

Figure 12. L-band BT emission of the Sun [16].

The so-called super-sampled Sun correction applied in L1OP v724 estimates the BT of multiple spots within the Sun disk, minimising the differences between the simulated signal and the BT observations in an area around the Sun and in the Sun tails [17].

$$V_{Sun} = \sum_i V_{1ki} T_i,\tag{14}$$

where V_{Sun} are the visibilities of the Sun as a function of the 1K visibilities (V_{1ki}) computed for each of the sub-sample points' times and T_i corresponds to the BT estimated for each of those points from the following equation:

$$
\begin{bmatrix}
F^{-1}(V-V_{G2})_1 - \langle F^{-1}(V-V_{G2})_0 \rangle \\
F^{-1}(V-V_{G2})_j - \langle F^{-1}(V-V_{G2})_0 \rangle \\
F^{-1}(V-V_{G2})_{n^*} - \langle F^{-1}(V-V_{G2})_0 \rangle \\
\cdots \\
\frac{T_0}{wnPOS} \\
\cdots \\
\frac{T_0}{w}
\end{bmatrix}
=
\begin{bmatrix}
F^{-1}(V_{1K1})_1 & F^{-1}(V_{1Ki})_1 & F^{-1}(V_{1KnPOS})_1 \\
F^{-1}(V_{1K1})_{n^*} & F^{-1}(V_{1Ki})_{n^*} & F^{-1}(V_{1KnPOS})_{n^*} \\
1/w & \cdots & 0 \\
\cdots & 1/w & \cdots \\
0 & 0 & 1/w \\
1/w & \cdots & 1/w
\end{bmatrix}
[T_i]
\tag{15}
$$

where:

- $F^{-1}(V-V_{G2})_1$ is the inverse Fourier transform of the difference between the measured visibilities and the Gibbs-2 synthetic visibilities;
- $\langle F^{-1}(V-V_{G2})_0 \rangle$ represents the average value over the clean 'o' pixels surrounding the sun disc and is calculated as the average of the inverse Fourier transform of the difference between the measured visibilities and the Gibbs-2 synthetic visibilities.
- T_0 is the first estimate of the Sun temperature, obtained assuming the Sun as a point source.
- w is a weight to be finely tuned to get the best compromise in condition number versus sensitivity (for now fixed at 10^6).
- $nPOS$ is the number of over-sampled points, fixed to 37 in the L1OP.
- 1,..,j,..,n is the number of points polluted by the solar radiation, including the disc of the sun and the tails.
- $V_{1K1}, \ldots, V_{1KnPOS}$ are the system response functions calculated over the over-sampled grid.
- $F^{-1}(V_{1Ki})_{n^*}$ is the inverse Fourier transform of V_{1K1} calculated over the polluted pixel.
- T_i is the set of estimated temperatures for the 37 points of the sun disc. It has been shown in [17] that the solution to this problem is explicit.

This correction succeeds in better reducing the residuals of solar radiations, as shown in Figure 13.

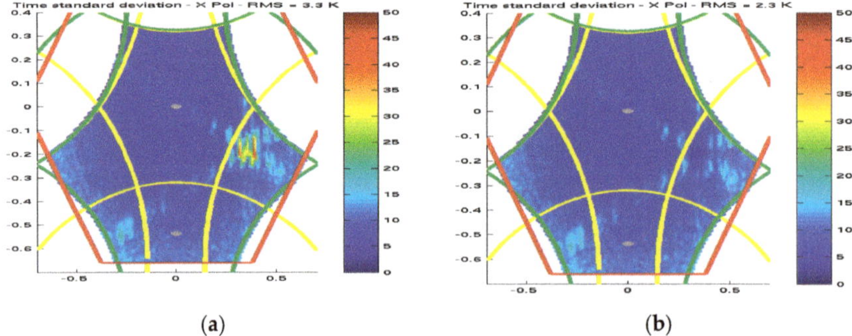

(a)   (b)

Figure 13. Time standard deviation of 100 snapshots over the ocean, with the Sun alias present at approximately [0.4, −0.15], with the old Sun correction (**a**) and the super-sampled correction (**b**).

Figure 13 shows that the new super-sampled Sun correction reduces the variations in the measurements along the Sun tails and within the Sun alias disk.

2.2.4. Sun Correction in the Back

SMOS measurements showed that the radiation coming from the Sun is observed even in the case the Sun is behind the antenna plane, through the antenna back-lobes [18].

The team considered that it was important to correct for this foreign source radiation and applied the Sun correction algorithm described in [19] and later modified in [20], even in the case the Sun was behind the antenna. Figure 14 shows the bias observed in the data before and after the extension of the Sun correction in the back.

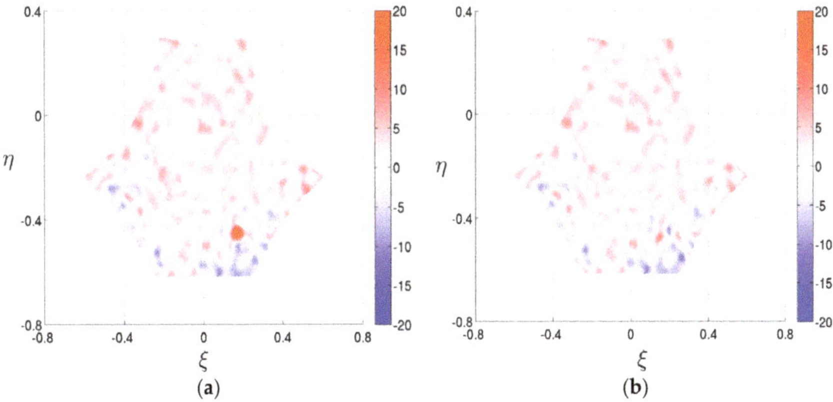

Figure 14. SMOS biases from a measurement in the Pacific after removing the expected forward model when the Sun is in the back of the instrument before (**a**) and after (**b**) the Sun BT correction in the back [18].

3. Results and Discussion

The changes applied in the v724 version of the processor have been analysed using a dedicated end-to-end processing campaign that involved 3 years of data. The results presented hereafter show the improvement of the data quality in different metrics, such as reduction of the spatial biases, improvement of stability, reduction of land–sea contamination biases for certain polarisations, better match to in-situ measurements and reduction of the χ^2 in the soil moisture retrievals.

3.1. Orbital and Seasonal Stability

The quality of the stability of the measurements is established by comparison with the ocean forward model. In this case, one of the metrics used by the SMOS team is the one provided by the Hovmoller plots showing the biases observed in the Pacific open ocean in time and latitude. This metric allows us to assess both the orbital stability (variation along the vertical axis) and the seasonal stability (variation along the horizontal axis). The analysis is done independently per polarisation and separately for ascending and descending passes.

Figure 15 shows an example of the stability of the measurements in the second mission reprocessing [21] and the expected behaviour for the third mission reprocessing, for both X and Y polarisations. Descending orbits suffer the most from larger instrument thermal dynamics and are always more prone to measurement instabilities.

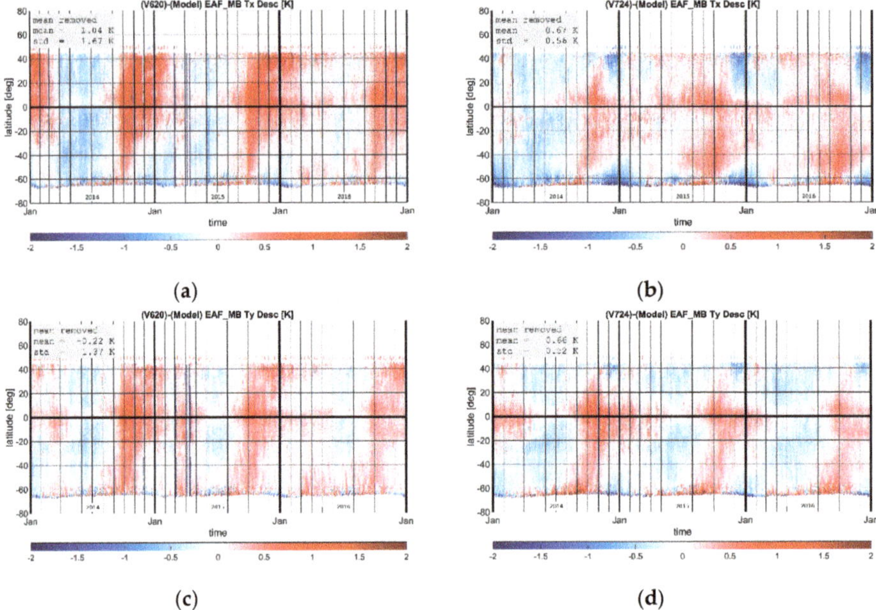

(a)

(b)

(c)

(d)

Figure 15. Hovmoller plots showing the bias between measurement and model averaged over the entire extended alias-free field of view for descending orbits for (**a**) v620 X polarization, (**b**) v724 X polarization, (**c**) v620 Y polarization and (**d**) v724 Y polarization.

Figure 15 shows that both the orbital and seasonal instabilities have been reduced in the third mission reprocessing. A further metric to assess the improvement in stability is the standard deviation of these Hovmoller plots. The values in Table 2 show an important reduction of the instabilities.

Table 2. Mean bias for the v620 data (second mission reprocessing) and the 724 data (third mission reprocessing), computed in the Pacific region between latitudes 45S and 10N.

Orbit Pass	Polarization	Second Mission Reprocessing	Third Mission Reprocessing
Ascending	X	1.73 K	0.57 K
	Y	1.41 K	0.35 K
Descending	X	1.67 K	0.56 K
	Y	1.37 K	0.52 K

3.2. Spatial Biases

Spatial biases have been one of the largest challenges in the SMOS image reconstruction process [21, 22]. The Level 2 Ocean Salinity Processor uses the ocean target transformation (OTT) technique to reduce them [23,24], but this technique only works well in scenes whose brightness temperature is roughly stable, such as measurements in the open ocean. Near the coastlines, or for any measurement over land, the technique does not work properly. Therefore, it is of utmost importance that the spatial biases are minimised at the image reconstruction level. The changes introduced in the v724 processor have considerably reduced the spatial biases in the extended alias-free field of view.

Figure 16 shows the spatial biases for X and Y polarisation. A very important aspect to note in the spatial bias improvement in Y polarisation is the reduction of a negative gradient from top to bottom of the OTT image.

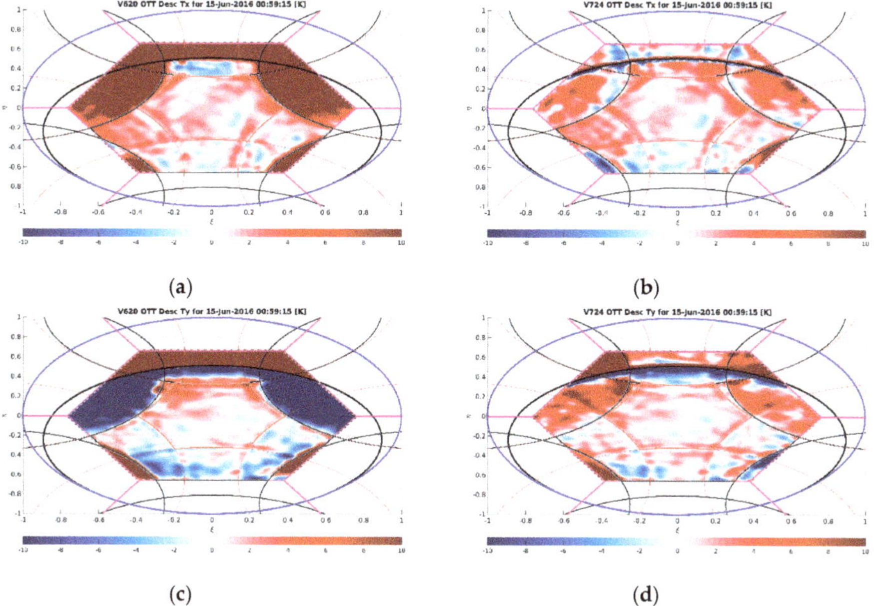

Figure 16. Spatial biases for descending orbits in (**a**) v620 X polarization, (**b**) v724 X polarization, (**c**) v620 Y polarization and (**d**) v724 Y polarization.

3.3. Land–Sea Contamination

Another critical aspect of the SMOS radiometric measurements is the land–sea contamination. This refers to the increase in the bias that occurs in the ocean measurements near land masses and vice versa. The adjustments in calibration and the new Gibbs-2 image reconstruction technique have achieved a significant improvement in land–sea contamination, even though this is still substantially present in the third mission reprocessing. Figure 17 shows the biases, over ocean, in the global maps for one particular month of data (June 2016) for the four polarisations for the second [21] and third mission reprocessing. The improvements are most noticeable in Y polarisation and in the fourth Stokes parameter. On the other hand, the contamination in Tx has changed but remains at similar levels, and similarly for the third Stokes parameter.

In must be noted that the Level 2 Ocean Salinity Processor includes an empirical correction of the land–sea contamination. Being able to reduce the original bias is an important aspect of the L_1 processor, but almost more important is that the residual bias remains constant, which can then be corrected empirically at Level 2. For this reason, another metric assesses the variation of the land–sea contamination bias at Level 1 by means of the standard deviation. Figure 17 shows this metric, over ocean, for Y polarisations only. Similarly to Figure 17, the land–sea contamination variation is substantially reduced, mainly for Y polarisation (Figure 18) and for the fourth Stokes parameter.

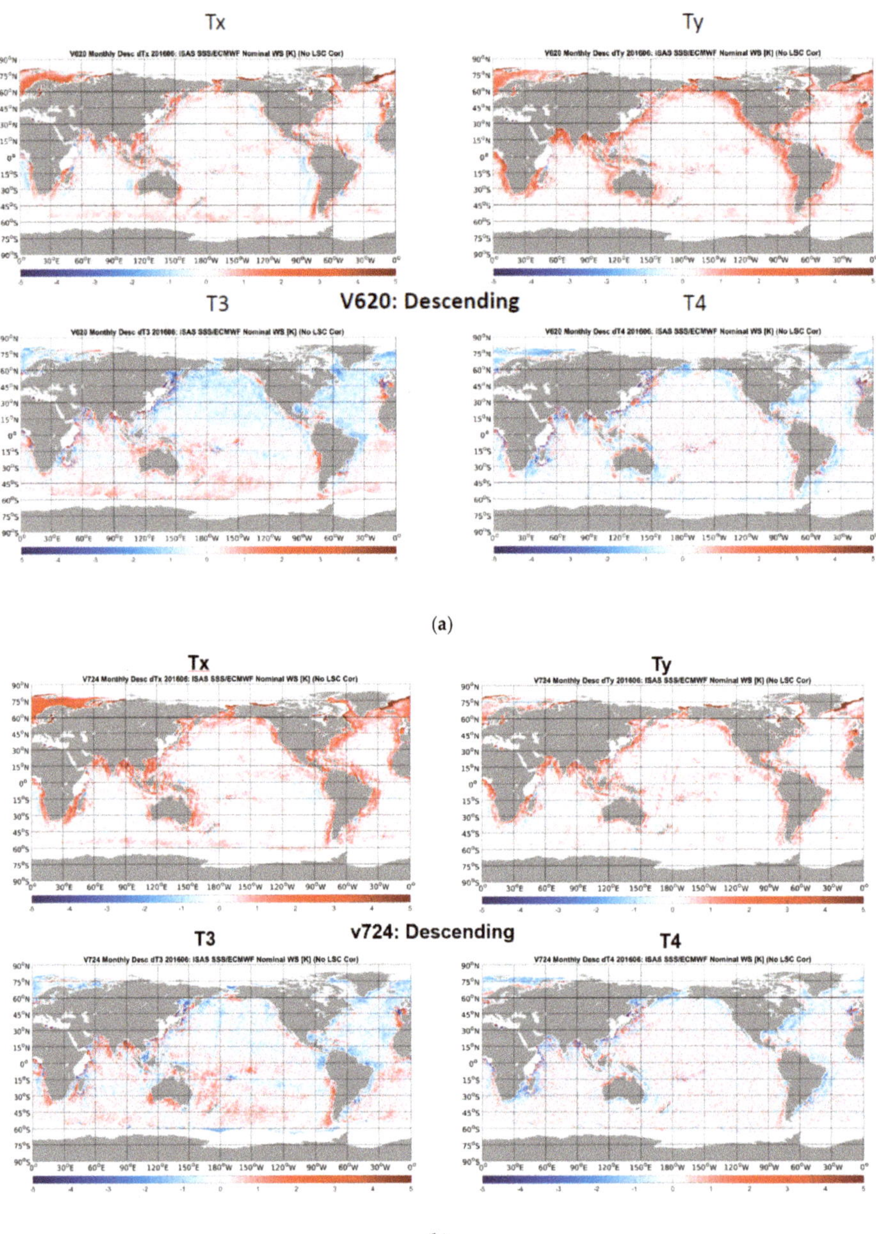

Figure 17. Maps of bias between SMOS measurements and the ocean forward model, showing an increase excess of bias in regions near the coast, known as land–sea contamination. (**a**) all four polarizations for v620 and (**b**) all four polarisations in v724.

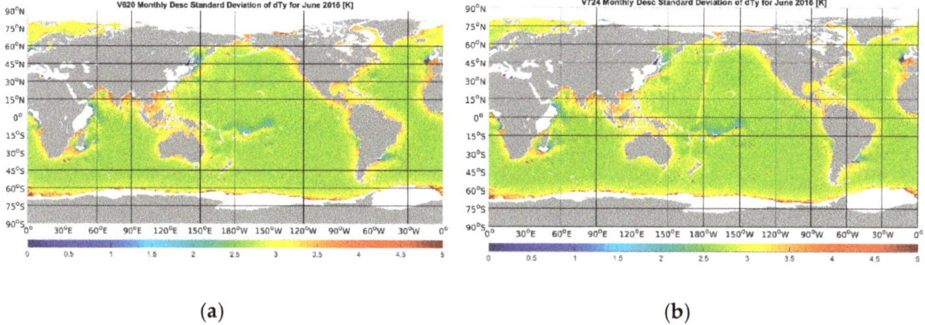

(a) **(b)**

Figure 18. Standard deviation of the SMOS measurements in June 2016 for v620 (**a**) and v724 (**b**).

3.4. Impact on Retrieved Soil Moisture and Vegetation Optical Depth (VOD)

As part of our standard metric protocol, new versions of L_1 data are systematically processed with the Level 2 soil moisture processor to assess the changes compared to the previous L_1 processor version. This assessment is made through spatial monthly maps showing the changes on retrieved soil moisture and retrieved opacity along with χ^2 changes. A second perspective is obtained through time series of retrieved soil moisture corresponding to a collection of in-situ time series of measured soil moisture for the two-year period 2011–2012 and provides quantitative metrics but for a limited number of grid points.

For the purposes of this analysis, the same Level 2 Soil Moisture v650 processor has been used in order to assess only the improvements in the L_1 processor.

3.4.1. Spatial Maps of Retrieved Soil Moisture and Opacity

Figure 19 displays the differences in soil moisture and opacity of v724 minus v620 for the month of June 2014, separated by ascending and descending orbit passes. The overall global change is rather neutral, with mean differences close to 0 but with a significant variability that appears very structured spatially. The significant changes correspond to specific areas, with contrasts between transition areas and forest in both retrieved soil moisture and opacity. Below dense forest v724, soil moisture and opacity tend to decrease, with patterns changing in position between ascending and descending orbits, e.g., the North American east coast, Amazonian forest, and African Congo forest. This is probably a signature of the Gibbs-2 correction, as the contrast of land/sea masses is not similar within the SMOS field of view for these locations depending on the orbit pass.

The L1C v724 data generate significant changes compared to the L1C v720, and the question whether those changes are in the right direction is addressed by the two following sections.

3.4.2. χ^2 Test

χ^2 is an important metric to assess the quality of the soil moisture retrievals and is used widely in many retrieval processes. It provides a measure of the agreement (best fit) between the geophysical modelling that resulted in the retrieved parameters and the L_1 data that were used accounting for the expected noise on the observed data. In this study, we considered rather the reduced χ_r^2 form, which is χ^2 divided by the number of degrees of freedom. Using χ_r^2 introduces a normalisation, which is preferable as the Level 2 processor includes L_1 data filtering that may result in slightly different numbers of degrees of freedom between the two L_1 datasets.

Figure 20 shows the changes of the χ_r^2 between v620 and v724, computed as the ratio χ_r^2 v724 divided by χ_r^2 v620 for June 2014 for ascending and descending orbits. Very similar maps are observed at other months of the year. Ratios >> 1 (toward red colours) indicate degraded (increased) v724 χ_r^2

with respect to v620 and ratios $\ll 1$ (toward blue colours) indicate improved (decreased) v724 χ_r^2 with respect to v620.

The team analysed the changes in χ_r^2 from v620 to v724 over land and concluded that χ_r^2 has improved (reduced) significantly over most of the globe. Most exceptions are either neutral (light blueish/reddish area) or are related to presence of radio frequency interference (RFI), especially in the Middle East region and South Asia. All maps report a significant improvement (decrease) in v724 χ_r^2 compared to v620 at global scale with distribution ratio modes marker ‣ always below 1 and strong negative asymmetry. Many continental areas show a deep blue colour, which indicates very significant improvements that also appear to be very stable in time for different seasons and different years. Similar to Figure 19, Figure 20 patterns show some differences between ascending and descending orbits. It is important to notice the good match of these blue spatial patterns in Figure 20 with the most significant change patterns in retrieved soil moisture and opacity reported in Figure 19; where v724 introduced the strongest changes in retrieved parameters is also where the best fit has improved the most with reduced χ_r^2. Finally, using the v724 data increases the number of successful retrievals by 2% to 3%.

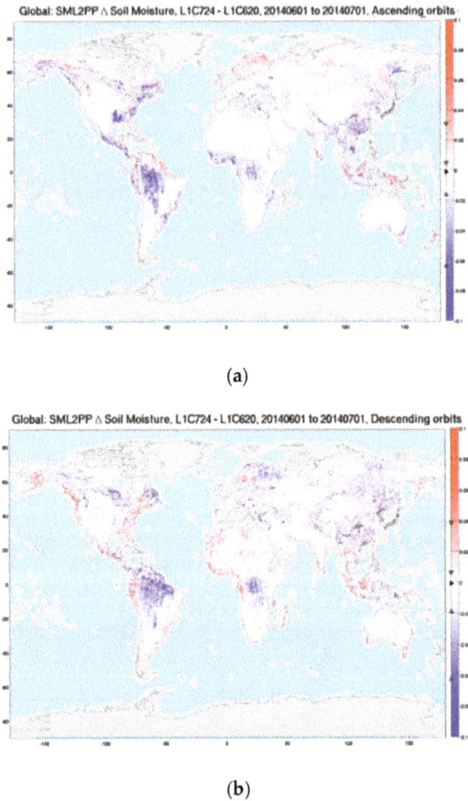

(a)

(b)

Figure 19. *Cont.*

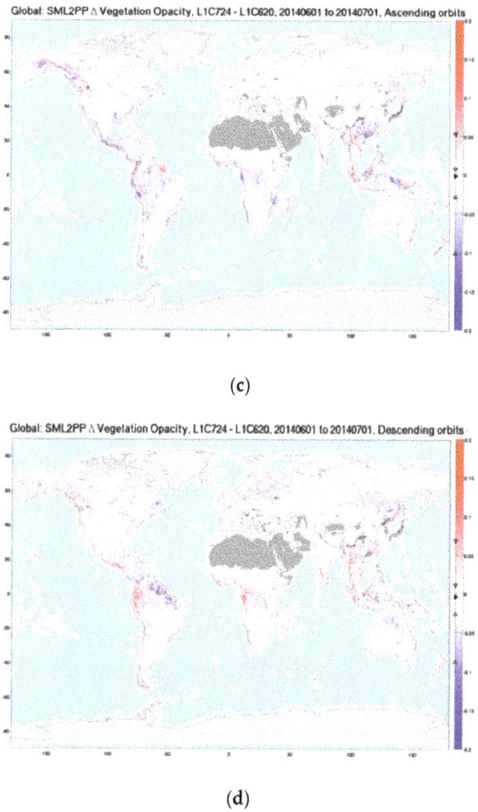

(c)

(d)

Figure 19. Maps of averaged difference, v724–v620, over the month of June 2014 in (**a**) soil moisture for ascending orbits, (**b**) soil moisture in descending orbits, (**c**) VOD in ascending orbits and (**d**) VOD in descending orbitsBlue (resp., red) indicates a decrease (resp., an increase) in V724 soil moisture, VOD compared to v620.

3.4.3. In-Situ Soil Moisture

Several networks of in-situ soil moisture measurements stations (ISMN) can be used to assess the quality of SMOS-retrieved soil moisture. SMOS coarse resolution observations and ultra-local in-situ measurement are not necessarily fully comparable, but become useful when assessing relative differences between two versions of processing of the same satellite data. SMOS soil moisture retrievals obtained from L1OP v620 data and from L1OP v724 data are compared against in-situ soil moisture time series of 250 validation sites taken from 11 in-situ soil moisture networks (Figure 21).

SMOS retrieval data and in-situ data are first co-located in space and time by taking the SMOS grid-points closest to the stations and by pairing in time SMOS and in-situ data of less than 7.5 min to a maximum of 30 min of absolute time difference, depending on the in-situ network temporal sampling characteristics. We denote the SMOS and in-situ collocated time series (S_t, I_t) and the associated difference time series $(\Delta_t = S_t - I_t)$.

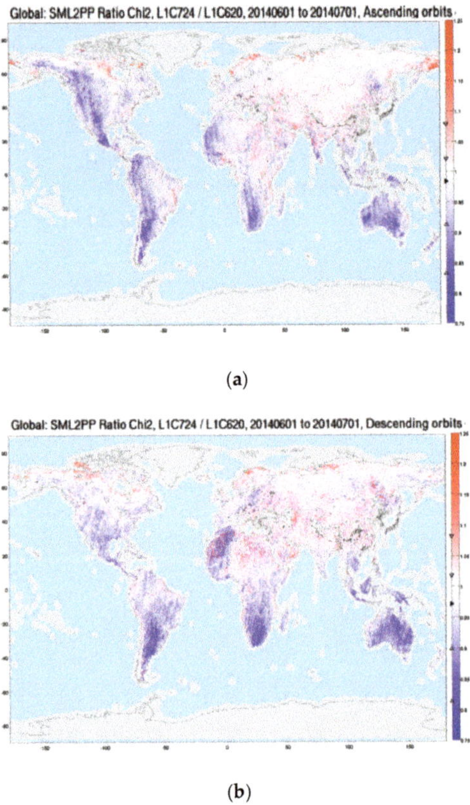

(**a**)

(**b**)

Figure 20. Maps of averaged χ^2_r ratios v724/V620 for the month of June 2014 for ascending orbit (**a**) and descending orbit (**b**). Blue indicates that v724 improves with lower χ^2 compared to v620′. The markers located on the colour bar show the distribution variables, the mode is represented by the ▸ marker. The 68.3% and 95.4% percentiles are shown by the inner and outer ▴ ▾ markers, respectively.

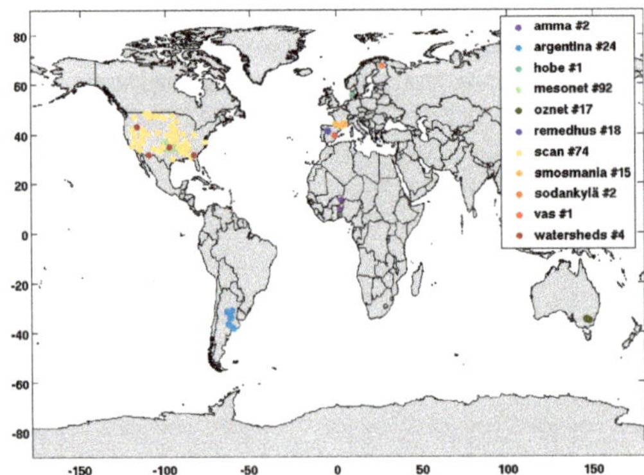

Figure 21. The 11 in-situ soil moisture networks and the position of their sites. The legend reports the network name, its associated colour and the number of the sites we considered for a total of 250 sites.

We computed the usual statistics and their 95% confidence intervals (CI95) obtained by bootstrap x to assess the two processor versions. For (S_t, I_t), we computed their means and standard deviations, μ_S and σ_S for SMOS and μ_I and σ_I for in-situ data and the correlation R. For the Δ_t differences, we computed the bias, the standard deviation (STDD) and the root mean square (RMSD).

The results vary from site to site, but most of them show a better correspondence between the in-situ measurements and the data than the v724 processor or similar performance to the v620 processor. This is reflected by the overall performances, which are obtained by computing the statistics on the concatenation of all sites' time series (S_t, I_t), which are reported in Tables 3 and 4 along with their graphic representation using Taylor diagrams (Figure 22).

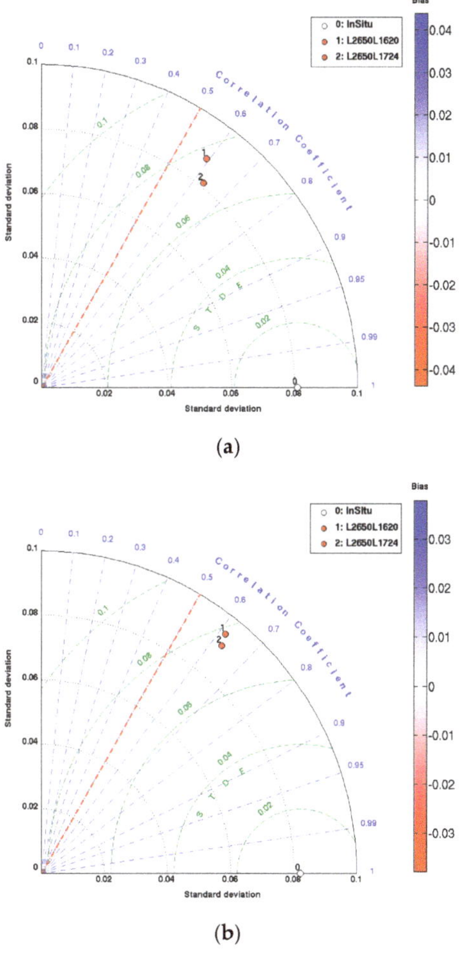

(a)

(b)

Figure 22. Taylor diagram of overall retrieved soil moisture time series with respect to in-situ measurement time series for ascending orbits (**a**) and descending orbits (**b**).

Table 3. Δ time series statistics and their CI95, 95% confidence intervals, given between parentheses.

L1OP	Ascending Orbits					Descending Orbits				
	R	bias	STDD	RMSD	#data	R	bias	STDD	RMSD	#data
v620	0.59 (0.016)	−0.031 (0.004)	0.076 (0.004)	0.083 (0.004)	38943	0.62 (0.017)	−0.034 (0.004)	0.078 (0.005)	0.085 (0.004)	42963
v724	0.063 (0.014)	−0.044 (0.004)	0.070 (0.004)	0.082 (0.004)	40108	0.63 (0.014)	−0.038 (0.004)	0.075 (0.004)	0.084 (0.004)	44099

Table 4. SMOS and in-situ time series statistics and their CI95, 95% confidence intervals, given between parentheses.

L1OP	Ascending Orbits				Descending Orbits			
	μ_S	μ_I	σ_S	σ_I	μ_S	μ_I	σ_S	σ_I
v620	0.167 (0.004)	0.198 (0.004)	0.088 (0.004)	0.081 (0.005)	0.171 (0.004)	0.205 (0.004)	0.094 (0.004)	0.082 (0.005)
v724	0.155 (0.004)	0.198 (0.004)	0.081 (0.004)	0.081 (0.004)	0.167 (0.004)	0.205 (0.004)	0.091 (0.004)	0.082 (0.004)

A Taylor diagram is a convenient 2D graphical representation focusing on the statistics R, σ and STDD, which are by nature debiased (mean-subtracted). The so-called 3D version given in Figure 22 makes the bias information available as a colour scale. Such a diagram is a polar coordinate representation of (σ, R). The standard deviations of series σ are used as the radius, and the correlation, R, with respect to a common reference is converted into an angle using acos(R). It is worth noting that the relation between the correlation and angle is highly non-linear; a 45° angle is already a 0.7 correlation. The reference data is always located at the x axis (correlation 1 with itself) and with a white marker (0 bias with itself) at the position σ_I, the reference being the in-situ data.

Figure 22 shows the performance of the overall retrieved soil moisture time series obtained from L1C V620 (1) and from L1C V724 (2) against the reference in-situ time series (0). Thanks to the concatenation, a large number of points (~40,000) allow computing reliable statistics, which result in a narrow CI95 that does not overlap for R, bias and STD making the separation of plots significant.

Compared to v620, v724 increases the correlation with respect to in-situ data and obtains an σ_S closer to σ_I. As usual, this is more prominent for ascending morning orbits, where Level 2 retrievals always perform better, with better thermodynamic equilibrium at the surface and a calmer ionosphere in the mornings than in the evenings. Different RFI contamination patterns are also likely playing a role.

These two results indicate an increase in signal-to-noise ratio for v724, generating less noisy retrieved soil moisture and possibly better long-term stability. However, for the latter, two years of data is probably too short, and it is necessary to wait for the full L_1 and L_2 10-year reprocessed data availability. Finally, similarly to the spatial maps, using the v724 data, provides here ~1% more successful retrievals in these time series.

4. Conclusions

The SMOS team has started a new reprocessing campaign, the third, after several improvements have been introduced in calibration and image reconstruction. In calibration, the changes mainly affect the NIR calibration parameters, the NIR antenna losses, the PMS sensitivities and the correction of the thermal coupling in one important thermistor. In image reconstruction, the changes focus on reducing the spatial biases induced by the dissimilarities of the antenna patterns, and on reducing the Sun effects in the image, which cannot be considered as a point source at L-band. These corrections improve the quality of the data, as indicated by several metrics that analyse spatial biases, measurement stability, and other image reconstruction errors, as well as by comparisons against in-situ measurements and χ^2

Remote Sens. **2020**, *12*, 1645

metrics from soil moisture retrievals. This reprocessing campaign comes just after SMOS has been in orbit for over 10 years.

Author Contributions: NIR calibration, R.O. and I.C.; Antenna losses, I.C., J.K. and R.O., PMS Sensitivities, J.C., I.C. and A.Z.; PMS Heater correction, J.C., I.C. and A.Z.; Antenna patch thermistor correction, J.K. and R.O.; Gibbs-2, A.K. and F.C.; Ice-sea mask, F.C.; Super-sample Sun correction, F.C.; Sun correction in the back, A.K., Software, J.B. and G.L.; Supervision, M.M.-N., R.O. and R.C.; Validation, J.T., P.R., R.O., R.D.-G., V.G.-G.; Writing—original draft, R.O., V.G.-G.; Writing—review & editing, I.C., J.K., J.C., P.R., J.T., J.B., G.L., R.C., M.M.-N., A.K., V.G.-G. All authors have read and agreed to the published version of the manuscript.

Funding: This work has been funded by the European Space Agency under the SMOS programme.

Conflicts of Interest: The authors declare no conflict of interest.

References

1. Kerr, Y.H.; Waldteufel, P.; Wigneron, J.P.; Delwart, S.; Cabot, F.; Boutin, J.; Escorihuela, M.J.; Font, J.; Reul, N.; Gruhier, C.; et al. The SMOS mission: New tool for monitoring key elements of the global water cycle. *Proc. IEEE* **2010**, *98*, 666–687. [CrossRef]
2. Font, J.; Camps, A.; Borges, A.; Martín-Neira, M.; Boutin, J.; Reul, N.; Kerr, Y.H.; Hahne, A.; Mecklenburg, S. SMOS: The challenging sea surface salinity measurement from space. *Proc. IEEE* **2010**, *98*, 649–665. [CrossRef]
3. Mecklenburg, S.; Drusch, M.; Kaleschke, L.; Rodriguez-Fernandez, N.; Reul, N.; Kerr, Y.; Font, J.; Martin-Neira, M.; Oliva, R.; Daganzo-Eusebio, E.; et al. ESA's Soil Moisture and Ocean Salinity mission: From science to operational applications. *Remote Sens. Environ.* **2016**, *180*, 3–18. [CrossRef]
4. McMullan, K.D.; Brown, M.A.; Martín-Neira, M.; Rits, W.; Ekholm, S.; Marti, J.; Lemanczyk, J. SMOS: The payload. *IEEE Trans. Geosci. Remote Sens.* **2008**, *46*, 594–605. [CrossRef]
5. Brown, M.A.; Torres, F.; Corbella, I.; Colliander, A. SMOS Calibration. *IEEE Trans. Geosci. Remote Sens.* **2008**, *46*, 646–658. [CrossRef]
6. Colliander, A.; Ruokokoski, L.; Suomela, J.; Veijola, K.; Kettunen, J.; Kangas, V.; Aalto, A.; Levander, M.; Greus, H.; Hallikainen, M.T.; et al. Development and Calibration of SMOS Reference Radiometer. *IEEE Trans. Geosci. Remote Sens.* **2007**, *45*, 1967–1977. [CrossRef]
7. SMOS L_2 OS Algorithm Theoretical Baseline Document, SO-TN-ARG-GS-0007, Issue 3.13. April 2016. Available online: https://earth.esa.int/documents/10174/1854519/SMOS_L2OS-ATBD (accessed on 5 March 2020).
8. Corbella, I.; Torres, F.; Duffo, N.; Durán, I.; Pablos, M.; Martín-Neira, M. Enhanced SMOS amplitude calibration using external target. In Proceedings of the IEEE International Geoscience and Remote Sensing Symposium, IGARSS 2012, Munich, Germany, 22–27 July 2012; IEEE: Munich, Germany, 2012; pp. 2868–2871.
9. Corbella, I.; Torres, F.; Duffo, N.; Durán, I.; González-Gambau, V.; Oliva, R.; Closa, J.; Martín-Neira, M. Calibration of the MIRAS Radiometers. *IEEE J. Sel. Top. Appl. Earth Obs. Remote Sens.* **2019**, *12*, 1633–1646. [CrossRef]
10. Corbella, I.; Torres, F.; Duffo, N.; Martín-Neira, M.; González-Gambau, V.; Camps, A.; Vall-Llossera, M. On-Ground Characterization of the SMOS Payload. *IEEE Trans. Geosci. Remote Sens.* **2009**, *47*, 3123–3133. [CrossRef]
11. Corbella, I.; Torres, F.; Duffo, N.; González-Gambau, V.; Pablos, M.; Duran, I.; Martín-Neira, M. MIRAS Calibration and Performance: Results from the SMOS In-Orbit Commissioning Phase. *IEEE Trans. Geosci. Remote Sens.* **2011**, *49*, 3147–3155. [CrossRef]
12. Kainulainen, J.; Oliva, R.; Closa, J.; Barbosa, J.; Martin-Neira, M. *In-Orbit Calibration of the Thermal Model of SMOS Noise Injection Radiometer to Account for Eclipse Time Bias*, Unpublished Work.
13. Duran, I.; Lin, W.; Corbella, I.; Torres, F.; Duffo, N.; Martín-Neira, M. SMOS floor error impact and migation on ocean imaging. In Proceedings of the 2015 IEEE International Geoscience and Remote Sensing Symposium (IGARSS), Milan, Italy, 26–31 July 2015; pp. 1437–1440. [CrossRef]
14. Corbella, I.; Torres, F.; Wu, L.; Duffo, N.; Duran, I.; Martín-Neira, M. Spatial biases analysis and mitigation methods in SMOS images. In Proceedings of the International Geoscience and Remote Sensing Symposium, IGARSS 2013, Melbourne, Australia, 21–26 July 2013; IEEE: Melbourne, Australia, 2013; pp. 3145–3418.

15. Khazâal, A.; Richaume, P.; Cabot, F.; Anterrieu, E.; Mialon, A.; Kerr, Y.H. Improving the Spatial Bias Correction Algorithm in SMOS Image Reconstruction Processor: Validation of Soil Moisture Retrievals with in Situ Data. *IEEE Trans. Geosci. Remote Sens.* **2019**, *57*, 277–290. [CrossRef]

16. Dulk, G.A.; Gary, D.E. The sun at 1.4 GHz: Intensity and polarization. *Astron. Astrophys.* **1983**, *124*, 103–107, ISSN 0004-6361.

17. Khazaal, A.; Cabot, F.; Anterrieu, E.; Kerr, Y.H. A new direct Sun correction algorithm for the Soil Moisture and Ocean Salinity Space Mission. *IEEE J. Sel. Top. Remote Sens. (JSTARS)* **2020**, *13*, 1164–1173. [CrossRef]

18. Khazaal, A.; Anterrieu, E.; Cabot, F.; Kerr, Y.H. Impact of Direct Solar Radiations Seen by the Back–Lobes Antenna Patterns of SMOS on the Retrieved Images. *IEEE J. Sel. Top. Appl. Earth Obs. Remote Sens.* **2017**, *10*, 3079–3086. [CrossRef]

19. Camps, A.; Vall-Llossera, M.; Duffo, N.; Zapata, M.; Corbella, I.; Torres, F.; Barrena, V. Sun effects in 2-D aperture synthesis radiometry imaging and their cancelation. *IEEE Trans. Geosci. Remote Sens.* **2004**, *42*, 1161–1167. [CrossRef]

20. Camps, A.; Vall-llossera, M.; Torres, F.; Corbella, I.; Duffo, N. *Sun Self-Estimation Algorithm*; Technical Report, SMOSP3-UPC-TN-0002 v1.0; Polytechnical University of Catalunya: Barcelona, Spain, 2007.

21. Martín-Neira, M.; Oliva, R.; Corbella, I.; Torres, F.; Duffo, N.; Durán, I.; Kainulainen, J.; Closa, J.; Zurita, A.; Cabot, F.; et al. SMOS instrument performance and calibration after six years in orbit. *Remote Sens. Environ.* **2016**, *180*, 19–39, ISSN 0034-4257. [CrossRef]

22. Oliva, R.; Martin-Neira, M.; Corbella, I.; Torres, F.; Kainulainen, J.; Tenerelli, J.E.; Cabot, F.; Martin-Porqueras, F. SMOS Calibration and Instrument Performance after One Year in Orbit. *IEEE Trans. Geosci. Remote Sens* **2013**, *51*, 654–670. [CrossRef]

23. Meirold-Mautner, I.; Mugerin, C.; Vergely, J.L.; Spurgeon, P.; Rouffi, F.; Meskini, N. SMOS ocean salinity performance and TB bias correction. In Proceedings of the EGU General Assembly, Vienna, Austria, 19–24 April 2009.

24. Tenerelli, J.; Reul, N. *Analysis of L1PP Calibration Approach Impacts in SMOS TB and 3-Days SSS Retrievals over the Pacific Using an Alternative Ocean Target Transformation Applied to L1OP Data*; Tech. Rep. IFREMER/CLS: Brest, France, 2010.

 *remote sensing* 

Article

Deriving VTEC Maps from SMOS Radiometric Data

Roselena Rubino [1],*, Nuria Duffo [1], Verónica González-Gambau [2], Ignasi Corbella [1], Francesc Torres [1], Israel Durán [1] and Manuel Martín-Neira [3]

[1] CommSensLab Research Group, Polytechnic University of Catalonia, c/Jordi Girona 1-3, 08034 Barcelona, Spain; duffo@tsc.upc.edu (N.D.); corbella@tsc.upc.edu (I.C.); xtorres@tsc.upc.edu (F.T.); israel.duran@tsc.upc.edu (I.D.)

[2] Physical Oceanography Department, Institute of Marine Sciences, ICM-CSIC and Barcelona Expert Center, Pg. Marítim de la Barceloneta 37-49, 08034 Barcelona, Spain; vgonzalez@icm.csic.es

[3] European Space Research and Technology Center, European Space Agency, 2200 AG Noordwijk, The Netherlands; manuel.martin-neira@esa.int

* Correspondence: roselena.rubino@tsc.upc.edu

Received: 27 March 2020; Accepted: 8 May 2020; Published: 18 May 2020

Abstract: In this work, a new methodology is proposed in order to derive vertical total electron content (VTEC) maps from the radiometric measurements of the Soil Moisture and Ocean Salinity (SMOS) mission as an alternative approach to those based on external databases and models. This approach uses spatiotemporal filtering techniques with optimized filters to be robust against the thermal noise and image reconstruction artifacts present in SMOS images. It is also possible to retrieve the Faraday rotation angle from the recovered VTEC maps in order to correct the effect that it causes in the SMOS brightness temperatures.

Keywords: faraday rotation angle (FRA); vertical total electron content (VTEC); L-band; radiometry; Interferometry; soil moisture; ocean salinity (SMOS)

1. Introduction

Over the past few years, the earth has been undergoing significant climate change and extreme weather events. Even though the water cycle is the most influential process in this situation, it is still relatively poorly understood. For this reason, its understanding remains a priority field under study by different research groups. Earth's water cycle and climate are intrinsically linked to some geophysical variables. Two of those variables are soil moisture and ocean salinity, which change constantly depending on the exchange of water between oceans, atmosphere, and landmasses [1]. Global measurements of both parameters were not available with a suitable temporal and spatial resolution until 2009.

In November 2009, The European Space Agency launched the SMOS (Soil Moisture and Ocean Salinity) mission to observe soil moisture over the earth's landmasses and salinity over the oceans at a global and frequent scale [2,3]. Its unique payload is MIRAS (Microwave Imaging Radiometer by Aperture Synthesis), a two-dimensional Y-shape synthetic aperture radiometer operating in the L-band (1.413 GHz) [4]. After more than 10 years in operation, MIRAS continues to provide good quality full polarimetric brightness temperature (TB) [5] to generate continuous and global maps of both geophysical variables.

MIRAS was designed to measure the radiation emitted from the earth (i.e., the brightness temperatures (TBs)). Each scene is measured with multi-incidence angles, which are taken into account when processing the data to construct an image (snapshot) over the extended alias-free field of view (EAF-FoV). A block of full polarimetric data per scene is obtained every 2.4 s [6].

As microwave radiation from Earth propagates through the ionosphere, the electromagnetic field components are rotated at an angle, called the Faraday rotation angle (FRA), which depends on the vertical total electron content (VTEC) of the ionosphere, the frequency, and the geomagnetic field. At the SMOS operating frequency (1.4135 GHz), the Faraday rotation is not negligible and must be compensated for to get accurate geophysical retrievals. It can be estimated using a classical formulation [7] that makes use of total electron content (TEC) and geomagnetic field data provided by external sources.

The Faraday rotation angle can alternatively be retrieved from the SMOS radiometric data. This is possible thanks to improvements in the image reconstruction algorithms developed in the last few years, particularly regarding the third and fourth Stokes parameters [6]. However, estimating the Faraday rotation from SMOS radiometric data per each pixel in the SMOS field of view is not straightforward because of the presence of spatial errors in SMOS images. Spatial ripples are due to calibration inaccuracies, image reconstruction artifacts, and antenna pattern uncertainties [8] that limit the quality of the retrieval. A previous work showed that the FRA can be dynamically retrieved at boresight per snapshot directly from SMOS full-polarization TB by applying filtering techniques [9]. The results show a good performance, but the FRA at boresight is not representative for the entire SMOS field of view as shown in [10].

The possibility of retrieving the Faraday rotation from SMOS radiometric data opens up the opportunity to estimate the total electron content of the ionosphere by using an inversion procedure from the measured rotation angle in the SMOS field of view. Currently, the SMOS ocean salinity team computes the VTEC over the ocean from the SMOS third Stokes measurements following the procedure detailed in [11]. These VTEC retrievals have been shown to improve the salinity retrievals. This methodology considers only the SMOS field of view region with the highest sensitivity of TB to VTEC, and it calculates the VTEC starting with a first order approximation initiated with the VTEC value from an external database. This value is then assigned to the entire SMOS field of view not taking into account the VTEC spatial variation within it. Therefore, this methodology is dependent not only on an external VTEC database but also on a forward radiative model.

The present work proposes a novel methodology to derive VTEC maps from SMOS radiometric data over the EAF-FoV. This methodology is expected to work independently on the target measured by the instrument. It applies spatiotemporal filtering techniques to overcome the issues involved. The structure of this paper is as follows: Section 2 details the different data sources and the methods used for deriving the VTEC maps from SMOS measurements. The results obtained with the novel methodology are shown and discussed in Section 3. Finally, conclusions are drawn in Section 4.

2. Data and Methods

2.1. Faraday Rotation

The rotation angle in the polarization of an electromagnetic field is directly proportional to the VTEC of the ionosphere, the geomagnetic field, and the sensor orientation, according to the following equation [7,12]:

$$\Omega_f = 1.355 * 10^4 * f^{-2} * B_0 * \cos \Theta_B * \sec \theta * VTEC, \tag{1}$$

where Ω_f represents the FRA in degrees; f, the frequency in GHz (1.4135 GHz); B_0, the geomagnetic field in Tesla; Θ_B, the angle between the magnetic field and the wave propagation direction; θ, the angle between the wave propagation direction and the vertical to the surface or so called incidence angle; and VTEC, the vertical total electron content in TEC Units (TECU) [$10^{16} electrons/m^2$]. Both the geomagnetic field and the VTEC are given at a geodetic altitude of 450 km.

The geomagnetic field is obtained from the data set of the International Geomagnetic Reference Field (IGRF) [13]. In the SMOS Level 2 operational processor, the vertical electron content used to correct the FRA is read from a SMOS auxiliary data field called "consolidated TEC" [14] (referred to hereafter as the VTEC database) for ascending and descending orbits over land, and only for ascending orbits over ocean. For measurements over the ocean in descending orbits, VTEC values are computed from SMOS TB measurements following the methodology detailed in [11], and the so derived VTEC values can be found in the OSDAP2 (Level 2 Ocean Salinity Data Analysis Product) [15].

The Faraday rotation can alternatively be retrieved directly from SMOS radiometric data using a different technique. At each spatial direction, Earth's radiation arrives at the instrument with a rotation equal to the addition of two angles. The first one is the geometric angle (φ) and it is given by the third Ludwing polarization definition [16] according to the instrument attitude and orientation with respect to the nominal ground-referenced horizontal (h) and vertical (v) polarizations. The second one is the FRA (Ω_f). Assuming that the h and v polarizations emitted by Earth are uncorrelated, the relationship between the brightness temperatures in full polarization at the ground and antenna levels can be expressed as follows [16]:

$$
\begin{bmatrix} T_B{}^{xx} \\ 2T_B{}^{xy} \\ T_B{}^{yy} \end{bmatrix} = \begin{bmatrix} \cos^2(\varphi + \Omega_f) & \sin^2(\varphi + \Omega_f) \\ -\sin 2(\varphi + \Omega_f) & \sin 2(\varphi + \Omega_f) \\ \sin^2(\varphi + \Omega_f) & \cos^2(\varphi + \Omega_f) \end{bmatrix} \begin{bmatrix} T_B{}^{hh} \\ T_B{}^{vv} \end{bmatrix} \tag{2}
$$

where the superscripts represent the polarization frames.

From Equation (2), the FRA can be calculated using full-pol radiometric data as follows (equivalent to Equation (22) of [12]):

$$
\Omega_f = -\varphi - \frac{1}{2}arctan\left(\frac{2\Re e(T_B{}^{xy})}{T_B{}^{xx} - T_B{}^{yy}}\right) \tag{3}
$$

where $2\Re e(T_B{}^{xy})$ corresponds to the third Stokes parameter at antenna level and $T_B{}^{xx}$ and $T_B{}^{yy}$ to the brightness temperatures in the x and y polarizations, respectively.

In a previous paper [9], the FRA was estimated from MIRAS data using spatiotemporal filtering techniques. A unique FRA value per snapshot (at boresight) was retrieved by averaging the FRA over a circle of radius 0.3 around the boresight in the $\xi-\eta$ plane, defining ξ and η as the director cosines with respect to the X and Y axes, respectively. Even though this methodology reproduces the natural variation of the Faraday rotation accurately, it is not enough to use one FRA value for all the EAF-FoV due to its spatial variation within the snapshot. Figure 1a shows the FRA over the EAF-FoV of one snapshot over a descending orbit in October 2011. The circle of radius 0.3 is drawn in black. The FRA was calculated using Equation (1) and reading the geomagnetic field and the VTEC from the external datasets [14,17], respectively. The error when the boresight FRA is considered for all pixels over the EAF-FoV is shown in Figure 1b.

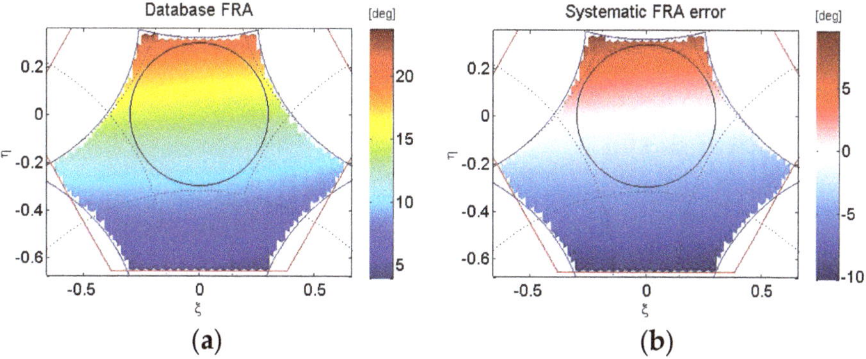

Figure 1. Faraday rotation angle (FRA) over the extended alias-free field of view (EAF-FoV): (**a**) database FRA, (**b**) systematic FRA error when considering the FRA value at boresight for the entire EAF-FoV.

It is known that the VTEC and therefore the FRA vary with solar activity, which can be assessed using sunspot numbers [18], and in a geographical and temporal way [7,19]. SMOS data is here considered within the 24th sun cycle, which started on January 4th, 2008, with each cycle lasting about 11 years. The sun's peak activity during this cycle was reached on March 2014 [18,19]. It is also important to note that the value of the FRA reaches its highest point during the year in the March equinox due to the sun's illumination geometry over the earth during that season.

The geographical variability of the FRA is presented in Figure 2. Equation (1) and the same mentioned external databases were used again to calculate the FRA at the coordinates of the SMOS boresight over a 3-day period. Two different time frames were used: one with high FRA (March 19th to 21st, 2014) and another with low FRA (January 14th to 16th, 2011). The FRA of both descending (DES) and ascending (ASC) orbits are shown for each period (be aware of the different scales in ASC/DES maps). Additionally, the latitude-time Hovmöller plots of the FRA for the entire mission are shown in Figure 3 for descending and ascending orbits to show the FRA temporal variability. The FRA was calculated over the eastern Pacific Ocean using also the database VTEC. These Hovmöller diagrams confirm the selected periods of high and low FRA.

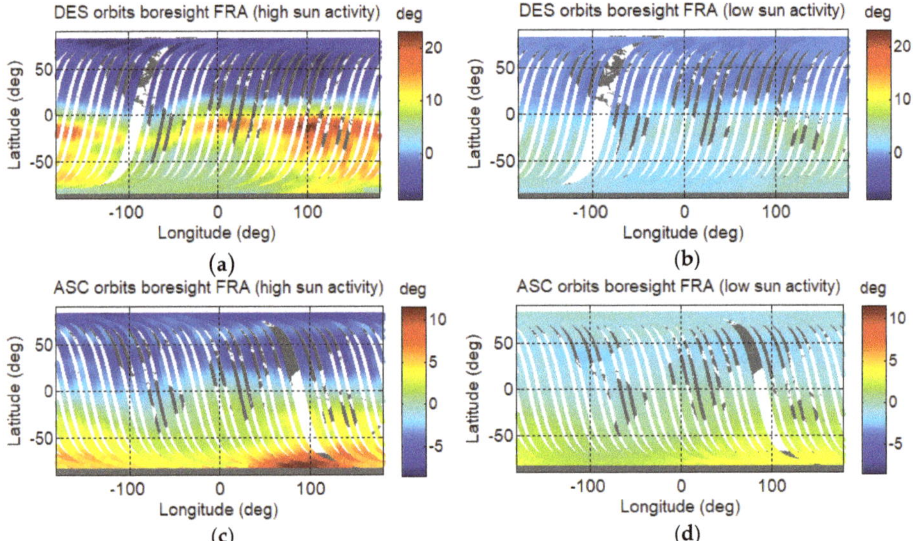

Figure 2. FRA in the Soil Moisture and Ocean Salinity (SMOS) boresight coordinates of 3 days in different periods: (**a**) descending orbits in March 2014 (high sun activity), (**b**) descending orbits in January 2011 (low sun activity), (**c**) ascending orbits in March 2014 (high sun activity), (**d**) ascending orbits in January 2011 (low sun activity).

From these figures, it is perceived that descending orbits present much higher FRA and higher dynamic ranges than ascending ones. In the afternoon local time, the surviving amount of VTEC generated by sunlight during the preceding hours is higher than in the morning local time [19] and because SMOS is in a 6 am—6 pm sun-synchronous orbit, the resulting FRA range is large. Consequently, in the first stage, the analysis was focused on descending orbits. A preliminary analysis was done during the March equinox of 2011 [20], but it was later decided to extend the study to the March equinox of 2014 as well, because, as can be perceived in Figure 3, the highest peak of FRA in the SMOS mission up until now corresponds to that period of time.

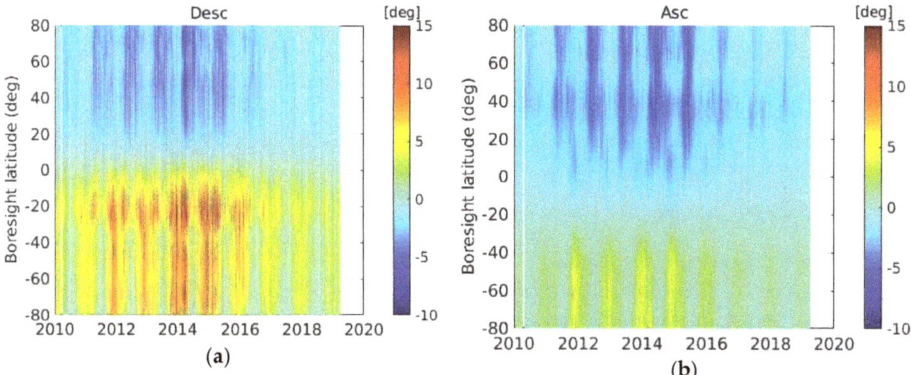

Figure 3. Latitude–time Hovmöller plots of the boresight FRA for the full mission for: (**a**) descending orbit, (**b**) ascending orbits.

2.2. Data Sources

2.2.1. SMOS Brightness Temperatures

MIRAS measures the brightness temperature of each scene providing a full-polarimetric block (Tx, Ty, and Txy) every 2.4 s [16]. The MIRAS Testing Software (MTS), developed by the Polytechnic University of Catalonia (UPC), is an independent processor used as a breadboard to test calibration and image reconstruction algorithms before their introduction into the SMOS Level 1 operational processor [21]. The SMOS brightness temperatures used in this work were processed by MTS from level 0 (raw data) up to level 1C (geolocated brightness temperatures).

2.2.2. Geomagnetic Field and the Consolidated VTEC Databases

The directly proportional relationship between the FRA and the VTEC is determined by Earth's electromagnetic field. Thus, a geomagnetic field dataset is needed. This dataset corresponds to the 12th Generation International Geomagnetic Reference Field (IGRF) [13], which is calculated by the International Association of Geomagnetism and Aeronomy (IAGA). It can be found in [17] and it has a geographical resolution of 5° in longitude and 2.5° in latitude for the entire globe as well as a daily temporal resolution.

SMOS level 1 data includes the consolidated VTEC. It is built by the SMOS Data Processing Ground Segment (DPGS) [14], and it can be obtained in the SMOS dissemination service website [15]. The data also has a geographical resolution of 5° in longitude and 2.5° in latitude for the entire globe but a temporal resolution of 2 h.

2.2.3. SMOS Level 2 VTEC (DTBXY Product)

The SMOS level 2 processor computes the VTEC with a methodology that uses the third Stokes parameter and an external database of VTEC. The VTEC value is calculated for the zone of the snapshot with highest sensitivity of TB to VTEC, which corresponds to pixels in the area around $\xi = 0$, $\eta = 0.2$ ($\xi = 0 \pm 0.025$, $\eta = 0.2 \pm 0.025$). Then, that value is used for the entire EAF-FoV. This VTEC is called A3TEC [11].

This A3TEC dataset is provided in the product called AUX_DTBXY (Delta TB) [15]. The latitude and the VTEC value per overpass are found in fields 32 and 34 respectively [22].

2.2.4. GPS VTEC

This third VTEC source consists on VTEC maps with a temporal resolution of 2 hours and a geographical resolution of 5° in longitude and 2.5° in latitude calculated from daily sets of GPS differential code bias values [23]. The IGS Ionosphere Working Group (Iono-WG) calculates the VTEC, which is named IGS Global Total Electron Content (IGSTEC). It is stored in IONEX format, and it can be downloaded from the official server (CDDISA at GSFC/NASA).

2.3. FRA End-to-End Simulator

In order to evaluate different approaches to retrieve VTEC maps, a first version of a FRA simulator was presented in [10]. This simulator was refined until it worked properly for the assessment. The improved FRA simulator is explained in this section.

For each snapshot, the ξ and η coordinates defined in the antenna frame (director cosine plane) are translated to Earth's surface coordinates. The incidence angle θ and the geometric rotation angle φ are computed using standard geometry [16]. Then, a simple geophysical model is used to simulate the Earth's emissions. Ocean TB are built by assuming a Fresnel model with a typical salinity value of 35 psu and a typical sea surface temperature of 294 K. Open ocean TB images per polarization, assuming a uniform ocean, are shown in the top row of Figure 4 (left: X-pol, middle: Y-pol, right: third Stokes parameter).

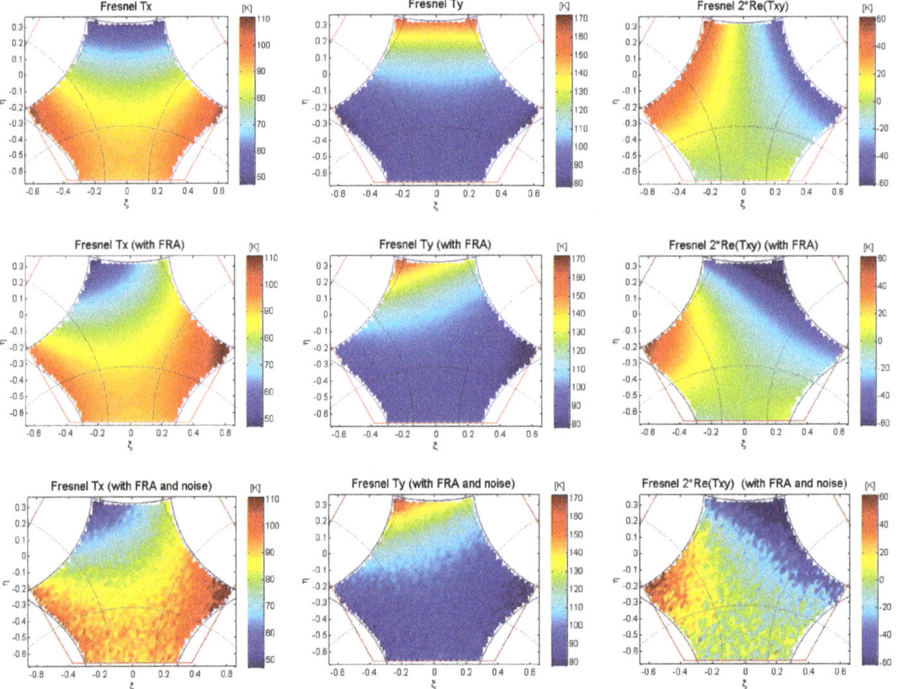

Figure 4. Typical open ocean Fresnel brightness temperature snapshots per polarization (left: X-pol, middle: Y-pol, right: third Stokes parameter). Top: Fresnel modeled brightness temperature (TB), middle: taking into account the FRA, bottom: adding the effect of noise in addition to the FRA.

To add the Faraday rotation for each pixel, Equation (1) is used together with both the VTEC and the geomagnetic databases. This Faraday rotation angle derived from external datasets shall be referred to as "database FRA" from now on in this paper. Once calculated, it is added to the geometrical angle

(φ) to obtain the total rotation per pixel when simulating Earth's emissions. Open ocean TB images taking into account the FRA are shown in Figure 4 (middle row).

Then, the effect of measurement noise is included in the simulator. To do so, first the radiometric sensitivity per polarization is calculated. The radiometric sensitivity corresponds to the smallest radiometric temperature that the instrument can detect. It is defined in the antenna reference frame as follows [24]:

$$\Delta T_{sens} = \Delta S \frac{T_{sys}}{\sqrt{B \tau_e}} \frac{\Omega_a}{|F_n(\xi, \eta)|^2} \sqrt{1 - \xi^2 - \eta^2} \alpha_w \sqrt{N_v} \tag{4}$$

where ΔS corresponds to the elementary area in the (ξ, η) grid defined by $\sqrt{3} d^2 / 2$; d refers to the distance between antennas normalized to the wavelength (d = 0.875 in SMOS); T_{sys} corresponds to the addition of the average antenna temperature (T_A) and the average receiver noise temperature at the antenna plane (T_R), both being different per polarization (see typical values for the open ocean in Table 1); B is the receiver noise equivalent bandwidth that in MIRAS equals 19 MHz; τ_e is the effective finite integration time ($\tau_e = \tau_i * Q$, where τ_i corresponds to the integration time that is 1.2 s for pure X and Y epochs and 0.4 s for mixed epochs and $Q = 0.552$ for SMOS 1 bit/2 level digital correlator [25]); Ω_a is the antenna equivalent solid angle that equals 1.4; $F_n(\xi, \eta)$ is the antenna pattern measured on the ground; α_w is the used window factor that in this case is a Blackman window ($\alpha_w = 0.45$); and N_v is the total number of visibilities samples that in MIRAS is 2791.

Table 1. Average antenna temperature, T_A, and average noise receiver temperature, T_R, at the antenna plane (typical values for open ocean).

Polarization	T_A [K]	T_R [K]
X	76.8	203
Y	95.5	206

The effect of noise is added to the TB with a normal distribution of zero mean and the standard deviation equal to the radiometric sensitivity of each pixel. Figure 4 (bottom row) shows typical open ocean TB images with FRA, including the effect of noise.

MIRAS measures sequentially for each polarization at different instants in time. When processing, one single instant is used for all polarizations. This introduces an error when translating the brightness temperatures from the $x - y$ polarization (antenna frame) to $h - v$ polarization (ground plane) that is unavoidable in the processing.

VTEC maps can be calculated with the FRA estimated using the TB. These maps can then be used in the correction of the FRA per pixel in the field of view (FoV) of every snapshot of the trace.

2.4. VTEC Retrieval from Radiometric Data

From the brightness temperature snapshots, the FRA can be retrieved by applying Equation (3). An indetermination emerges when both the numerator and the denominator tend to 0 ($T_B^{xx} \approx T_B^{yy}$ and $2\mathcal{R}e(T_B^{xy}) \approx 0$); that occurs at low incidence angles [26]. To avoid it, pixels with incidence angles lower than 25° are discarded. Figure 5a shows a database VTEC snapshot from a descending SMOS overpass over the Pacific Ocean on March 20th, 2014. Figure 5b shows the retrieved VTEC snapshot where some pixels are affected by the indetermination of Equation (3). Figure 5c shows the retrieval once pixels causing the indetermination of Equation (3) are rejected with the chosen threshold.

Once the FRA is retrieved, the VTEC is calculated using Equation (1). This equation presents an indetermination when the geomagnetic field is orthogonal to the wave propagation direction, which occurs close to the Equator, in a zone where the FRA vanishes. To avoid this indetermination, pixels accomplishing $\Theta_B \approx \pi/2$ are rejected by using an appropriate threshold established empirically ($\cos \Theta_B < 0.27$). Figure 5d shows another database VTEC snapshot from a descending SMOS overpass over the Pacific Ocean on the same date, March 20th, 2014, in order to compare it with its retrieval

(Figure 5e). The error introduced by the indetermination is noticeable. Figure 5f shows the retrieval once pixels causing the indetermination of Equation (1) are rejected with the chosen threshold (pixels causing indetermination of Equation (3) were rejected previously). It is important to remark that this threshold is only used when the geomagnetic field is orthogonal to the signal path.

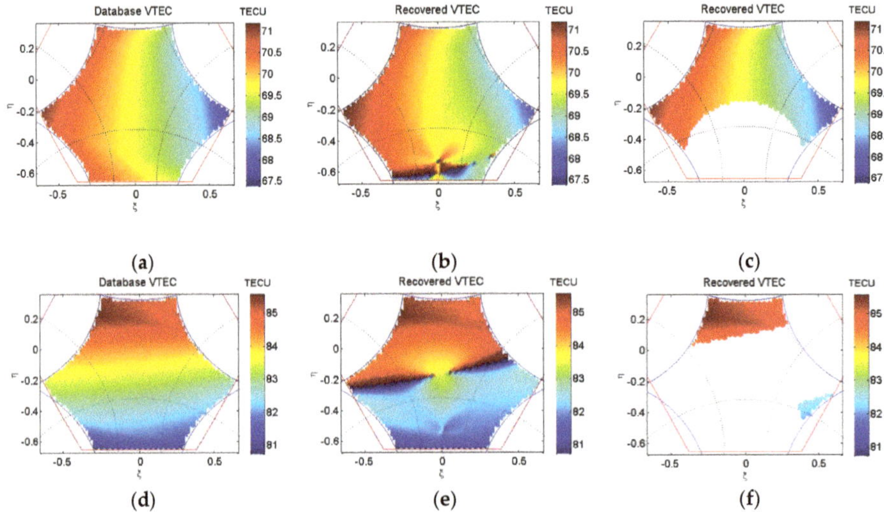

Figure 5. Vertical total electron content (VTEC) snapshots: (**a**) database VTEC, (**b**) its retrieved VTEC snapshot with pixels affected by the indetermination of Equation 3, (**c**) its retrieved VTEC filtering affected pixels, (**d**) another database VTEC snapshot, (**e**) its retrieved VTEC snapshot with pixels affected by the indetermination of Equation (1), (**f**) its retrieved VTEC filtering affected pixels.

After retrieving VTEC snapshots, the geolocation is done at an altitude of 450 km in an ETOPO5 grid (1/12°).

Figure 6 shows the VTEC maps of a descending SMOS overpass in the Pacific Ocean (March 20th, 2014) processed with the simulator (taking into account the effect of the noise). In order to have a reference, the database VTEC can be seen in the left (Figure 6a). Because both the geographical (5° in longitude and 2.5° in latitude) and temporal resolution (2 h) of the "Consolidated TEC" is coarse, a spatiotemporal interpolation is done in order to obtain the VTEC in that SMOS overpass in an ETOPO5 grid. In Figure 6b, the retrieved VTEC is shown, where it is noticeable how the effect of noise introduces errors in the retrieval.

To reduce the effect of noise and artifacts in the retrieved VTEC from SMOS data, spatiotemporal filtering techniques are required [9]. The sizes of both filters were optimized with the simulator that takes into account the effect of noise. To do so, the size of the spatial filter was set to 0.179 in the director cosine plane (10 times the minimum $\Delta\xi = 0.0179$) in TB snapshots and the length of the temporal filter was varied from 15 to 83 snapshots with a step of 4 snapshots. Figure 7a shows the root mean square error (RMSE) of the deviations with respect to the database VTEC of this optimization. The temporal filter is an averaging triangular window considering the current snapshot with the highest weight. Its size was then fixed to 43 snapshots (optimum value from Figure 7a) and the size of the spatial filter was optimized by varying its radius from 0.1253 to 0.2506 with a step of 0.0179. The RMSE of this optimization is shown in Figure 7b, where it can be seen that the optimum size of the spatial filter corresponds to a radius of 0.1969 in the cosine plane. Finally, a fine tuning was done for the size of both filters to select which ones to use. The optimum temporal filter corresponds to 43 snapshots, again, and the spatial filter to 0.189 in the $\xi - \eta$ plane (Figure 7c). It is important to remark that the temporal filter is applied to TB snapshots, and the spatial filter to VTEC snapshots at the antenna frame. The spatial filter was also tested over the ground instead of at the antenna reference frame. To do so, the window size of the spatial filter was calculated to be equivalent to the spatial filter at the antenna, which corresponds to a radius of approximately 190 km over the ground. There was not a clear improvement in the retrieval, but there was an important difference in the execution time, with the calculation over the ground being much slower. Therefore, the spatial filter was applied at the antenna reference frame.

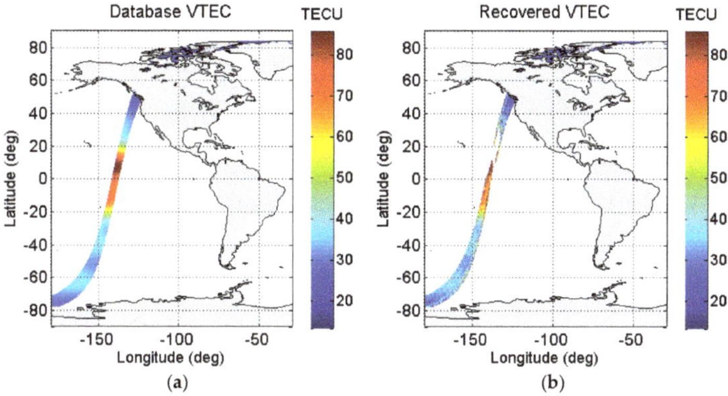

Figure 6. VTEC of a descendent orbit over the Pacific Ocean, March 20th, 2014: (**a**) database VTEC and (**b**) simulated VTEC retrieval.

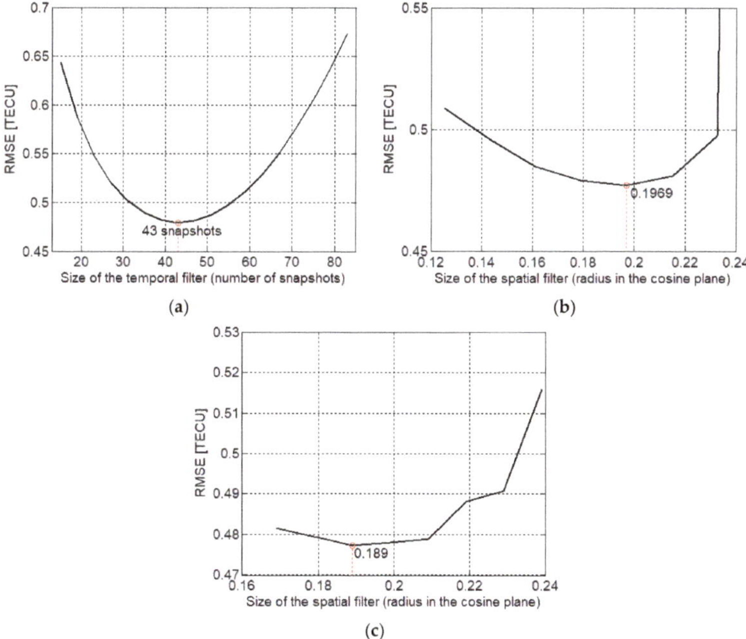

Figure 7. Root mean square error of the retrieved VTEC with respect to the database VTEC when optimizing (**a**) the size of the temporal filter with a coarse binning, (**b**) the size of the spatial filter with a coarse binning, setting an optimum temporal filter size, and (**c**) the size of the spatial filter with a fine binning, setting the temporal filter with the most optimum temporal filter size.

Hence, the methodology consists of:

1. Applying a temporal filter as a triangle filter with a window of 43 TB snapshots.
2. Computing the FRA from the TB using Equation (3), rejecting pixels with incidence angles lower than 25° in order to avoid the indetermination in pixels with $T_B{}^{xx} \approx T_B{}^{yy}$ and $2\Re e(T_B{}^{xy}) \approx 0$.
3. Computing the VTEC from the retrieved FRA using Equation (1), rejecting pixels with a threshold of $\cos \Theta_B < 0.27$ to avoid the indetermination that emerges from that equation.
4. Applying a spatial filter with a radius of 0.189 in the director cosine plane of VTEC snapshots.
5. Generating VTEC maps in an ETOPO5 grid at 450 km of altitude.

2.5. Recovered VTEC Maps with Simulated Data

The descending SMOS overpass in the Pacific Ocean on March 20th, 2014 was processed in the simulator using the methodology described and the results are shown in Figure 8. The recovered VTEC is shown at the left and the error of the retrieval with respect to the database is shown at the right. The retrieved VTEC follows the variation of the database VTEC with a RMSE of 0.48 TEC units. It can be seen that the error in the retrieval does not follow any systematic pattern. The highest error is found close to where the geomagnetic field is orthogonal to the wave propagation direction (indetermination in Equation (1)).

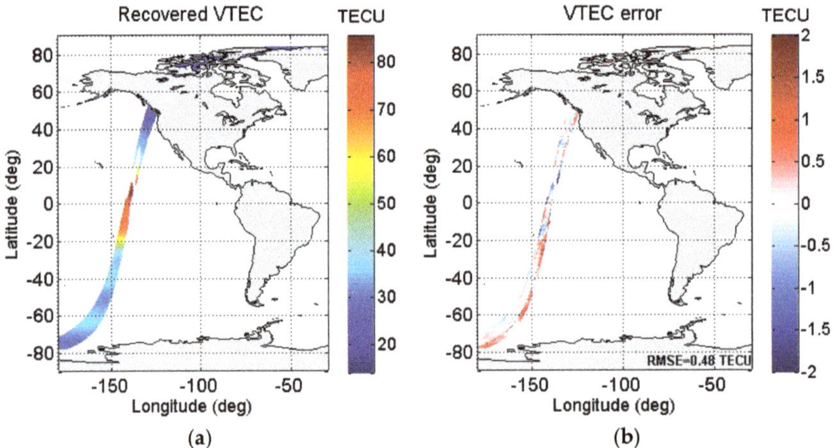

Figure 8. VTEC of a descendent orbit over the Pacific Ocean, March 21st, 2011 processed with the simulator applying the proposed methodology: (**a**) recovered VTEC, (**b**) VTEC error with respect to the database VTEC.

By applying Equation (1) with the database VTEC, an FRA reference can be obtained. Likewise, the FRA can be computed from the retrieved VTEC for all the pixels in the EAF-FoV. Figure 9 shows the database FRA (red) and the simulated retrieved FRA (green) as a function of latitude of a pixel with $\xi = 0$ and $\eta = 0.2$, as well as the error of the FRA retrieval with respect to the database FRA. The gap in the retrieval comes from the rejected pixels in the zone of the orthogonality that is between the geomagnetic field and the wave propagation direction (incidence angle), where it can be seen how the FRA vanishes.

The methodology is able to recover the FRA following the geophysical and temporal variation with a negligible error (with a RMSE of 0.07°), showing good performance.

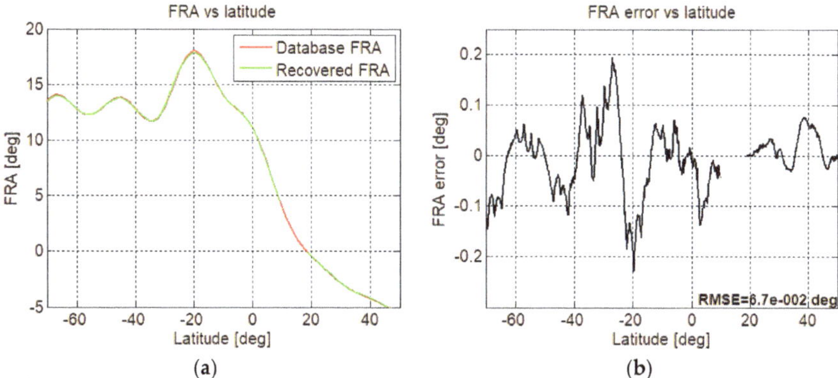

Figure 9. FRA vs. latitude of a pixel along the descending orbit: (a) database FRA (red) and retrieved simulated FRA (green), (b) error of the retrieved simulated FRA with respect to the database.

3. Results and Discussion

Considering the promising results obtained when assessing the methodology with the simulated data, SMOS radiometric data were processed to derive VTEC maps.

3.1. VTEC Retrievals from SMOS Data

Results of the retrieved VTEC with SMOS radiometric data, and the difference between the retrieved VTEC and the database VTEC (used here as a reference) are shown in Figure 10 (top). The recovered VTEC presents a systematic pattern with higher differences in the edges of the swath. This pattern did not appear in the retrieved VTEC from simulated TB. It needs to be characterized at some point in future research.

In order to mitigate the effect on the swath laterals, some empirical approaches were assessed. The first approach used only the alias free-field of view (AF-FoV) instead of using the EAF-FoV region. By doing so, the retrieval of FRA could only be performed over a much narrower swath after a complete SMOS overpass. Hence, a second attempt was based on assigning the average VTEC of the AF-FoV to the entire EAF-FoV in each snapshot [9], disregarding the TEC variability along the snapshot. The third and selected approach consists of extending the value of the VTEC in the pixels of the AF-FoV closest to the EAF-FoV to the latter. The processed orbit using that approach and its difference with respect to the database VTEC are shown in Figure 10 (bottom).

The lateral bands of the southern hemisphere become softened. In the northern hemisphere, a similar softening happens, though not as noticeably as in the southern hemisphere. Additionally, the retrieved VTEC is generally lower than the database VTEC, something that in the simulation does not happen. Table 2 shows the root mean square of the difference (RMSD) between the retrieved and the database VTEC in the EAF-FoV and the difference between the retrieved VTEC in the AF-FoV extended to the EAF-FoV with respect to the database VTEC. For a reference, the statistics of the simulated retrieval are also presented. The statistics are calculated in a range of latitudes between 60° N and 60° S.

Table 2. Statistics of the VTEC retrieval with respect to the database VTEC: (a) with simulated data, (b) with the retrieval in the EAF-FoV, (c) with the retrieval in the AF-FoV extended to the EAF-FoV.

Retrieval with	RMSD [TECU]
Simulated data	0.48
Retrieval in the EAF-FoV	15.69
Retrieval in AF-FoV and extension to the EAF-FoV	10.81

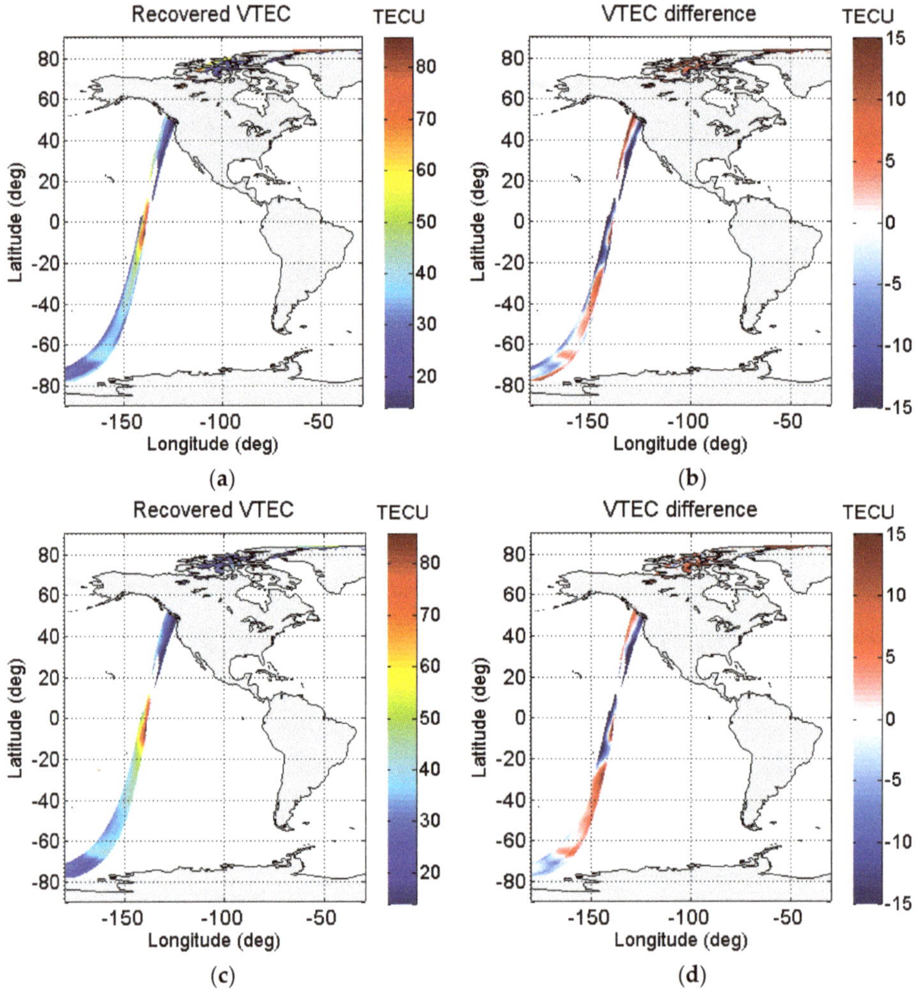

Figure 10. VTEC of a descendent orbit over the Pacific Ocean, March 20th, 2014 obtained from SMOS radiometric data: (**a**) retrieved VTEC, (**b**) VTEC difference with respect to the database VTEC, (**c**) retrieved VTEC with the refined methodology (extension of alias free-field of view (AF-FoV) to the laterals), (**d**) difference of the retrieved VTEC with the refined methodology and the database VTEC.

Figure 11 shows the retrieved FRA (from the VTEC shown in Figure 10c) as a function of the latitude. The database FRA (red) is compared against the retrieved FRA (green) at a pixel in the center of the swath ($\xi = 0$ and $\eta x 0.2$) and its difference is shown in Figure 11b. When processing SMOS radiometric data with the proposed methodology, even though it is possible to recover the FRA geophysical and temporal variation, there is a difference with respect to the database FRA.

Greater differences are perceived in the southern hemisphere. In the northern hemisphere and up to 10° S, both FRAs are very similar, a scenario that does not happen when analyzing pixels in the laterals of the overpass (not shown). This is noticeable in Figure 10d. Still, the root mean square of the difference between the retrieved FRA and the FRA database (Figure 11b) is 1.5°, which represents only 6.50% of the dynamic range of the database FRA.

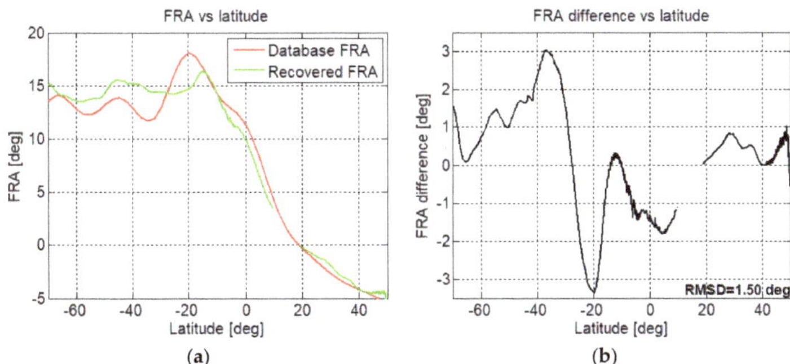

Figure 11. FRA vs. latitude of a pixel along the descending orbit: (**a**) database FRA (red) and retrieved FRA with SMOS radiometric data (green), and (**b**) retrieved FRA difference with respect to the database.

3.2. Comparison of Retrieved VTEC from SMOS with Other External VTEC Sources

In this section, a comparison of the VTEC retrieved with the proposed methodology and that from other sources shown in Section 2 is presented in Figure 12 shows the VTEC of the middle pixel of the swath as a function of the latitude provided by different sources. The line in red corresponds to the database VTEC, the green line to the recovered VTEC using the proposed methodology, the magenta line to the A3TEC (VTEC from the DTBXY product), and the blue line to the IONEX VTEC coming from GPS data [23].

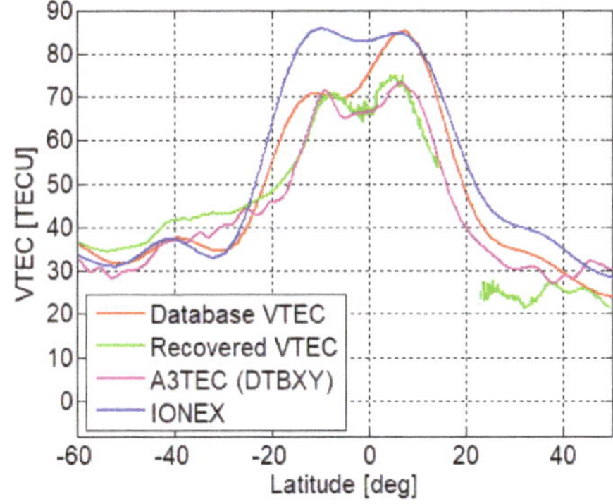

Figure 12. Comparison of the VTEC coming from different sources.

The retrieval with the proposed methodology (green line) follows the temporal-geophysical variation that the A3TEC (magenta) presents. Both of them come from SMOS radiometric data. The gap around 20° N corresponds to the zone where the geomagnetic field is orthogonal to the wave propagation direction causing the Faraday rotation to vanish. The A3TEC provides data in that zone but tends to the value of the database VTEC, which is expected, because in the procedure to retrieve it, that auxiliary database is used as a first guess, and in that zone, the sensitivity of the TB to TEC is very low. The retrieved VTEC with the proposed methodology has fewer ripples than the A3TEC. It was found that the origin of those ripples is due to the remaining noise as reported in [11]. IONEX is always above the VTEC value retrieved from SMOS data, both using the methodology proposed in this paper and the VTEC from DTBXY products, which is in agreement to [11]. The presented methodology proposes an alternative to the current methodology used in order to eliminate the dependency on any external database VTEC.

3.3. Impact of RFI Contamination in the Retrieved VTEC

In order to analyze the performance of the proposed methodology on a global scale, the analysis was extended to all the descending orbits over the ocean on March 21st, 2011. This particular year was chosen because it opened up the possibility to evaluate the impact in the presence of radio-frequency interference (RFI). The RFI contaminates the TB, which has an impact in the recovered VTEC. The retrieved VTEC with the proposed methodology and the one provided by the database VTEC (used as a reference) are shown in Figure 13 as well as the difference between them.

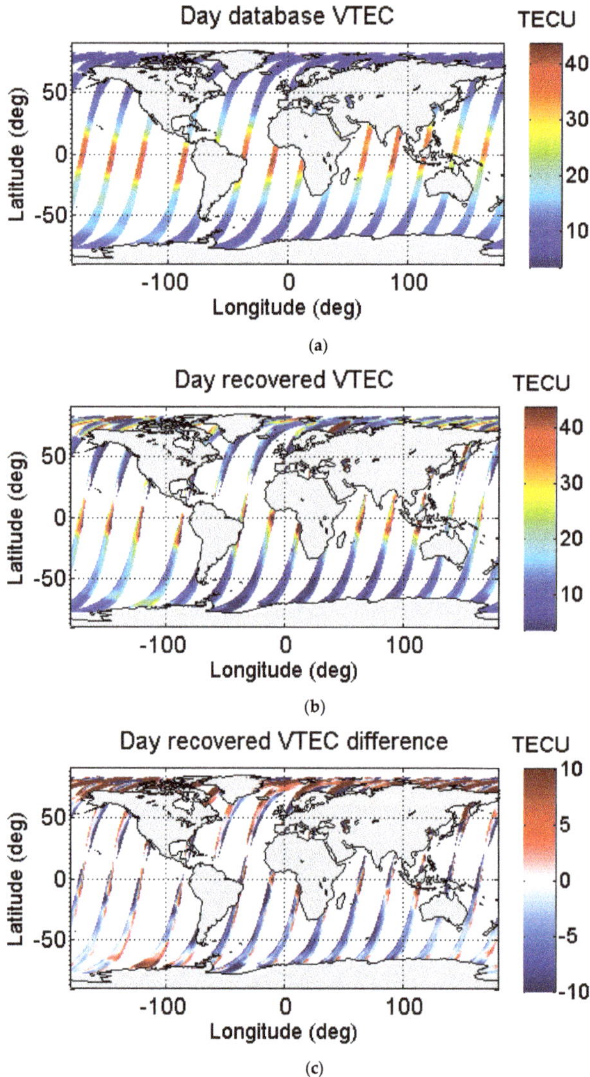

Figure 13. VTEC of all descending orbits on March 20th, 2011: (**a**) database VTEC, (**b**) retrieved VTEC using radiometric SMOS data with the proposed methodology, and (**c**) differences between the retrieved VTEC and the database VTEC with a RMSD of 17.84 total electron content units (TECU).

Greater differences between the recovered and the database VTEC are concentrated at northern, high latitudes, close to ice edges. There were significant differences over the Bering, the Beaufort, and the Barents Seas on that date (see zoomed in portions of Figure 14a,c, respectively). We analyzed whether this was related to TB contamination by RFIs (radio-frequency interferences) sources. In 2011, there were RFIs affecting the Bering Sea, which were switched off in 2012 [27]. The VTEC retrieval of a descending orbit over the same region on March 20th, 2012 was processed and it is shown in Figure 14b. When the RFI was shut down, it was possible to recover a VTEC that was less affected by errors. Similarly, a descending orbit over the Barents Sea on March 22nd, 2019 was processed (when the RFI source was already switched off) and it is shown in Figure 14c. Once again, it was confirmed that RFIs were affecting the VTEC retrieval in 2011 (Figure 14c).

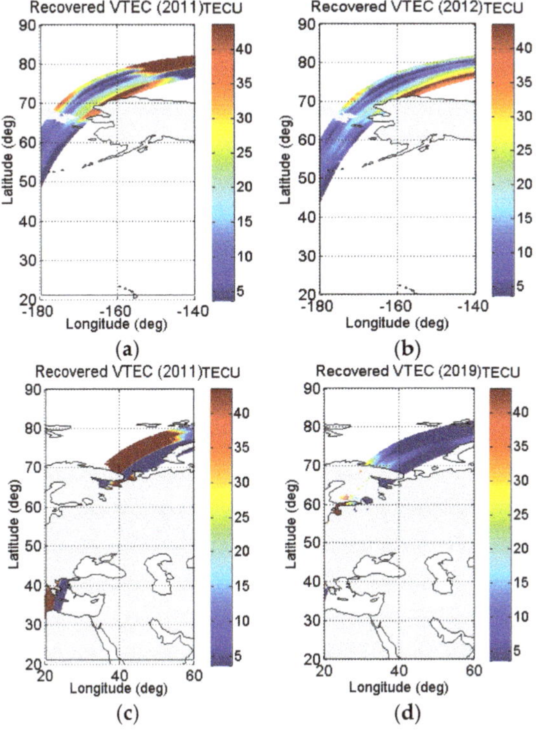

Figure 14. Recovered VTEC of descending orbits over (**a**) the Bering Sea in March 21st, 2011, (**b**) the Bering Sea on March 20th, 2012, (**c**) the Barents Sea on March 21st, 2011, and (**d**) Barents Sea on March 22nd, 2019.

4. Conclusions

Measuring the Faraday rotation from radiometric data allows for the estimation of the total electron content of the ionosphere by using an inversion procedure. This allows for the possibility of creating a VTEC product from SMOS data. Eventually, this product can then be re-ingested in the SMOS level 2 processor in order to improve the geophysical retrievals.

The proposed methodology works independently of the target seen by the instrument. This is an important improvement with respect to the current methodology that also derives VTEC maps from SMOS radiometric data [11], which use a forward radiative model and are focused only over the ocean because SMOS's main purpose is to improve salinity measurements. Moreover, the developed methodology estimates the VTEC for all the pixels in the EAF-FoV with information from different incidence angles, instead of using only the FoV region with the highest sensitivity to TEC, and extending this value across the FoV as does the methodology detailed in [8].

The analysis of the retrieved VTEC maps has been focused over the ocean, where the impact of ionospheric corrections is stronger. These maps have been inter-compared with the database VTEC, the IONEX GPS data, and the A3TEC products. The retrieved VTEC maps provide values generally lower than those of the external VTEC database and the IONEX GPS data, which is in agreement with the differences found when comparing the other SMOS-derived product (A3TEC) with the same two external data sources [8]. However, SMOS-derived VTEC products cannot be fully validated by comparing them with the external VTEC database and the IONEX GPS data, since the spatial resolution of the latter ones is much coarser than that provided by the SMOS products. The comparison between both SMOS-derived VTEC products reveals greater differences in the northern hemisphere. The origin of these discrepancies needs to be investigated.

Further work is needed to evaluate the feasibility of providing global SMOS-derived VTEC maps, including ocean, land, and ice. The main challenge is to obtain accurate TEC retrievals over land areas in (i) regions where SMOS TB measurements are degraded by strong RFI contamination and (ii) regions where TB at horizontal and vertical polarizations are very similar (such as in dense forests), making the TEC retrieval ill-conditioned. A dedicated study of retrieved VTEC over land to assess the performance of the proposed method on a global scale is currently on-going. Besides, the Faraday rotation vanishes in regions of the earth where the geomagnetic field is orthogonal with the signal path. Therefore, the retrieval of VTEC in these regions is not possible and maps will present data gaps. Improvements over the ocean also need to be addressed. The retrieved VTEC maps present a remaining systematic pattern (more noticeable in the northern hemisphere, as shown in Figure 14c) that might be introduced by the instrument when measuring the Faraday rotation (not present in the simulation experiments). Ongoing work is focused on characterizing this FRA systematic pattern in a region and for a period with very low FRA values, so the measurement can be assumed as a systematic error of the instrument, and a correction can be built upon that.

The recovered VTEC maps could be used in the SMOS Level-2 processor to correct the Faraday rotation, which could potentially improve geophysical retrievals, as reported when using the A3TEC method [11]. As a preceding step to analyze the impact of using these VTEC maps on salinity retrievals, the computation of the OTT (ocean target transformation) will be evaluated and used in order to correct the spatial bias presented in the TB as is done by the SMOS ocean salinity team [28]. If an improvement in the stability of the OTT is achieved, more accurate salinity retrievals by using these VTEC maps would be expected.

Author Contributions: Conceptualization, F.T., Investigation, R.R., N.D., V.G.-G., and I.C., methodology, R.R., N.D., V.G.-G., I.C., and I.D., supervision, N.D., V.G.-G., and I.C., writing—original draft, R.R., writing—review an editing, N.D., V.G.-G., I.G., and M.M.-N. All authors have read and agreed to the published version of the manuscript.

Funding: This research was supported by the European Space Agency and Deimos Engenharia (Portugal), SMOS P7 Subcontract DME CP12 no. 2015-005; ERDF (European Regional Development Fund); by the Spanish public funds, projects TEC2017-88850-R and ESP2015-67549-C3-1-R; and through the award "Unidad de Excelencia María de Maeztu" MDM-2016-0600, financed by the "Agencia Estatal de Investigación" (Spain) and by the European Regional Development Fund (ERDF).

Conflicts of Interest: The authors declare no conflict of interest.

References

1. Esa Earth's Water Cycle. Available online: https://www.esa.int/Our_Activities/Observing_the_Earth/SMOS/Earth_s_water_cycle (accessed on 14 February 2019).
2. Kerr, Y.H.; Waldteufel, P.; Wigneron, J.-P.; Delwart, S.; Cabot, F.; Boutin, J.; Escorihuela, M.-J.; Font, J.; Reul, N.; Gruhier, C.; et al. The SMOS Mission: New Tool for Monitoring Key Elements of the Global Water Cycle. *Proc. IEEE* **2010**, *98*, 666–687. [CrossRef]

3. Font, J.; Camps, A.; Borges, A.; Martín-Neira, M.; Boutin, J.; Reul, N.; Kerr, Y.H.; Hahne, A.; Mecklenburg, S. SMOS: The Challenging Sea Surface Salinity Measurement From Space. *Proc. IEEE* **2010**, *98*, 649–665. [CrossRef]

4. McMullan, K.D.; Brown, M.A.; Martin-Neira, M.; Rits, W.; Ekholm, S.; Marti, J.; Lemanczyk, J. SMOS: The Payload. *IEEE Trans. Geosci. Remote Sens.* **2008**, *46*, 594–605. [CrossRef]

5. Martín-Neira, M.; Oliva, R.; Corbella, I.; Torres, F.; Duffo, N.; Durán, I.; Kainulainen, J.; Closa, J.; Zurita, A.; Cabot, F.; et al. SMOS instrument performance and calibration after six years in orbit. *Remote Sens. Environ.* **2016**, *180*, 19–39. [CrossRef]

6. Wu, L.; Torres, F.; Corbella, I.; Duffo, N.; Duran, I.; Vall-llossera, M.; Camps, A.; Delwart, S.; Martin-Neira, M. Radiometric Performance of SMOS Full Polarimetric Imaging. *IEEE Geosci. Remote Sens. Lett.* **2013**, *10*, 1454–1458. [CrossRef]

7. Le Vine, D.M.; Abraham, S. The effect of the ionosphere on remote sensing of sea surface salinity from space: Absorption and emission at L band. *IEEE Trans. Geosci. Remote Sens.* **2002**, *40*, 771–782. [CrossRef]

8. Corbella, I.; Torres, F.; Wu, L.; Duffo, N.; Duran, I.; Martin-Neira, M. Spatial biases analysis and mitigation methods in SMOS images. In Proceedings of the 2013 IEEE International Geoscience and Remote Sensing Symposium—IGARSS, Melbourne, Australia, 21–26 July 2013; pp. 3415–3418.

9. Corbella, I.; Wu, L.; Torres, F.; Duffo, N.; Martin-Neira, M. Faraday Rotation Retrieval Using SMOS Radiometric Data. *IEEE Geosci. Remote Sens. Lett.* **2015**, *12*, 458–461. [CrossRef]

10. Rubino, R.; Torres, F.; Duffo, N.; Gonzalez-Gambau, V.; Corbella, I.; Martin-Neira, M. Direct Faraday rotation angle retrieval in SMOS field of view. In Proceedings of the 2017 IEEE International Geoscience and Remote Sensing Symposium (IGARSS), Fort Worth, TX, USA, 23–28 July 2017; pp. 697–698.

11. Vergely, J.-L.; Waldteufel, P.; Boutin, J.; Yin, X.; Spurgeon, P.; Delwart, S. New total electron content retrieval improves SMOS sea surface salinity. *J. Geophys. Res. Oceans* **2014**, *119*, 7295–7307. [CrossRef]

12. Yueh, S.H. Estimates of Faraday rotation with passive microwave polarimetry for microwave remote sensing of Earth surfaces. *IEEE Trans. Geosci. Remote Sens.* **2000**, *38*, 2434–2438. [CrossRef]

13. Alken, P.; Maus, S.; Chulliat, A.; Manoj, C. NOAA/NGDC candidate models for the 12th Generation International Geomagnetic Reference Field. *Earth Planets Space 2015* **2015**, *67*. [CrossRef]

14. Barbosa, J. *SMOS Level 1 and Auxiliary Data Products Specifications*; Indra Sistemas: Alcobendas, Spain, 2014.

15. ESA SMOS Online Dissemination. Available online: https://smos-diss.eo.esa.int/oads/access/ (accessed on 22 February 2019).

16. Martin-Neira, M.; Ribo, S.; Martin-Polegre, A.J. Polarimetric mode of MIRAS. *IEEE Trans. Geosci. Remote Sens.* **2002**, *40*, 1755–1768. [CrossRef]

17. IAGA V-MOD Geomagnetic Field Modeling: International Geomagnetic Reference Field IGRF-12. Available online: https://www.ngdc.noaa.gov/IAGA/vmod/igrf.html (accessed on 22 February 2019).

18. Daily and Monthly Sunspot Number (Last 13 Years) | SILSO. Available online: http://www.sidc.be/silso/dayssnplot (accessed on 13 March 2019).

19. Kakoti, G.; Bhuyan, P.K.; Hazarika, R. Seasonal and solar cycle effects on TEC at 95°E in the ascending half (2009–2014) of the subdued solar cycle 24: Consistent underestimation by IRI 2012. *Adv. Space Res.* **2017**, *60*, 257–275. [CrossRef]

20. Rubino, R.; Duffo, N.; González-Gambau, V.; Corbella, I.; Durán, I.; Torres, F. Refining the Methodology to Correct the Faraday Rotation Angle from SMOS Measurements. In Proceedings of the IGARSS 2019—2019 IEEE International Geoscience and Remote Sensing Symposium (Under Review), Yokohama, Japan, 28 July–2 August 2019.

21. Corbella, I.; Torres, F.; Duffo, N.; Gonzalez, V.; Camps, A.; Vall-llossera, M. Fast Processing Tool for SMOS Data. In Proceedings of the IGARSS 2008—2008 IEEE International Geoscience and Remote Sensing Symposium, Boston, MA, USA, 7–11 July 2008; pp. II-1152–II-1155.

22. Bengoa, B. *SMOS Level 2 and Auxiliary Data Products Specifications*; Indra Sistemas: Alcobendas, Spain, 2017.

23. Hernández-Pajares, M. *IGS Ionosphere WG Status Report: Performance of IGS Ionosphere TEC Maps-Position Paper*; IGS Workshop: Bern, Switzerland, 2004.

24. Camps, A.; Corbella, I.; Bara, J.; Torres, F. Radiometric sensitivity computation in aperture synthesis interferometric radiometry. *IEEE Trans. Geosci. Remote Sens.* **1998**, *36*, 680–685. [CrossRef]

25. Corbella, I.; Torres, F.; Camps, A.; Bara, J.; Duffo, N.; Vall-Ilossera, M. L-band aperture synthesis radiometry: Hardware requirements and system performance. In Proceedings of the IGARSS 2000. IEEE 2000 International Geoscience and Remote Sensing Symposium. Taking the Pulse of the Planet: The Role of Remote Sensing in Managing the Environment. Proceedings (Cat. No.00CH37120), Honolulu, HI, USA, 24–28 July 2000; Volume 7, pp. 2975–2977.

26. Barre, H.M.J.P.; Duesmann, B.; Kerr, Y.H. SMOS: The Mission and the System. *IEEE Trans. Geosci. Remote Sens.* **2008**, *46*, 587–593. [CrossRef]

27. Oliva, R.; Daganzo, E.; Richaume, P.; Kerr, Y.; Cabot, F.; Soldo, Y.; Anterrieu, E.; Reul, N.; Gutierrez, A.; Barbosa, J.; et al. Status of Radio Frequency Interference (RFI) in the 1400–1427 MHz passive band based on six years of SMOS mission. *Remote Sens. Environ.* **2016**, *180*, 64–75. [CrossRef]

28. Tenerelli, J.; Reul, N. *Analysis of L1PP Calibration Approach Impacts in SMOS TB and 3-Days SSS Retrievals over the PACIFIC Using an Alternative Ocean Target Transformation Applied to L1OP Data*; Tech. Rep.; IFREMER/CL: Paris, France, 2010.

 *remote sensing*

Article

Copernicus Imaging Microwave Radiometer (CIMR) Benefits for the Copernicus Level 4 Sea-Surface Salinity Processing Chain

Daniele Ciani [1,*], Rosalia Santoleri [1], Gian Luigi Liberti [1], Catherine Prigent [2], Craig Donlon [3] and Bruno Buongiorno Nardelli [4]

[1] Consiglio Nazionale delle Ricerche, Istituto di Scienze Marine (CNR-ISMAR), 00133 Rome, Italy
[2] Laboratoire d'Études du Rayonnement et de la Matière en Astrophysique et Atmosphères (LERMA), 75014 Paris, France
[3] European Space Agency, ESA-ESTEC, 2201 AZ Noordwijk, The Netherlands
[4] Consiglio Nazionale delle Ricerche, Istituto di Scienze Marine (CNR-ISMAR), 80133 Naples, Italy
[*] Correspondence: daniele.ciani@cnr.it

Received: 11 July 2019; Accepted: 31 July 2019; Published: 3 August 2019

Abstract: We present a study on the potential of the Copernicus Imaging Microwave Radiometer (CIMR) mission for the global monitoring of Sea-Surface Salinity (SSS) using Level-4 (gap-free) analysis processing. Space-based SSS are currently provided by the Soil Moisture and Ocean Salinity (SMOS) and Soil Moisture Active Passive (SMAP) satellites. However, there are no planned missions to guarantee continuity in the remote SSS measurements for the near future. The CIMR mission is in a preparatory phase with an expected launch in 2026. CIMR is focused on the provision of global coverage, high resolution sea-surface temperature (SST), SSS and sea-ice concentration observations. In this paper, we evaluate the mission impact within the Copernicus Marine Environment Monitoring Service (CMEMS) SSS processing chain. The CMEMS SSS operational products are based on a combination of in situ and satellite (SMOS) SSS and high-resolution SST information through a multivariate optimal interpolation. We demonstrate the potential of CIMR within the CMEMS SSS operational production after the SMOS era. For this purpose, we implemented an Observing System Simulation Experiment (OSSE) based on the CMEMS MERCATOR global operational model. The MERCATOR SSSs were used to generate synthetic in situ and CIMR SSS and, at the same time, they provided a reference gap-free SSS field. Using the optimal interpolation algorithm, we demonstrated that the combined use of in situ and CIMR observations improves the global SSS retrieval compared to a processing where only in situ observations are ingested. The improvements are observed in the 60% and 70% of the global ocean surface for the reconstruction of the SSS and of the SSS spatial gradients, respectively. Moreover, the study highlights the CIMR-based salinity patterns are more accurate both in the open ocean and in coastal areas. We conclude that CIMR can guarantee continuity for accurate monitoring of the ocean surface salinity from space.

Keywords: sea surface salinity; microwave remote sensing; CIMR; copernicus marine service

---

1. Introduction

The salinity of the ocean is a crucial parameter to investigate the water cycle, the ocean dynamics and the marine biogeochemistry from the global to the regional scale. It was classified as an Essential Climate Variable in the context of the Global Climate Observing System (GCOS) programme. Indeed, salinity affects both the sea water density and the marine carbonate chemistry (alkalinity), making it a fundamental variable to investigate the thermohaline global circulation, the local surface and deep circulation, the water mass transformation and the uptake of carbon by the ocean including ocean acidification, e.g., [1–4].

In the past, salinity has suffered from poor observational coverage, being uniquely sampled via point-wise in situ observations, impacting the accurate assessment of its spatial variability and dynamics, in particular for the construction of gap-free optimally interpolated fields, e.g., the in situ-Based Reanalysis of the Global Ocean Temperature and Salinity (ISAS) [5]. The three-dimensional ISAS dataset was recently improved using quasi-global ARGO floats observations [6]. In addition, the characterization of the sea-surface salinity (SSS) has benefited from satellite observations and on their synergy with in situ measurements, e.g., [7–9].

In the framework of the Copernicus Marine Environment Monitoring Service (CMEMS), a daily, mesoscale resolving SSS multi-year gap-free Level-4 analysis (L4 hereinafter) product was developed by Buongiorno Nardelli et al. 2016 [10] and Droghei et al. 2018 [11]. The approach of [10,11], unlike [5–9], relies on a multi-dimensional (multivariate) optimal interpolation (OI) algorithm that combines both Soil Moisture Ocean Salinity (SMOS) satellite retrievals and in situ salinity measurements with satellite sea-surface temperature information. This product is distributed operationally in near real time in late 2018. Despite the success of this product, there are no missions planned to secure continuity of satellite SSS measurements after the SMOS and Soil Moisture Active Passive (SMAP) missions, which could impact both the accuracy and the effective resolution of the CMEMS SSS operational products.

The Copernicus Imaging Microwave Radiometer (CIMR) satellite mission [12] is being developed as a High Priority candidate Mission (HPCM) in the context of the European Copernicus Expansion Programme. To address the needs of Copernicus and the Integrated European Policy for the Arctic, the CIMR mission will carry a wide-swath (>1900 km) conically scanning multi-frequency microwave radiometer. CIMR measurements will be made over a forward scan arc followed $\simeq$260 s later by a second measurement over a backward scan arc. Polarised (H and V) channels centered at 1.414, 6.925, 10.65, 18.7 and 36.5 GHz are included in the mission design under study (full Stokes parameters are foreseen). The real-aperture resolution of the 6.925/10.65 GHz channels is <15 km and 5/4 km for the 18.7/36.5 GHz channels, respectively. The 1.414 GHz channel will have a real-aperture resolution of $\simeq$60 km (fundamentally limited by the size of the $\simeq$8 m deployable mesh reflector). However, all channels will be oversampled allowing gridded products to be generated at much better spatial resolution. Channel NEdT is 0.2–0.8 K with a goal absolute radiometric accuracy of $\simeq$0.5 K. Radio Frequency Interference (RFI) will be mitigated on-board the satellite using dedicated processors. CIMR will fly in a dawn-dusk orbit providing $\simeq$95% global all weather coverage every day with one satellite and complete (no hole-at-the-pole) sub-daily coverage of the polar regions. CIMR will operate in synergy with the EUMETSAT MetOp-SG(B) mission so that over the polar regions (>60°N and 60°S) collocated and contemporaneous measurements between CIMR and MetOp Microwave Imager and SCA scatterometer measurements will be available within ±10 min. Thus, CIMR is designed to provide global mesoscale-to-submesoscale resolving observations of sea-surface temperature, sea-surface salinity and sea-ice concentration. The availability of the CIMR SSSs with global coverage would represent an opportunity to continue ingesting satellite SSS measurements in the CMEMS multivariate OI, potentially allowing to improve the interpolated product effective spatial resolution and accuracy with respect to L4 analyses built from in situ observations alone. The present study is thus focused at demonstrating this potential.

We will implement an Observing System Simulation Experiment (OSSE) based on the CMEMS MERCATOR Global Operational model outputs [13] and relying on the present-day version of the CMEMS SSS processing chain. In particular, the multivariate OI algorithm will be first run ingesting only synthetic in situ SSS extracted from the MERCATOR outputs. In a second run, the OI processing will rely on both synthetic in situ and CIMR observations, the latter also built from the MERCATOR model outputs accounting parametrically for the expected CIMR accuracy. Finally, the L4 SSS resulting from the two versions of the OI algorithm will be validated against the original MERCATOR global

SSS. The impact of the future CIMR SSS in the OI processing will be demonstrated and quantified throughout the paper.

In Section 2, we present the datasets and methods used to carry out the study. This section illustrates both the principles of the OI algorithm as well as the details on the input-data processing and the structure of the OSSE. We go on presenting the main results of the OSSE in Section 3. In the end, Section 4 will discuss the main results and illustrate the main conclusions and perspectives of our study.

2. Materials and Methods

This section contains the description of the dataset used to carry out the study, the principles of the CMEMS L4 SSS processing chain and the structure of the OSSE for evaluating the potential of the future CIMR SSS remote measurements.

2.1. Data

- One year of daily SSS and SST data were extracted from the CMEMS MERCATOR global operational model, [13]. The model is based on the NEMO hydrodynamical framework and assimilates satellite SST, sea ice concentration, sea surface height and in situ thermohaline vertical profiles.
  We focused on the year 2016 that is also compatible with the other datasets used in this study. The MERCATOR outputs are mapped on a $1/12°$ regular grid for the global ocean. The SSS and SST timeseries were used for several purposes, like generating the synthetic in situ and future CIMR SSS observations, running the OI algorithm and finally assessing the quality of the L4 SSS maps given by the OI processing chain (see Sections 2.2, 2.4 and 3).
- One year (2016) of daily SST, Ocean Wind Speed (OWS), Total Cloud Liquid Water (TCLW) and Total Cloud Water Vapour (TCWV) were obtained from the Second Advanced Microwave Scanning Radiometer (AMSR-2) observations, distributed by Remote Sensing Systems [14]. These data are mapped on a $1/4°$ regular grid for the global ocean. They are distributed as L3U (uncollated) data, i.e., reporting measurements for both the ascending and the descending satellite orbits for each grid box, including the boxes where swath overlapping occurs. The AMSR-2 data were used in combination with the MERCATOR SSS to optimize the generation of the synthetic CIMR L3 SSS (see Section 2.4).

2.2. Processing Chain Description

The CMEMS SSS operational chain combines in situ and satellite-derived observations using the multidimensional OI technique originally introduced by Buongiorno Nardelli 2012 [15] and based on the findings of Bretherton et al. 2016 [16]. Presently, this processing chain provides global L4 weekly SSS mapped over a $1/4°$ regular grid. The algorithm is based on the assumption that SSS and SST covary at spatial scales smaller than those characterizing atmospheric fluxes. This allows us to extract useful dynamical information from the high-pass filtered satellite L4 SSTs for use with the SSS fields.The L4 SSTs are generally obtained by merging microwave and infrared observations, the latter having resolutions $\mathcal{O}(1\ km)$ [17]. and provide guidance to the OI system when interpolating SSS fields resulting in an enhanced effective resolution compared to simple space-time interpolation approaches. The technique, originally developed to interpolate in situ SSS, was successively adapted to ingest satellite observations from SMOS and finally calibrated to compute dynamically consistent SSS/SSD datasets [10,11,18]. The theoretical framework of the SSS multivariate OI is briefly illustrated below, referring the reader to [10,11,15,18] for more details.

The optimal SSS analysis ($\overline{SSS}_{analysis}$) is given by a weighted sum of the SSS observations anomalies ($\overline{SSS}_{obs}$) with respect to a first guess background field ($\overline{SSS}_{background}$). The weights provide an unbiased estimate (i.e., it has the same mean as the true field) with the minimum expected estimate error (in a least squares sense):

$$\overline{\text{SSS}}_{\text{analysis}} = \overline{\text{SSS}}_{\text{background}} + C(R+C)^{-1}(\overline{\text{SSS}}_{\text{obs}} - \overline{\text{SSS}}_{\text{background}}). \tag{1}$$

In Equation (1), C is the background error covariance matrix and R, assumed diagonal, is the observations error covariance matrix (here defined by constant values per each observation type/platform). In the implementation described by [10,11], the background field is provided by an analysis built from in situ observations alone through a classical OI. More recently, in the framework of CMEMS, the background field estimate was modified by computing a first round of space-time OI based on in situ input data relying on a monthly climatology. The structure of the background error covariance matrix, C, is given by Equation (2):

$$C(\Delta r, \Delta t, \Delta \text{SST}) = e^{-(\Delta r/L)^2} e^{-(\Delta t/\tau)^2} e^{-(\Delta \text{SST}/T)^2}, \tag{2}$$

where:

- Δr, Δt, and ΔSST respectively indicate the spatial, temporal, and thermal separations;
- L, τ, and T are the spatial, temporal, and thermal decorrelation lengths. Their values have been defined by previous studies [15] and are L = 500 km, τ =7 days, T = 2.75 K.
- The SST L4 data are high-pass-filtered (cut-off at 1000 km).

This covariance model approximates a multivariate approach that includes both SSS and SST in the state vector used to build the observation matrix (that is then used to estimate the covariance matrix C). In practice, the multivariate covariance model gives more weight to observations found on the same isothermal of the interpolation point compared to observations found at the same spatial and temporal separation but characterized by different SST values. The L4 SSTs, due to their high spatial resolution, accurately describe the thermal signature of the ocean mesoscale features at global scale. Their ingestion in the OI algorithm has already shown to improve L4 SSS effective resolution (e.g., [11]). Moreover, the processing chain relies on a dataset of pseudo in situ SSS observations to overcome the sparseness of avaliable in situ SSS. The pseudo-observations are extracted from the background field (1 every 16 grid boxes) and are used as additional input. In coastal areas (at distances < 200 km from the coast), the pseudo-observations are taken from the climatological background. This strategy guarantees the homogeneity of the L4 SSS spatial resolution in case of prolonged (in time) or extended (in space) input data gaps during the interpolation process.

In the present study, the input data for the OI processing, i.e., the in situ, satellite, pseudo SSS and high resolution L4 SST, are all derived from the MERCATOR numerical simulations (see also Section 2.3). In Equation (1), the constant values included in the observation error covariance matrix R are divided by the signal variance, thus representing the different noise-to-signal levels for each type of input data. These values have been found through dedicated tuning experiments and are 0.5 for the pseudo observations, 0.1 for the synthetic CIMR satellite SSS and 0.05 for the synthetic in situ observations. The mean value used for CIMR may appear too optimistic (only twice the error associated with synthetic in situ) and could be optimized by considering the expected latitudinal dependence of CIMR error. This dependence, however, is not presently handled by the CMEMS processing chain and is thus left to future developments.

2.3. OSSE Description

Selecting one year (2016) of daily SSS data from the CMEMS MERCATOR global operational model, we subsampled the SSS fields in two ways, according to the following purposes:

1. the simulation of synthetic in situ SSS observations;
2. the simulation of the expected future CIMR satellite observations, taking into account the number of satellite passes over Earth and the expected uncertainty on the SSS retrieval.

The processing chain is run in two different configurations: once ingesting the synthetic in situ SSS alone and, in a second run, the in situ plus the CIMR observations. In both cases, the result of the processing, i.e., the multivariate OI algorithm, is a global map of L4 SSS. Finally, the SSS maps obtained in the two configurations is compared with the original MERCATOR SSS fields, our benchmark. The metrics of the comparison are the root mean square error (RMSE) and the power spectral density (PSD). A flowchart of the OSSE is provided in Figure 1. The details on the simulation of the synthetic data are given in Section 2.4.

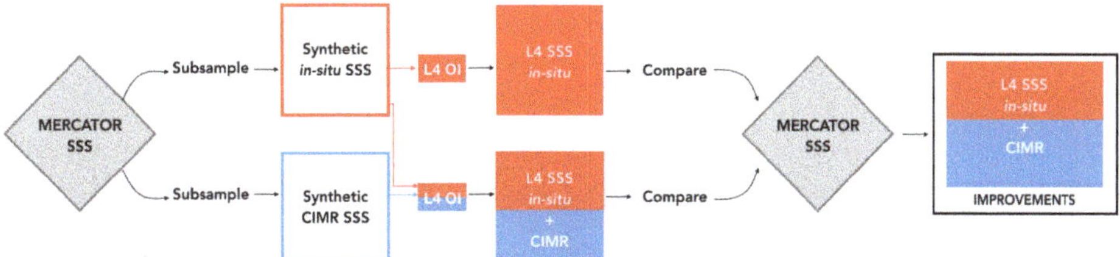

Figure 1. Workflow of the observing system simulation experiment.

2.4. Input Data Preparation: Simulating the SSS Observations

Here, we discuss the generation of the synthetic in situ and CIMR SSS observations.

1. We generated synthetic in situ SSS observations from the MERCATOR simulations. This was achieved by colocating the MERCATOR SSS with the quality controlled in situ observations from ARGO floats and CTD casts ingested by the in situ Analysis System (ISAS) [5,6]. Such in situ data were previously binned on daily basis over a regular 1/4° grid. An example of daily in situ SSS is provided in Figure 5b. In the figure, the distribution of the pseudo observations mentioned in Section 2.2 is also given;

2. A one year long time series of SSS and SST was extracted from the CMEMS MERCATOR global operational model. In order to mimic the CIMR SSS observations given by the 1.4 GHz measurements (see Table 1, and [12,19] for the CIMR measurements frequencies), we low-pass filtered the SSSs with a 55 km cut-off wavelength and we remapped them onto a regular 1/4° grid, i.e., exactly the same as the output of the present-day CMEMS SSS L4 processing chain. The more recent release of the CIMR mission requirements document [12] indicates a 1.4 GHz real aperture resolution less than 60 km, which is consistent with the 55 km of [19] also used in the present study. The MERCATOR SSTs were simply remapped over a 1/4° regular grid in order to simulate the L4 SST to be ingested in the SSS OI processing (see Section 2.2).

3. We then generate synthetic CIMR-SSS starting from the low-pass filtered MERCATOR SSS described at point 2. This includes both the expected uncertainties on the CIMR SSS and the expected satellite coverage. The SSS retrieval uncertainty (hereinafter referred to as σSSS) was provided by the theoretical estimates of Kilic et al. 2018 [19] (see, e.g., Figure 7 in [19]). The authors derived a Look Up Table (LUT) containing the σSSS as a function of the local SSS, SST, OWS, TCLW and TCWV. The ranges of variability of the aforementioned parameters are given by Table 2.

Table 1. CIMR instrument measurement frequencies.

Frequency (GHz)	Spatial Resolution (Km)
1.4	55
6.9	15
10.65	15
18.7	5
36.5	5

Table 2. Variability range of the LUT quantities used in [19].

Variable	Range
SST (K)	271–303
OWS (m/s)	0–25
SSS (PSU)	0–38
TCWV (kg·m^{-2})	4–40
TCLW (g·m^{-2})	0–500
σSSS (PSU)	0.27–0.99

Based on the LUT estimates, we could generate a time series of daily L4 σSSS. This was achieved relying on the daily SST-OWS-TCLW and TCWV observations provided by AMSR-2 [14] plus the daily MERCATOR SSS. In order to fill the gaps in the AMSR-2 observations, we used a linear interpolator, leading to a gap-free proxy of the σSSS behaviour. However, in order to rely on higher quality data, the areas relying solely on interpolation were not taken into account when running the OI algorithm. This is further justified at the end of this section and schematically represented in Figures 4 and 5. Checking different SSS-SST-OWS-TCLW-TCWV combinations, we found that the local low SST and high OWS are primarily responsible for the σSSS increase, consistently with [19]. An example of the σSSS is provided in Figure 2.

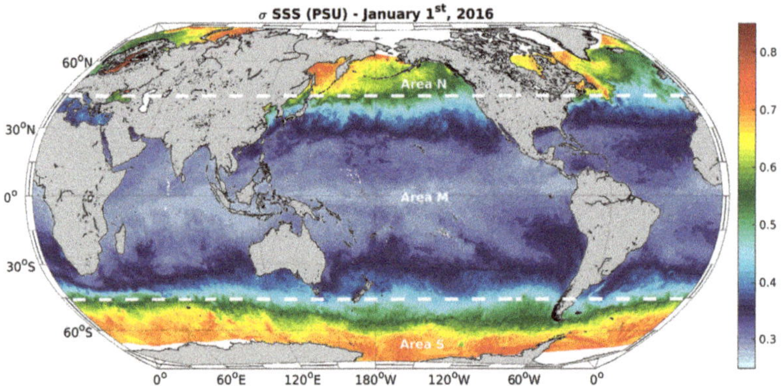

Figure 2. σSSS computed according to [19], example from 1 January 2016. The additional information in white are referenced in Section 3.2.1.

Based on these results, we added a white random Gaussian noise to the MERCATOR-SSS according to Equation (3):

$$\text{SSS}_{\text{noise}} = \text{SSS} + \text{WGN}(\sigma\text{SSS}), \tag{3}$$

where WGN(σSSS) stands for a σSSS dependent White Gaussian Noise. Figure 3 shows an example of the MERCATOR SSS after addition of the white noise. In the present work, only white noise has been taken into consideration. While actual errors would likely include instruments drift or other variations with time, these effects are still in evaluation [12]. For the moment, the error estimates (as also stated by [19]) are mostly based on the local ocean-atmosphere conditions at the measurement site. The impact of the CIMR radiometric and orbital stability on the global L4 SSS estimates maps is thus left for future investigations.

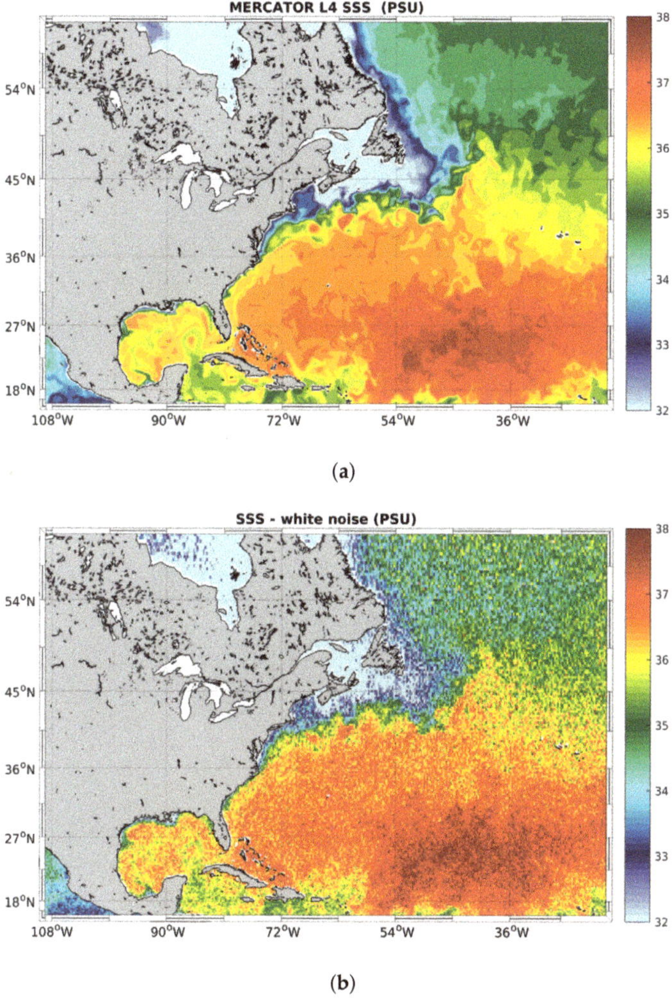

(a)

(b)

Figure 3. (**a**) MERCATOR SSS, 1 January 2016, Gulf Stream area; (**b**) MERCATOR SSS with addition of white noise according to Equation (3); 1 January 2016, Gulf Stream area.

In order to simulate the CIMR coverage, we used the 28-days cycle of the CIMR satellite overpasses over Earth provided by the European Space Agency, remapped onto a regular 1/4° grid. Such 28 days cycle was applied to our time series of MERCATOR SSS, arbitrarily assuming that the first day corresponds to 1 January 2016 and repeating the cycle throughout the year. The number of overpasses (NVIS) per day is between 0 (in a few small areas of the tropics) and 11 (in polar regions) (Figure 4a). In all the areas where NVIS exceeded 1, we used an averaged SSS obtained as follows: we oversampled the MERCATOR SSS according to NVIS (adding each time a different random Gaussian noise) and then computed the mean SSS in each grid box.

As a final step, in order to make the synthetic satellite-derived SSS more realistic, we derived a sea-ice mask, land mask and rainfall observations mask (i.e., a no observation mask) using the REMSS AMSR-2 SST L3U observations [14]. This is also schematically represented in Figure 4a–c. Except for extreme cases, L-band brightness temperatures are not significantly affected by precipitation, remaining within the CIMR radiometric accuracy [12,20]. In principle, this means a SSS retrieval is possible. However, in these cases, since higher frequency channels are significantly affected by precipitation, the retrieval would require ancillary information on relevant geophysical variables (e.g., SST and OWS). This would require an independent evaluation of the expected uncertainty on the SSS retrieval.

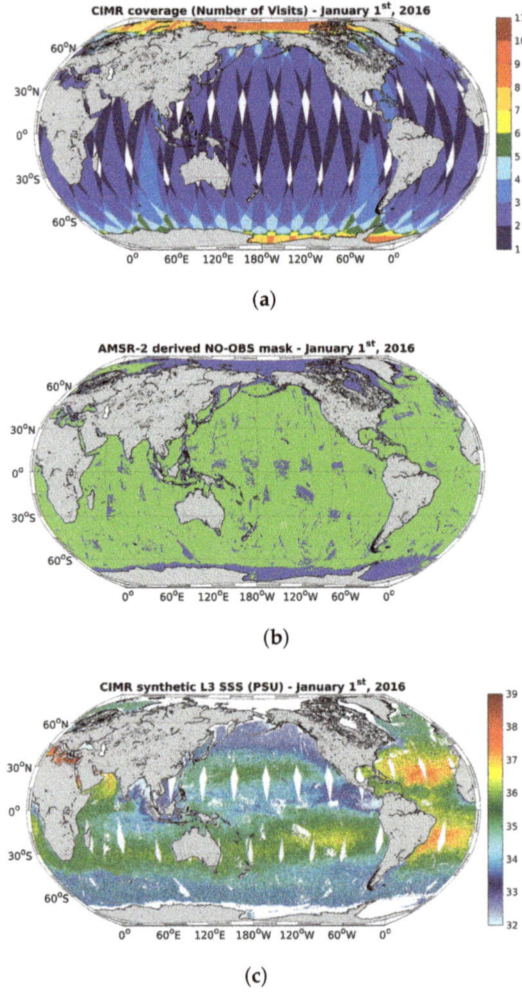

Figure 4. Simulating the CIMR observations from the MERCATOR SSS. (**a**) expected CIMR coverage; (**b**) daily mask for land, sea-ice and precipitation from AMSR-2 SST observations (blue and green respectively stand for available and unavailable observations); (**c**) synthetic CIMR observations obtained combining the information on the CIMR overpasses, the AMSR-2 observations and the noise. All of the figures are mapped onto a regular 1/4° grid (the same as the present-day CMEMS L4 SSS) and refer to 1 January 2016.

3. Results

In this section, we present the results of the OSSE described in Section 2.3 and summarized by Figure 1. Two different configurations of the CMEMS L4 SSS processing chain (including and not including the CIMR synthetic observations) are qualitatively and quantitatively validated against the

MERCATOR SSS, constituting the true SSS field. In the following, the L4 SSS given by the OI of in situ observations and the ones obtained combining in situ plus the CIMR estimates will be referred to as IL4 and CIL4, respectively.

3.1. Qualitative Validation

Observing the IL4 (Figure 5a) and the CIL4 (Figure 5c), a qualitative validation can be carried out using the original MERCATOR SSS as a benchmark (Figure 5e). We present a case study on one of the most dynamically active areas of the global ocean: the Gulf Stream (1 January 2016). In general, when the CIMR observations are not included, the resulting OI SSS misrepresents the salinity values as well as the mesoscale activity found in the benchmark salinity field. Indeed, the IL4 are given by a smooth field, close to a climatological estimate, with mesoscale activity appearing only in proximity of the in situ observations, where the multivariate algorithm can account for the spatial, temporal and thermal decorrelation (given by the L4 SST field) as indicated in Equation (2). This statement is confirmed by visual inspection of Figure 5a–f.

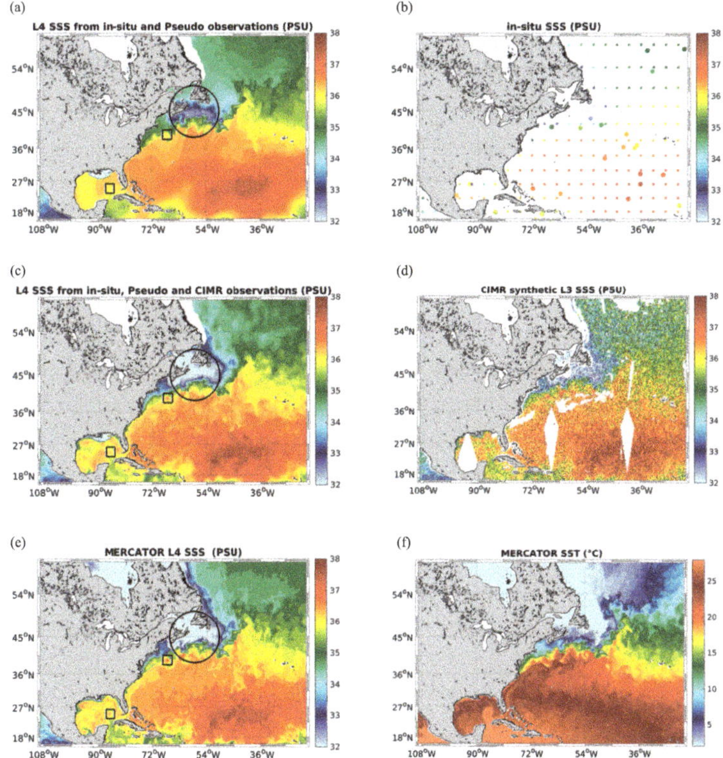

Figure 5. (**a**) L4 SSS from in situ observations (IL4); (**b**) extraction of in situ SSS from the MERCATOR SSS, squares and circles, respectively, stand for pseudo and in situ observations; (**c**) L4 SSS from the combination of in situ and CIMR observations (CIL4); (**d**) simulated CIMR L3 SSS; (**e**) MERCATOR SSS (benchmark); (**f**) MERCATOR SST. All figures refer to 1 January 2016, in the Gulf Stream Area.

Moreover, in Figure 5, we highlighted the basin south of Newfoundland and New Scotland using a black circle. Here, the IL4 underestimates the true SSS by about 1 to 1.5 PSU. When the CIMR observations are ingested in the OI processing, the salinity values are corrected and agree with the reference SSS. Finally, we discuss the CIMR performances in resolving the signature of two eddies located off New Jersey and in the Gulf of Mexico. In Figure 5a,c,e, these eddies are highlighted by two black squares. If the CIMR observations were not used in the OI processing, their signatures in the SSS field would either disappear or only partially be resolved.

3.2. Quantitative Validation

The potential of the future CIMR SSS is here demonstrated through quantitative analyses. The metrics of the validation are based on the computation of the RMSE and PSD.

3.2.1. Temporal Variability of the CIMR Impact in the CMEMS SSS

We computed the time series of the RMSE between the outputs of the CMEMS L4 processing chain and the true SSS field. Such statistics are based on weekly data for the year 2016. The main results of the validation are summarized by Figure 6a–c. The statistics have been computed in three latitudinal bands: 90°S to 45°S (referred to as Area S), 45°S to 45°N (referred to as Area M) and 45°N to 90°N (referred to as Area N). This choice is due to the behavior of the average σSSS, whose map is well approximated by Figure 2, where these areas have been highlighted. The 45°S/N latitudes correspond to the areas where the σSSS reaches half of its maximum magnitude, i.e., $\simeq$0.45 PSU, and then rapidly increases up to a maximum of $\simeq$0.9 PSU moving towards the polar regions. On the other hand, in the 45°S to 45°N latitudinal band, the average σSSS is mostly around 0.3 PSU.

As a general comment, the quantitative validations of the L4 SSS show that CIMR SSS will undoubtedly bring benefits for the CMEMS SSS operational products. The CIMR SSSs guarantee to reconstruct L4 salinity maps that systematically reduce the RMSE with respect to the true SSS, compared to products relying on in situ observations alone. In the Area M, the RMSE exhibits the largest improvements, whose magnitude is around 50% throughout the whole year 2016. The improvement is evaluated according to Equation (4) [21]:

$$\text{IMPROVE} = 100 \times \left[1 - \left(\frac{RMSE_{CIL4}}{RMSE_{IL4}} \right)^2 \right]. \tag{4}$$

In the Area N, the improvements brought by CIMR vary between 20% and 40%, with the largest values observed during summertime. The RMSE time series of both the CIL4 and the IL4 exhibit a seasonal behaviour with enhanced values during summertime ($\simeq$3 PSU), which is a known behaviour for the CMEMS SSS, also discussed by Xie et al. 2019 [22]. The Area S is the only region exhibiting slight degradation using CIMR observations within CMEMS. Here, we could observe improvements reaching 30% during the austral Summer and Fall but from June to November, the CIL4 RMSE increases by about 10^{-2} PSU compared to the IL4 reconstruction, mostly indicating that CIMR is not bringing a useful contribution to the SSS reconstruction in these areas during the austral Spring and Winter.

3.2.2. Spatial Variability of the CIMR Impact in the CMEMS SSS

In order to quantify the spatial variability of the CIMR improvements, we compared the local temporal RMSE between the CIL4 and the IL4. This was achieved by means of Equation (5), using the MERCATOR SSS as a reference:

$$\Delta\text{RMSE} = \text{RMSE}(\text{SSS}^{CIL4}) - \text{RMSE}(\text{SSS}^{IL4}). \tag{5}$$

The negative ΔRMSE values indicate an RMSE reduction with respect to the true SSS, hence, an improved SSS retrieval given by CIMR. Figure 7a indicates that the CIMR observations improve the SSS retrieval in 63% of the world ocean, exhibiting better performances in the Area M (45°S to 45°N) and in coastal regions, which are characterized by the main upwelling systems and the larger river inputs. This is easily explained considering that the density of in situ observations is lower in coastal regions than in the open ocean, as confirmed by Figure 7b. In the figure, the white halo located in correspondence of the coastal waters indicates the absence of in situ SSS estimates. Thus, progressively approaching the coastline, the relative contribution of the CIMR observations improve the SSS variability of the L4 interpolated fields. However, the CIMR sensor in itself is not expected to perform better in costal zones (because of a large measurement footprint). This is confirmed by the analyses reported in Figure 8, where both the $RMSE^{CIL4}$ and $RMSE^{IL4}$ generally increase as the coastline is approached. In the future, provided the characteristics of the receiving antenna, an estimate of the expected land contamination will be possible. This will enable a more realistic evaluation of the CIMR performances within the CMEMS SSS in coastal areas.

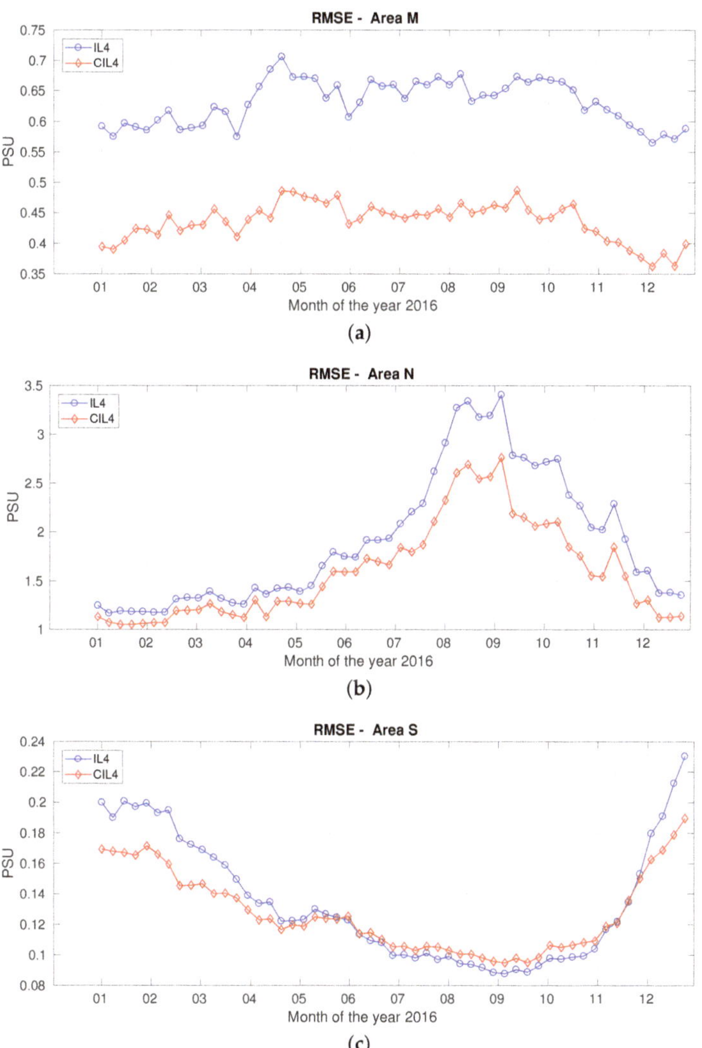

Figure 6. (**a**) RMSE between the OI L4 SSS and the MERCATOR outputs. Blue and red, respectively, stand for IL4 and CIL4 reconstructions. The statistics are referred to the 45°S to the 45°N latitudinal band (Area M); (**b**) analyses referred to the the 45°N to the 90°N latitudinal band (Area N); (**c**) analyses referred to the 90°S to the 45°S latitudinal band (Area S).

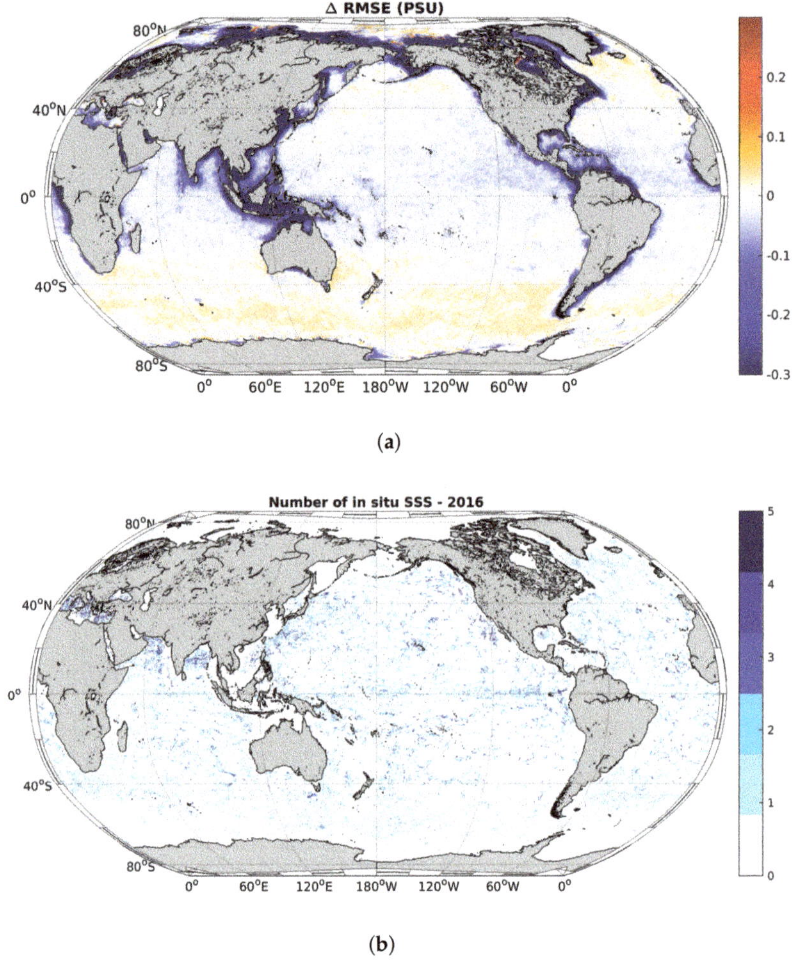

Figure 7. (**a**) ΔRMSE based on weekly data, year 2016; (**b**) density of in situ SSS for the year 2016. The maximum number of in situ observations is ≃140 (in the North Atlantic). The colorbar is saturated to 5 in order to facilitate the visualization of the measurement sites at a global scale.

An overall CIL4 degradation is observed in the Area S and in the 60°W–10°W zone of Area N, where the $\mathrm{RMSE}^{\mathrm{CIL4}}$ exceeds the $\mathrm{RMSE}^{\mathrm{IL4}}$ by 0.01 PSU on average, which is also confirmed by Figure 8a,b. This behaviour is discussed in Section 3.3 in more detail. As an additional analysis, we compared the SSS gradients magnitude found in the CIL4 and IL4 reconstructions. This was performed in a similar fashion as for the SSS fields, i.e., computing the $\Delta\mathrm{RMSE}_\nabla$:

$$\Delta\mathrm{RMSE}_\nabla = \mathrm{RMSE}(|\nabla\mathrm{SSS}^{\mathrm{CIL4}}|) - \mathrm{RMSE}(|\nabla\mathrm{SSS}^{\mathrm{IL4}}|) \tag{6}$$

with $|\nabla\mathrm{SSS}| = \sqrt{(\partial_x\mathrm{SSS})^2 + (\partial_y\mathrm{SSS})^2}$, where the quantities ∂_x and ∂_y are estimated via a centered finite differences numerical scheme and the subscripts "x,y" respectively stand for the zonal and meridional directions.

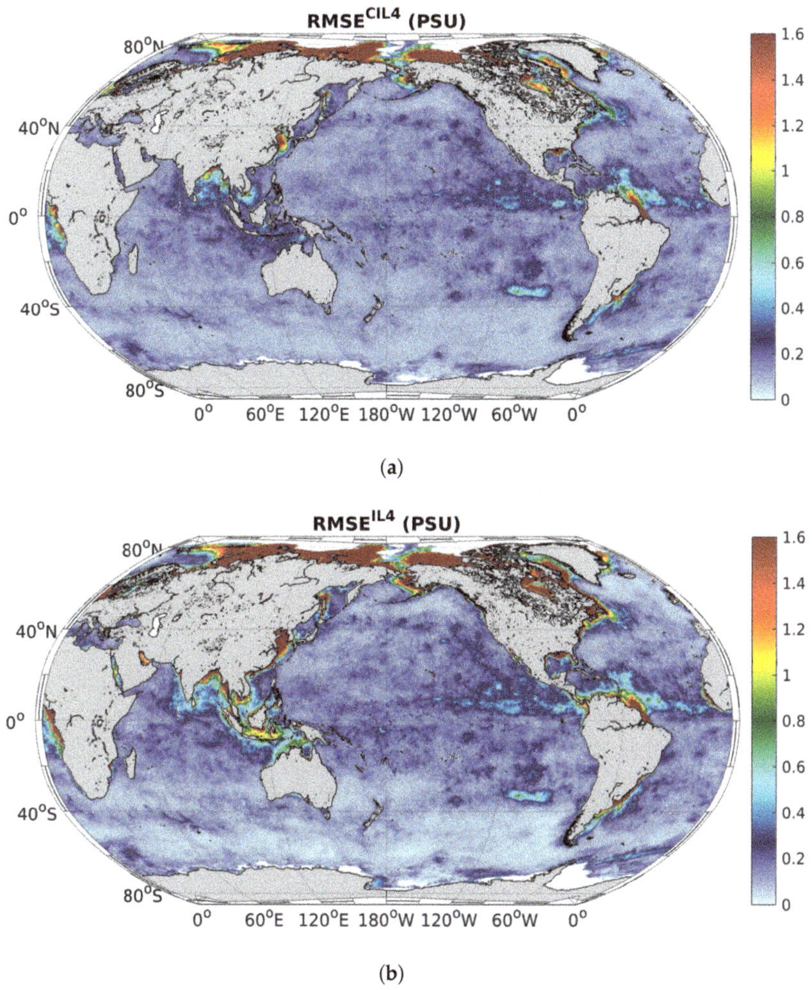

Figure 8. (a) RMSE^{CIL4}; (b) RMSE^{IL4}.

Accurate estimates of the SSS gradients in L4 products is crucial from a physical point of view. Combined with the information on SST, it gives access to the patterns of the surface density gradients, allowing for diagnosing the global ocean surface dynamics. In addition, the surface density gradients allow for predicting the subsurface circulation from surface observations [23], which justifies the interest in evaluating the CIMR contribution for the monitoring of this variable. CIMR itself will provide global SST fields at the same time as SSS based on the use of 6.9 GHz channel data where the real aperture of the CIMR channel is $\simeq$15 km. According to Figure 9, CIMR improves the SSS gradients retrieval in 70% of the world ocean. This is more evident in the Area M and in coastal waters. As for the previous analysis, the Areas S and N show reduced performance, where the averaged RMSE($|\nabla$SSS$^{CIL4}|$) exceeds by about 0.04 PSU·m^{-1} the one based on in situ observations alone. At latitudes exceeding 75°N, the ΔRMSE$_\nabla$ shows alternating patterns of large improvement and degradation. This is not in agreement with the behavior of the ΔRMSE, where an overall improvement of the SSS values is observed. This indicates that, in this region, the CIMR SSS contributes to accurately describe the temporal variability of the SSS but only locally improves the estimate of the SSS gradients with respect to a climatological field.

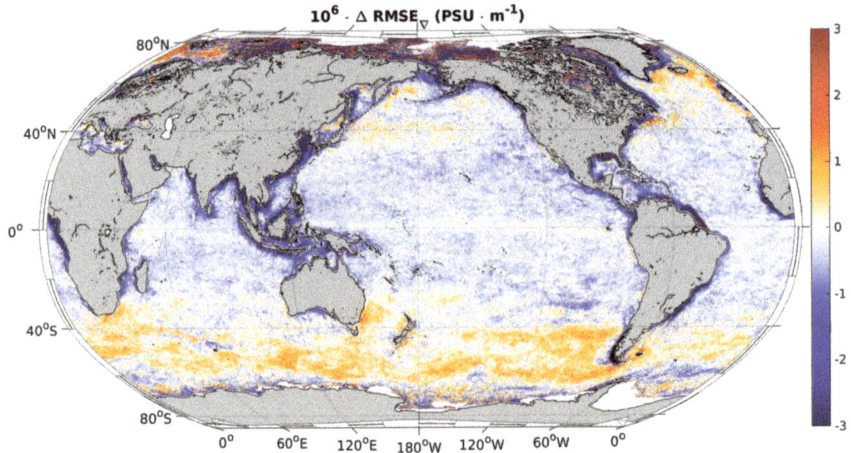

Figure 9. ΔRMSE$_\nabla$ based on weekly data, year 2016.

The statistical significance of the results shown in Figures 7–9 was evaluated via a bootstrap resampling technique. The analysis confirmed that both the ΔRMSE and ΔRMSE$_\nabla$ are in the 95% confidence interval.

3.3. Further Insights on the Spatial-Temporal Variability of the CIMR Performances

The spatial-temporal variability of the CIMR performances is consistent with the average physical conditions of the ocean surface throughout the year 2016. The evolution of the daily $\langle\sigma\text{SSS}\rangle$, i.e., the CIMR spatially averaged measurement uncertainty confirms that the Area S exhibits the largest values throughout the year 2016 (Figure 10a). This is mostly due to the persistent low mean SST ($\simeq$5 °C) and high OWS ($\simeq$10.5 m·s^{-1}) in this area [19] indicating a well mixed surface layer. This was obtained using weekly AMSR-2 observations. Moreover, during the austral Spring and Winter, the $\langle\sigma\text{SSS}\rangle$ exhibits a further increase which results in the degradation of the CIL4 estimates illustrated in Figure 6c.

Following the same logic, the fairly constant CIL4 improvements observed in the Area M are also explained. Indeed, in the 45°S to 45°N latitudinal band, the $\langle\sigma\text{SSS}\rangle$ is permanently around 0.35, guaranteeing an optimal SSS retrieval from CIMR. The CIMR measurement uncertainty of the Area N enables generally improving the CIL4 estimates compared to the IL4 reconstruction. Here, the $\langle\sigma\text{SSS}\rangle$ also exhibits a larger seasonal cycle than in the Area S, due to the enhanced SST and OWS variability of the northern hemisphere, in agreement with the results of Dunstan et al. 2018 [24]. This explains the time dependence of the CIL4 improvements found in the Area N, yielding a more precise reconstruction during the Summer.

The behaviour of the $\langle\sigma\text{SSS}\rangle$ also explains the spatial variability of the CIMR benefits within the CMEMS SSS operational products, summarized by Figures 7a and 9. These figures suggest that CIMR will improve the SSS and SSS gradients estimates in the 60% and 70% of the world ocean, respectively. Nevertheless, the Area S is the main degradation zone for both the CIL4-SSS and CIL4-SSS spatial gradients. The large values of $\langle\sigma\text{SSS}\rangle$ during April to November 2016 are most likely the responsible of this degradation, whose signature emerges in the ΔRMSE and ΔRMSE$_\nabla$.

The present-day version of the CMEMS SSS processing chain may also be responsible for the partial degradation observed in the Area S. In future studies, we plan to further tune of the operational chain in this area. This will be achieved considering a different weighting of the CIMR SSS in the multivariate OI algorithm, given their decreased accuracy during the austral Spring and Winter.

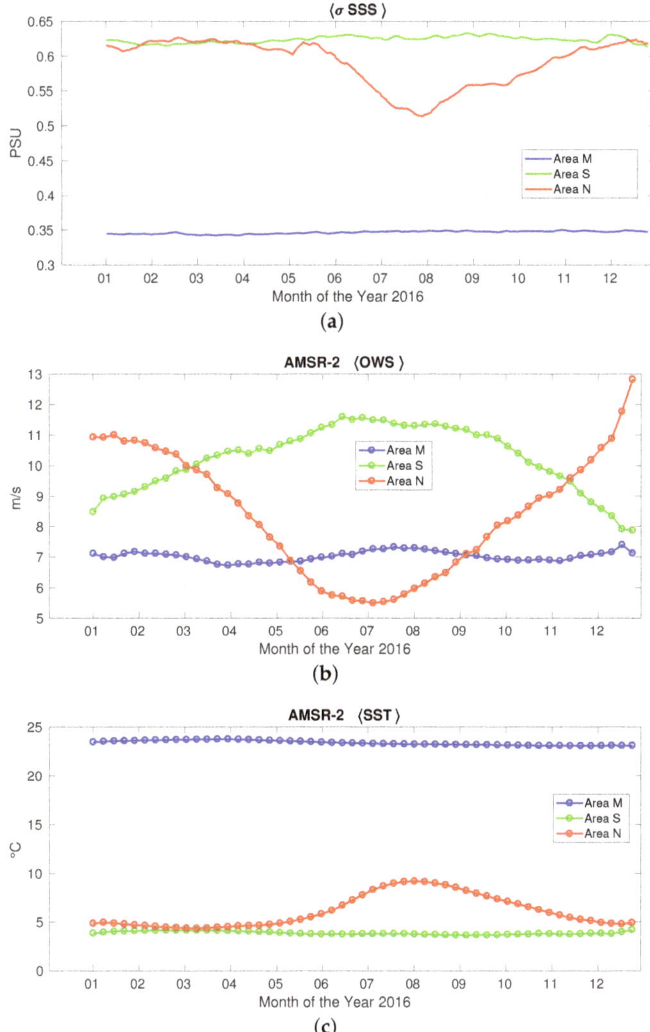

Figure 10. (**a**) blue line: $\langle\sigma SSS\rangle$ in the 45°S to 45°N latitudinal band (Area M). Red line: $\langle\sigma SSS\rangle$ from 45°N to 90°N (Area N) and from 45°S to 90°S (Area S); (**b**) same analysis for the AMSR-2 derived OWS; (**c**) same analysis for the AMSR-2 derived SST.

3.4. Spectral Content of IL4 and CIL4

Here, we describe the capability of CIMR to retrieve the signatures of mesoscale activity in the CMEMS SSS. This will be assessed via a spectral analysis of the IL4 and CIL4 compared to the reference SSS provided by MERCATOR. We perform the spectral analyses in five land-free areas of the world ocean: the North and South Atlantic, the North and South Pacific and the Indian Ocean. The SSS power spectral density (PSD) computation is compliant with [11] and is based on Fast Fourier Transform with the Blackman–Harris window for the reduction of spectral leakage.

The spectra are computed over the entire 2016 using weekly data for both the L4 reconstructions. The time average of the mean zonal spectra is presented in Figure 11 as a function of the spatial wavenumber, indicating spectral properties of the IL4, CIL4 and MERCATOR SSS that agree in all the aforementioned investigation areas. Compared to the true SSS, the IL4 spectra (red lines of Figure 11) show a PSD drop around 0.1 degrees^{-1}. On the other hand, the CIMR SSS allows for building interpolated SSS maps with spectral properties that follow the true SSS, except for the addition of noise at scales below 0.8 degrees^{-1} (on average), where the CIL4 spectra begin to flatten.

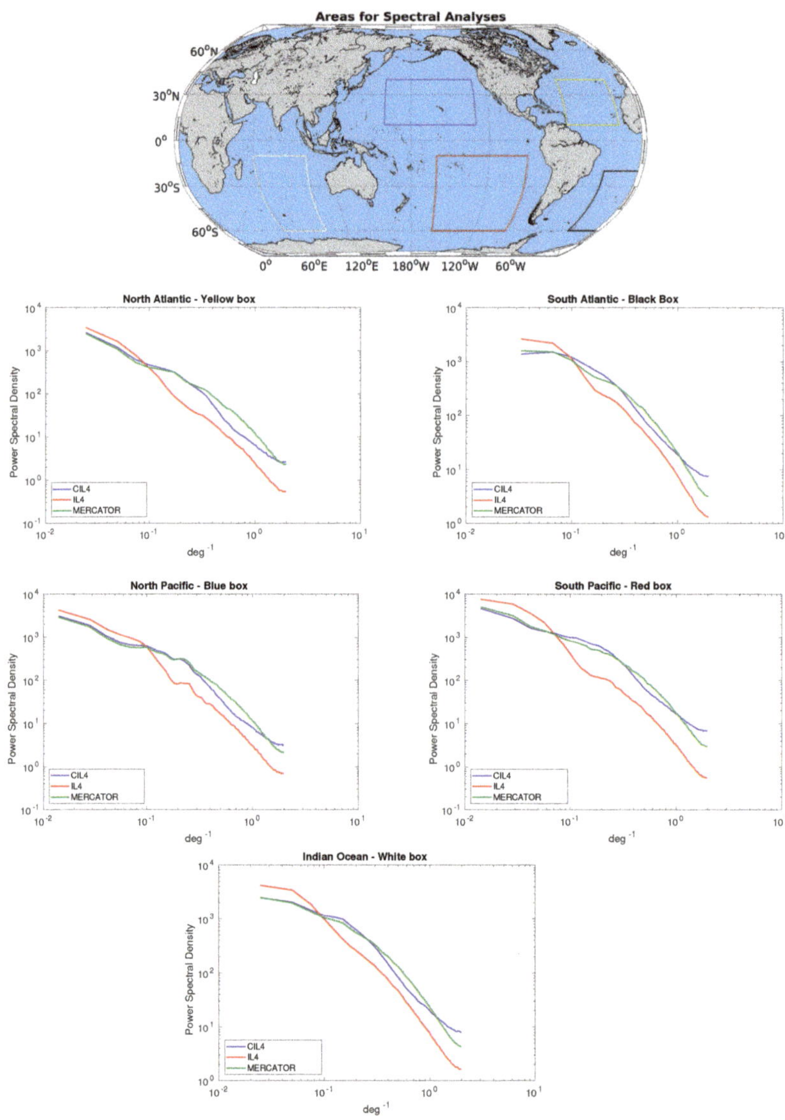

Figure 11. Time average of the mean zonal spectra of the MERCATOR SSS (green line), the IL4 (red line) and the CIL4 (blue line). The time average is based on weekly data for the year 2016. The spectra are computed in five different areas of the global ocean, referenced in the top panel of the figure.

The spectral analysis presented here is not fully rigorous from a physical point of view. Indeed, our maps are presented over a regular $1/4°$ grid and we are computing the spectra over latitudinal extents exceeding $40°$, thus considering spatial grid separations ranging from approximately 27 to 14 km, e.g., going from the equator up to $60°$N/S. Nevertheless, the results presented here are still proving that the CIMR SSS are expected to enhance the effective spatial resolution of the CMEMS SSS with respect to reconstructions based on in situ observations. In order to quantify the spectral properties from a physical point of view, we performed the spectral analysis in a $20°$ box centered over the equator ($\pm 10°$ around the equator) in the Pacific and the Atlantic Oceans, i.e., guaranteeing grid separations around 27 km on average. The behaviour of the spectra are similar to the ones presented in Figure 11. The CIL4 reconstruction spectrum evolves in agreement with the one of the MERCATOR SSS until scales of 1.4 deg^{-1} where the effect of noise becomes evident. This additional analysis indicates that the CIL4 are dynamically consistent with the MERCATOR fields until scales of approximately 70 km.

4. Discussion and Conclusions

The Copernicus Marine Environment Monitoring Service is presently serving a wide community of users with the distribution, amongst others, of global optimally interpolated L4 SSS (and sea-surface density) mapped on a $1/4°$ regular grid, updated weekly and based on a combination of in situ, satellite SSS and high-resolution SST data. The CMEMS SSS are obtained combining ARGO and CTD observations with SMOS SSS measurements [10,11,18]. After the SMOS era, the potential availability of the CIMR SSS could guarantee the accuracy and effective spatial resolution of the CMEMS SSS datasets. This was shown throughout this paper via a synthetic CMEMS SSS processing chain relying on simulated CIMR SSS retrievals. The core of our investigation was the CMEMS MERCATOR operational model, which was used to simulate the input SSS for the optimal interpolation processing as well as for assessing the accuracy of the CIMR-based L4 SSS. We focused on the year 2016. The main results of our evaluation study are summarized and discussed here.

CIMR will guarantee the retrieval of improved L4 SSS estimates (CIL4) with respect to optimally interpolated products based on in situ observations alone (IL4). Using the MERCATOR outputs as a benchmark, we obtained that the CIL4 RMSE is systematically reduced throughout the year in the latitudinal band $45°$S to $90°$N (Areas M and N), showing improvements up to 50% (based on Equation (4)) compared to the IL4 reconstructions. In the Area S, the CIMR benefits were confirmed during half of the year 2016, exhibiting a slight criticism during the austral Spring and Winter.

In past studies [11], the multivariate algorithm for the production of the L4 SSS was shown to perform much better in the open ocean. Offshore, the high pass filtered SST and SSS are generally more correlated and the assumptions made to derive the multidimensional covariance function are more strictly valid. This is more evident when SSS observations in coastal waters are too sparse, preventing an accurate mapping of the salinity changes related to groundwater fluxes or when the SST patterns are modified by localized heat fluxes (e.g., wind interactions with highly variable coastal orography). In the present study, we showed that the availability of remotely sensed CIMR SSS not only proves useful to monitor salinity changes associated with mesoscale-to-large scale processes in the open ocean, but also significantly improves our capability to describe salinity patterns in coastal areas.

Moreover, the use of the CIMR SSS will enable improving the effective spatial resolution of the global CMEMS L4 SSS, compared to L4 SSS obtained interpolating in situ observations alone. In our OSSE, the CIL4 reconstruction showed spectral properties in agreement with the true SSS, i.e., the MERCATOR model outputs. The optimal interpolation scheme only adds noise to scales smaller than 80 km, according to the tests performed in the Pacific and Atlantic equatorial bands.

These results indicate that the benefits of the potential SSS observations from CIMR will go even beyond the operational requirements within CMEMS. Their application to scientific and societal studies will be wide. The global L4 SSS obtained with CIMR will enable capturing the signatures of the major mesoscale dynamical features, e.g., the main Gulf Stream or Agulhas Rings [25], guaranteeing the monitoring of their spatial distribution and migration pathways. This will contribute to evaluating the global scale SSS distribution and budget. The monitoring of the global SSS is also a key element in studies of water cycle, oceanic water formation and ocean-atmosphere coupled dynamics [26]. For example, Ballabrera-Poy et al. 2002 [27] pointed out that SSS can be crucial in predicting the El Niño Southern Oscillation (ENSO) dynamics over time scales of 6 to 12 months. Indeed, the positive SSS anomalies in proximity of the Pacific equatorial band can modulate ENSO via their impact on the subsurface oceanic stratification. Quite interestingly, CIMR showed optimal measurement performances in the tropical Pacific area. Moreover, the accurate SSS estimate is useful for applications of three-dimensional fields reconstruction from surface information, as pointed out by [7,23,28] for the reconstruction of the three-dimensional horizontal and vertical oceanic motions and tracers. In conclusion, the expected performance of the CIMR mission confirmed the importance of ingesting the CIMR SSS within the framework of the CMEMS SSS data production. The future loss of the SMOS and SMAP missions fully justifies the high priority of the CIMR mission development within the framework of Copernicus.

Author Contributions: Conceptualization: D.C., B.B.N., R.S. and G.L.L. Formal analysis: D.C. and B.B.N. Funding acquisition: R.S. Investigation: D.C., B.B.N., G.L.L., and R.S. Supervision: B.B.N., G.L.L., R.S., C.P., and C.D. Validation: D.C. and B.B.N. Writing of original draft: D.C. and B.B.N. Writing—review and editing, B.B.N., C.D., G.L.L. and C.P.

Funding: This study was financed by the CIMR-Apps Mission Application Study, Contract 4000125189/18/NL/AI.

Acknowledgments: The authors wish to acknowledge the three anonymous Reviewers for providing constructive comments on the manuscript. Moreover, the authors thank the Barcelona Expert Center team and Kayla Jia for taking care of the reviewing process.

Conflicts of Interest: The authors declare no conflict of interest. The funders had no role in the design of the study; in the collection, analyses, or interpretation of data; in the writing of the manuscript, or in the decision to publish the results.

References

1. Richardson, P.; Bower, A.; Zenk, W. A census of Meddies tracked by floats. *Prog. Oceanogr.* **2000**, *45*, 209–250. [CrossRef]

2. Bower, A.S.; Hunt, H.D.; Price, J.F. Character and dynamics of the Red Sea and Persian Gulf outflows. *J. Geophys. Res.* **2000**, *105*, 6387–6414. [CrossRef]

3. Kolodziejczyk, N.; Hernandez, O.; Boutin, J.; Reverdin, G. SMOS salinity in the subtropical North Atlantic salinity maximum: 2. Two-dimensional horizontal thermohaline variability. *J. Geophys. Res. Oceans* **2015**, *120*, 972–987. [CrossRef]

4. Land, P.E.; Shutler, J.D.; Findlay, H.S.; Girard-Ardhuin, F.; Sabia, R.; Reul, N.; Piolle, J.F.; Chapron, B.; Quilfen, Y.; Salisbury, J.; et al. Salinity from space unlocks satellite-based assessment of ocean Acidification. *Environ. Sci. Technol.* **2015**. [CrossRef] [PubMed]

5. Gaillard, F.; Brion, E.; Charraudeau, R. ISAS–V5: Description of the method and user manual. In *IFREMER Rapport LPO*; LPO: Brest, France, 2009; Volume 9, p. 34.

6. Gaillard, F.; Reynaud, T.; Thierry, V.; Kolodziejczyk, N.; Von Schuckmann, K. In situ–based reanalysis of the global ocean temperature and salinity with ISAS: Variability of the heat content and steric height. *J. Clim.* **2016**, *29*, 1305–1323. [CrossRef]

7. Guinehut, S.; Dhomps, A.L.; Larnicol, G.; Le Traon, P.Y. High resolution 3D temperature and salinity fields derived from in situ and satellite observations. *Ocean Sci.* **2012**, *8*, 845–857. [CrossRef]

8. Umbert, M.; Hoareau, N.; Turiel, A.; Ballabrera-Poy, J. New blending algorithm to synergize ocean variables: The case of SMOS sea surface salinity maps. *Remote Sens. Environ.* **2014**, *146*, 172–187. [CrossRef]

9. Olmedo, E.; Taupier-Letage, I.; Turiel, A.; Alvera-Azcárate, A. Improving SMOS Sea Surface Salinity in the Western Mediterranean Sea through Multivariate and Multifractal Analysis. *Remote Sens.* **2018**, *10*, 485. [CrossRef]

10. Buongiorno Nardelli, B.; Droghei, R.; Santoleri, R. Multi-dimensional interpolation of SMOS sea surface salinity with surface temperature and in situ salinity data. *Remote Sens. Environ.* **2016**, *180*, 392–402. [CrossRef]

11. Droghei, R.; Buongiorno Nardelli, B.; Santoleri, R. A New Global Sea Surface Salinity and Density Dataset From Multivariate Observations (1993–2016). *Front. Mar. Sci.* **2018**, *5*, 84. [CrossRef]

12. Donlon, C.J. *Copernicus Imaging Microwave Radiometer (CIMR) Mission Requirements Document*; Version 2.0; ESA-ESTEC Noordwijk 2201 AZ, The Netherlands, 2019; Available online: http://esamultimedia.esa.int/docs/EarthObservation/Copernicus_CIMR_MRD_v2.0_Issued_20190305.pdf (accessed on 3 August 2019).

13. Nouel, L. *Global Ocean 1/12 of a Degree Physics Analysis and Forecast Updated Daily—Product User Manual (CMEMS-GLO-PUM-001-024)*; Issue 1.4; E.U. Copernicus: Brussels, Belgium, 2019.

14. Wentz, F.; Meissner, T.; Gentemann, C.; Hilburn, K.; Scott, J. Remote Sensing Systems GCOM-W1 AMSR2 Daily data, Environmental Suite on 0.25 Degrees Grid, Version V.8 2014. Available online: www.remss.com/missions/amsr (accessed on October 2018).

15. Buongiorno Nardelli, B. A novel approach for the high-resolution interpolation of in situ sea surface salinity. *J. Atmos. Ocean. Technol.* **2012**, *29*, 867–879. [CrossRef]

16. Bretherton, F.P.; Davis, R.E.; Fandry, C. A technique for objective analysis and design of oceanographic experiments applied to MODE-73. *Deep. Sea Res. Oceanogr. Abstr.* **1976**, *23*, 559–582. [CrossRef]

17. Robinson, I.S. *Measuring the Oceans from Space: The Principles and Methods of Satellite Oceanography*; Springer Science & Business Media: Berlin/Heidelberg, Germany, 2004.

18. Droghei, R.; Buongiorno Nardelli, B.; Santoleri, R. Combining in situ and satellite observations to retrieve salinity and density at the ocean surface. *J. Atmos. Ocean. Technol.* **2016**, *33*, 1211–1223. [CrossRef]

19. Kilic, L.; Prigent, C.; Aires, F.; Boutin, J.; Heygster, G.; Tonboe, R.T.; Roquet, H.; Jimenez, C.; Donlon, C. Expected Performances of the Copernicus Imaging Microwave Radiometer (CIMR) for an All-Weather and High Spatial Resolution Estimation of Ocean and Sea Ice Parameters. *J. Geophys. Res. Oceans* **2018**, *123*, 7564–7580. [CrossRef]

20. Skou, N.; Hoffman-Bang, D. L-band radiometers measuring salinity from space: Atmospheric propagation effects. *IEEE Trans. Geosci. Remote Sens.* **2005**, *43*, 2210–2217. [CrossRef]

21. Rio, M.H.; Santoleri, R. Improved global surface currents from the merging of altimetry and Sea Surface Temperature data. *Remote Sens. Environ.* **2018**, *216*, 770–785. [CrossRef]

22. Xie, J.; Raj, R.P.; Bertino, L.; Samuelsen, A.; Wakamatsu, T. Evaluation of Arctic Ocean surface salinities from SMOS and two CMEMS reanalyses against in situ data sets. *Ocean Sci. Discuss.* **2019**. [CrossRef]

23. Isern-Fontanet, J.; Lapeyre, G.; Klein, P.; Chapron, B.; Hecht, M. Three-dimensional reconstruction of oceanic mesoscale currents from surface information. *J. Geophys. Res. Oceans* **2008**, *113*, 153–169. [CrossRef]

24. Dunstan, P.K.; Foster, S.D.; King, E.; Risbey, J.; O'Kane, T.J.; Monselesan, D.; Hobday, A.J.; Hartog, J.R.; Thompson, P.A. Global patterns of change and variation in sea surface temperature and chlorophyll a. *Sci. Rep.* **2018**, *8*, 14624. [CrossRef]

25. Carton, X. Hydrodynamical Modeling of Oceanic Vortices. *Surv. Geophys.* **2001**, *22*, 179–263. [CrossRef]

26. Font, J.; Camps, A.; Borges, A.; Martín-Neira, M.; Boutin, J.; Reul, N.; Kerr, Y.H.; Hahne, A.; Mecklenburg, S. SMOS: The challenging sea surface salinity measurement from space. *Proc. IEEE* **2009**, *98*, 649–665. [CrossRef]

27. Ballabrera-Poy, J.; Murtugudde, R.; Busalacchi, A. On the potential impact of sea surface salinity observations on ENSO predictions. *J. Geophys. Res. Oceans* **2002**, *107*, SRF–8. [CrossRef]

28. Buongiorno Nardelli, B.; Mulet, S.; Iudicone, D. Three-Dimensional Ageostrophic Motion and Water Mass Subduction in the Southern Ocean. *J. Geophys. Res. Oceans* **2018**, *123*, 1533–1562. [CrossRef]

Article

Assimilation of Satellite Salinity for Modelling the Congo River Plume

Luke Phillipson * and Ralf Toumi

Space and Atmospheric Physics Group, Department of Physics, Imperial College London, London SW7 2AZ, UK; r.toumi@imperial.ac.uk
* Correspondence: l.phillipson14@imperial.ac.uk

Received:13 November 2019; Accepted:15 December 2019; Published: 18 December 2019

Abstract: Satellite salinity data from the Soil Moisture and Ocean Salinity (SMOS) mission was recently enhanced, increasing the spatial extent near the coast that eluded earlier versions. In a pilot attempt we assimilate this data into a coastal ocean model (ROMS) using variational assimilation and, for the first time, investigate the impact on the simulation of a major river plume (the Congo River). Four experiments were undertaken consisting of a control (without data assimilation) and the assimilation of either sea surface height (SSH), SMOS and the combination of both, SMOS SSH. Several metrics specific to the plume were utilised, including the area of the plume, distance to the centre of mass, orientation and average salinity. The assimilation of SMOS and combined SMOS SSH consistently produced the best results in the plume analysis. Argo float salinity profiles provided independent verification of the forecast. The SMOS or SMOS SSH forecast produced the closest agreement for Argo profiles over the whole domain (outside and inside the plume) for three of four months analysed, improving over the control and a persistence baseline. The number of samples of Argo floats determined to be inside the plume were limited. Nevertheless, for the limited plume-detected floats the largest improvements were found for the SMOS or SMOS SSH forecast for two of the four months.

Keywords: SMOS; data assimilation; 4D-Var; Congo River plume; satellite salinity; Angola Basin; ROMS

---

1. Introduction

The ten largest rivers transport a combined 40% of the freshwater and particulate material into the oceans [1–4]. Some larger rivers can generate offshore plumes of significant distance [4,5] that can support high levels of biological productivity, due to the vast amount of supplied nutrients [6,7]. Therefore, the accuracy of modelling these river plumes is valuable for fishing communities. Furthermore, terrestrial material, both suspended and dissolved, are transported via river plumes, affecting sediment and pollutant distributions and the biogeochemistry of carbon [3].

The second-largest river (in terms of the annual mean daily discharge) is the Congo in West Africa, discharging an average 39,866 m^3/s of freshwater per month into the Angola Basin. Such extensive outflow produces a plume as large as 800 km from the river mouth [4,5], influencing a substantial section of the basin. Yankovsky and Chapman [8] noted that the plume could be classified as surface-advected; with identifiable near-and far-field regions. Variations in the near-field have been identified with the speed of outflow, the orientation of the estuary mouth and local coastal currents [9,10]. For the far-field, a more complex situation arises through the interaction between the wind stress, ocean circulation patterns, tidal currents and river discharge variations that contribute to different scenarios of the Congo River plume dispersion [11]. Denamiel et al. [10] performed the first numerical simulation of the Congo River plume where they found the plume had a northward extension for the majority of the year except during February-March where the plume had a sizeable westward expansion (up to 800 km). Subsequent model studies have focused on the plume's buoyancy-driven dynamics [12,13], the effect of the Congo on ocean temperatures [14] and a complete simulation of the Congo river-to-sea continuum with a multi-scale unstructured mesh model [15].

Very few studies have focused on improving forecasting river plumes. Liu et al. [16] produced a realistic hind-cast of the Columbia River estuarine-plume-shelf circulation and used a skill score to quantify the five dynamical regions (estuary, near- and far-field plume, near-surface and deep layers). Data assimilation (DA) is a powerful tool that produces an optimal estimate of the initial conditions (typically for a forecast) using observations and information from the dynamical model [17]. Until recently, DA could not easily be utilised in the context of river plume modelling. While the more traditional Argo floats provide relatively accurate in-situ salinity profiles, their lack of spatial coverage (typically 3° by 3°) is a significant limitation. In 2009 the European Space Agency (ESA) launched the first satellite to monitor surface sea salinity under the Soil-Moisture-Ocean Salinity (SMOS) mission [18] providing global coverage of surface salinity. Köhl et al. [19] assimilated SMOS but found no benefits to the global model salinity. Conversely, Lu et al. [18] found that SMOS data plays a complementary role in model salinity simulations with an Ensemble Optimal Interpolation DA scheme (EnOI). A new version of SMOS that dramatically improved the land/sea interference and coastal coverage has been available from 2017 [20]. The Congo River plume can now be fully analysed on a 9-daily scale instead of monthly, expanding the scope of the data for use in coastal assimilation (Figure 1).

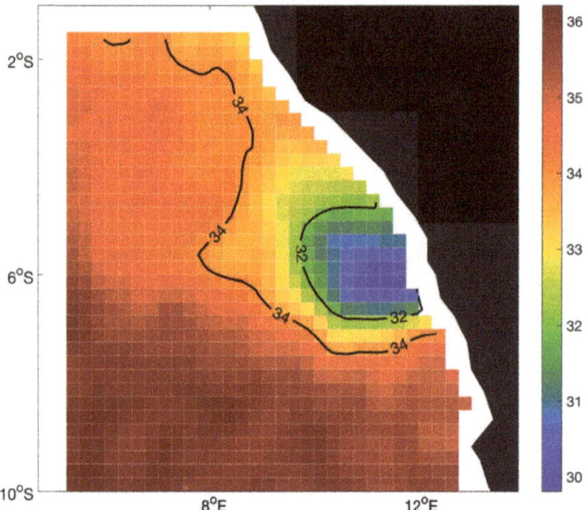

Figure 1. A snapshot of the 2017 Soil Moisture and Ocean Salinity (SMOS) data from the Barcelona Expert Centre (BEC) [20] for January 2013 with filled salinity (PSU), and the 32 PSU and 34 PSU isohaline (lines of constant salinity) in black.

A recent study by Mu et al. [21] was the first to show that the assimilation of this newer data could improve simulations of the upper ocean salinity. However, prior to the assimilation, Mu et al. [21] applied an additional bias correction, specific to their study region (South China Sea) using a Generalised Regression Neural Network (GRNN). The assimilation of this bias-corrected GRNN SMOS data set improved results compared to the original data set without such bias correction.

In this paper, the SMOS data as originally produced by Olmedo et al. [20] is assimilated to study the impact on the simulation of the Congo River plume. This will be the first time that satellite salinity has been used with DA specifically for river plumes and the first time a Congo River plume forecast has been attempted. Satellite altimetry sea surface height (SSH) is also assimilated individually and in combination with SMOS, to explore whether these two data streams could compliment each other.

The rest of this paper is structured as follows: the materials and method sections introduces the numerical model, observations, assimilation scheme and experiment design. The results then follows split into an assessment of the assimilation analysis and forecast. The paper closes with a discussion on the results and a final conclusion.

2. Materials and Methods

2.1. Numerical Model

ROMS is a hydrostatic, primitive equation, Boussinesq ocean general circulation model [22]. Previous studies have utilised various versions of this ROMS model of the Angola Basin in understanding the Congo River plume dynamics [10], effects on ocean temperature [14] and more recently on the impact of DA on the ocean current predictability [23]. The model domain extends between 1°S–21°S and 3.7°E–13.8°E with a 10 km resolution and 40 terrain-following vertical levels. Lateral boundary and initial conditions for temperature, salinity, ocean current velocities and sea surface height were obtained from the HYCOM reanalysis [24]. Atmospheric forcing at the surface for downward radiative surface fluxes, sea level pressure, 2 m specific humidity, 2 m air temperature, 10 m winds, and total precipitation were obtained from European Centre for Medium-Range Weather Forecasts (ECMWF) reanalysis (ERA)-Interim reanalysis data (ERA-I) [25].

Seven rivers (Nyanga, Kouilou, Kwanza (Cuanza), Kuene, and the Congo) are incorporated into ROMS as boundary conditions. For each river channel, several source points are assigned a unique outflow rate with a constant temperature (15 degrees) and near-zero salinity (0.1 PSU). Following White and Toumi [14], river flow rates were obtained from various sources, including the RivDIS v1.1 database [26], the Global Environmental Monitoring System/Global River Inputs (GEMS/GLORI) database [27], The Office of Industrial Studies Renewable Energies and Environment (BEI ERE) and the University of Brazzaville available at (http://hmf.enseeiht.fr/travaux/CD0809/bei/beiere/groupe5/node/53). The Congo represents the largest of the rivers in the model domain with a channel of 20 km (width) × 60 km (length) (2 × 6 grid points) containing eight source points. The average annual cycle of the Congo River discharge computed from the observed monthly mean between 1902 and 2005 is used for the simulations presented here (Figure 2). The bimodal signature has been suggested to be part of the seasonal migration of the Intertropical Convergence Zone (ITCZ) [28] and, Northern and Southern Africa Easterly Jets [29] enhancing convection across central Africa.

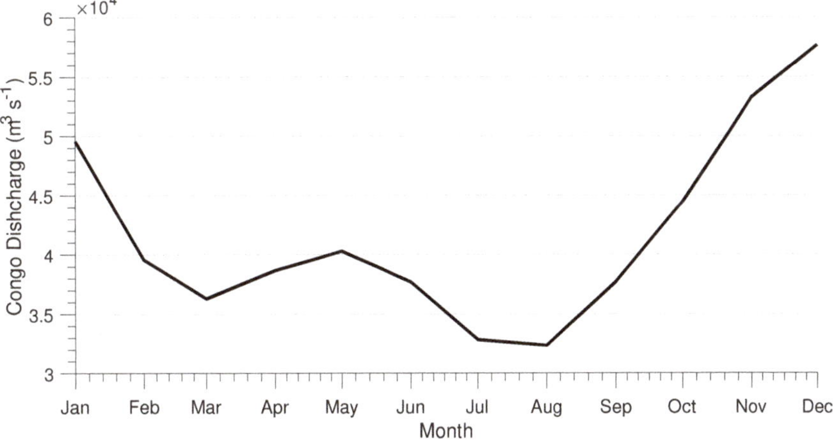

Figure 2. The average annual cycle of the Congo River discharge used within the model. The average is calculated from the observed monthly mean discharge between 1902 and 2005 provided by the BEI ERE in collaboration with the University of Brazzaville. Adapted from Phillipson [30].

2.2. Observations

The Barcelona Expert Centre (BEC) released an updated version of SMOS in 2017 (available at http://bec.icm.csic.es) which corrected for systematic biases created by landmasses (land contamination) and radio interference, and also reduced, significant data gaps due to the non-convergence of the retrieval algorithm [20]. The resulting daily gridded product has a spatial resolution of $0.25°$ and time-averaging window of 9 days. For the tropics, the validation with in-situ data (Argo floats) provided by BEC shows that between 2011–2016 the root mean square error (RMSE) is 0.24 PSU [31].

Gridded sea surface height (SSH) observations consisted of a merged Ssalto/Duacs dataset (TOPEX/Poseidon, Jason-1&2, Envisat, ERS-1&2, and GFO measurements) distributed by Aviso with support from the Centre National d'Etudes Spatiales. ROMS does not resolve the global steric signal. This signal was removed from SSH using a database provided by Willis [32]. Furthermore, SSH was calibrated to ensure ROMS and AVISO dynamic topography were spatially and temporally equal on a long-term average. We remove the nearest 50 km to the coastline. This represents a typical length scale of land contamination for the Aviso data.

In situ salinity profile observations from Argo floats were obtained from the EN4 dataset provided by the Met Office Hadley Centre [33].

2.3. Assimilation Scheme

For this study, incremental, strong constraint four-dimensional variational data assimilation (IS4D-Var) adjusting the initial conditions, surface forcing, and boundary conditions were utilised within ROMS [22]. Before initialising the IS4D-Var process, several DA parameters are required to be specified. The choices of each are described in-depth within Phillipson [30].

The most important of these choices relate to the background error covariance matrix for the initial condition. This is formulated following [34] as in a combination of multivariate balance relationships, the standard deviations of the model using a long climatology simulation, and a diffusion operator [22]. The standard deviation field represents a spatially varying parameter controlling the relative weighting of the short forecast (prior) in the data assimilation system. A single assigned observational error then controls the relative weighting of the observations. Both are kept constant throughout the entire DA cycle.

During some initial assimilation tests (not shown) it became apparent that the standard deviation used in the background error covariance for the initial condition of salinity required specific tuning with regards to river plumes [30]. The large difference in the estimated standard deviation in salinity for the open ocean (0.2–0.5 PSU) as compared to that of the Congo plume (of up to 10 PSU) caused issues.

Previous studies assimilating SMOS [18,21] have used an observation error of between 0.1–0.4 PSU. However, these studies did not focus on river plumes. As far as the authors are aware, no studies of DA for such extensive river plumes exist, and so this challenge is unique to modelling the Congo River (as also likely for Amazon River). The initial tests of the assimilation of SMOS using a 'typical' error of 0.1 PSU and the original estimated standard deviation from the climatology run drastically over-fit the observations near the river plume (0.1 PSU $\ll$ 10 PSU). Moore et al. [35] noted a similar issue in developing an operational ROMS 4D-Var analysis system for the California current. Moore et al. [35] suggested that the salinity standard deviation could be capped at a certain level to account for this. Therefore following [35] the salinity standard deviation within the background error covariance was capped at 0.6 PSU. This cap was chosen in order to isolate the area of the plume, i.e., most of north of 12 degrees longitude. After further testing (not shown), an observational error of 1.2 PSU for the SMOS observations were assigned (double that of the applied salinity cap). Lower errors ($<$1.2 PSU) resulted in overconfidence in the observations near the river plume mouth where the variability in the model is at it's largest. Therefore, an error of 1.2 PSU enabled the SMOS observations to have the most substantial impact near the plume without over-fitting. Although this error is much larger than the typical error of SMOS in the open ocean (a quality report from BEC noted an error of 0.24 PSU in the tropics SMOS-BEC Team [31]), this was adopted as a pilot approach for the assimilation of SMOS in river plume modelling [30].

The observation error for satellite altimetry measuring sea surface height is assigned as 0.04 m as in [23].

2.4. Data Assimilation Experiments

Assimilation was performed sequentially (four-day assimilation window) using the Angola Basin ROMS with IS4D-Var for four experiments during January, April, July and August 2013. During each month, four cycles of DA was performed (16 days) with the initial conditions for each new cycle obtained from the final posterior analysis from the previous cycle. The first cycle was initialised with a minimum one-year spin-up of the model without assimilation. Following the final cycle, a 16-day forecast was performed. Although simulating the whole year would be preferable to capture the whole season of the Congo River plume, four months was regarded as adequate to examine the impact of assimilating SMOS. The four months cover a range of discharges. January represents a month of high discharge and April a transitional month into lower discharges during July and August.

Four experiments were undertaken: a control run (CNTRL) without any assimilation; the assimilation of daily gridded satellite altimetry (SSH); daily gridded satellite sea surface salinity (SMOS) and the combination of both (SMOS SSH). A regional domain RMSE and correlation coefficient (R) was estimated to assess the quality of the DA algorithm. Specific to the validating the Congo river plume, the 34 PSU isohaline (line of constant salinity) was used as an upper bound of the river plume, a similar choice to Kang et al. [3]. With this definition the plume area bounded by the 34 PSU isohaline, the distance to the centre of mass from the source, orientation and average salinity within can be determined. Several Argo floats are present in Angola Basin providing independent data.

A total of 195 Argo float samples were taken in the Angola Basin over the entire 4 months analysed, providing independent data of salinity. However, none were located within the defined 34 PSU isohaline. Therefore in order to determine if some Argo floats were present within the Congo River plume, independent of a defined isohaline, a plume detection algorithm was used. Here, we follow Hopkins et al. [4], who used two main criteria (stratification and mean salinity less than the open ocean) to determine plume-detected Argo floats (Figure 3). Twenty-three samples (12%) were determined to be inside the plume (Pink lines and circles in Figure 3). By definition, stratification is present for all plume-detected Argo floats, meaning the comparison with SMOS sea surface salinity is robust. This small sample size is a partially a consequence of the expense of running a 4D-Var model system, and while more months and samples would have been ideal, the limited choice of months analysed was a computational restraint.

A persistence forecast was also included as a baseline forecast. An additional metric denoted the salinity skill score; $ss = 1 - (AE\_model/AE\_persist)$, where AE_model (AE_persist) is the absolute error for the model (persistence) forecast, was computed to highlight improvements ($ss>0$) or degradation ($ss<0$) against persistence.

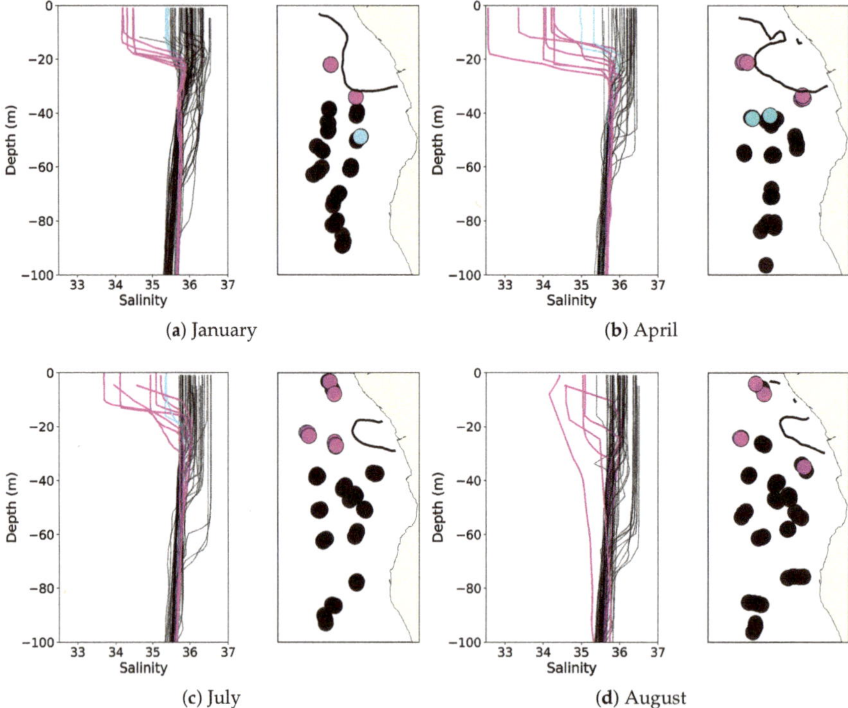

Figure 3. Argo float salinity profiles for top 100 meters (left) and their associated locations (right) during (**a**) January, (**b**) April, (**c**) July and (**d**) August. Black profiles and circles represent Argo floats determined to be outside the plume according to the plume detection algorithm. Purple profiles and circles represent Argo floats determined to be inside the plume. Blue profiles and circles represent Argo floats that only meet one criteria from the plume detection algorithm. Overlaid in black are the average 34 PSU isohalines for each respective month.

3. Results

3.1. Assimilation Analysis

The SMOS RMSE reduced for each experiment assimilating SMOS during the assimilation analysis as expected (Table 1). Interestingly, the combined assimilation SMOS SSH experiment had the smallest SMOS RMSE for three of the four months analysed, at 12–22% less than the SMOS experiment. The SMOS correlation coefficient (R) mirrored this result, exhibiting small improvements for SMOS SSH as compared to the SMOS only assimilation. Similarly, the SMOS RMSE and R also reduced and increased, respectively, for the SSH experiment as compared to the control for the majority of the study period. This result supports the benefits of altimetry assimilation in modelling the regional salinity field. Following this systemic reduction of SMOS errors, the data assimilation algorithm appears to be successfully fitting the salinity observations within the model.

Table 1. The root mean squared error (RMSE) and mean correlation coefficient (R) for each experiment (CNTRL, SSH, SMOS, SMOS SSH) computed over each four day analysis cycle (4) during (**a**) January, (**b**) April, (**c**) July and (**d**) August. Adapted from Phillipson [30].

(a) January			(b) April		
Exp. Name	RMSE (PSU)	R	Exp. Name	RMSE (PSU)	R
CNTRL	2.76	0.54	CNTRL	2.33	0.76
SSH	2.61	0.71	SSH	2.35	0.79
SMOS	0.61	0.92	SMOS	0.38	0.94
SMOS SSH	0.48	0.95	SMOS SSH	0.43	0.93
(c) July			(d) August		
Exp. Name	RMSE (PSU)	R	Exp. Name	RMSE (PSU)	R
CNTRL	0.88	0.70	CNTRL	1.4	0.51
SSH	0.84	0.71	SSH	1.0	0.61
SMOS	0.25	0.95	SMOS	0.31	0.94
SMOS SSH	0.22	0.96	SMOS SSH	0.28	0.95

The SMOS plume specific statistics (area, distance to the centre of mass, orientation and salinity within) also exhibits improvements for each month within the analysis (Table 2). Note this improvement is expected since we are assimilating and comparing against the same SMOS observations, and therefore solely representing whether the DA system can replicate the SMOS plume within the ROMS model.

Table 2. The SMOS plume specific statistics for each experiment (CNTRL, SSH, SMOS, SMOS SSH) assimilation analysis over each study month. Statistics include plume area mean absolute error (MAE), distance to the plume centre of mass from source (COM Dist.) MAE, plume orientation (angle) MAE, inside plume salinity MAE and a metric mean % improvement over the CNTRL.

Exp. Name	Area MAE (km^2)	COM Dist. MAE (km)	Angle MAE ($^\circ$)	Inside Salinity MAE (PSU)	Mean % Improv.
January					
CNTRL	27,454	78	2.0	2.53	-
SSH	74,515	75	1.2	2.37	−30%
SMOS	8084	6	1.9	1.06	56%
SMOS SSH	8754	5	1.1	0.95	68%
April					
CNTRL	144,409	13	10.1	1.78	-
SSH	160,850	9	11.5	1.78	1%
SMOS	5291	11	0.4	0.54	68%
SMOS SSH	7936	9	0.7	0.55	72%
July					
CNTRL	18,654	39	7.6	3.59	-
SSH	10,582	36	13.2	3.26	−4%
SMOS	7425	13	3.8	2.42	52%
SMOS SSH	6154	12	2.4	2.53	58%
August					
CNTRL	26,897	118	39.9	4.99	-
SSH	23,309	73	6.4	4.37	37%
SMOS	4053	7	0.9	2.22	83%
SMOS SSH	4882	3	0.9	2.73	81%

In the CNTRL experiment without assimilation, the plume experiences substantial biases for the majority of statistics (Table 2). For the area of the plume, the CNTRL is often vastly different than observed from SMOS. Figure 4 illustrates this bias for January (a month of high discharge) with an area mean absolute error (MAE) of around $+27{,}000$ km^2. Further discrepancies in the structure (stretched along the coast), orientation ($2°$ MAE), distance to the centre of mass (80 km MAE) and average salinity within (2.5 PSU MAE) are present.

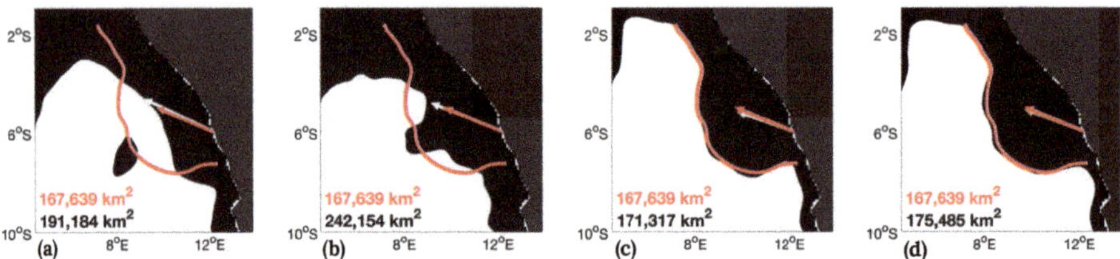

Figure 4. The 34 PSU plume area for (**a**) the control (CNTRL), (**b**) sea surface height (SSH), (**c**) SMOS and (**d**) SMOS SSH assimilation analysis (black filled) and SMOS data (red outlines) averaged for the January analysis. Overlaid are arrows denoting the orientation and distance to centre of mass for each experiment (white) and the SMOS data (red). The bottom left numbers denote the area of the plume for each experiment (black) and SMOS data (red).

Contrary to improvements in the SMOS RMSE and R (Table 1), the subsequent assimilation of SSH does not always aid in significantly reducing this bias consistently across all months, sometimes enhancing it. For example, the January and April average plume area exhibit increased biases of up to $48{,}000$ km^2 (160%).

Only with the assimilation of SMOS (either alone or in combination with SSH) is the most systematic improvement over the CNTRL found across all months for the analysis. Most notably improving the area in the analysis by as much as 95% (visually represented in Figure 4). Here, the structure from SMOS is well replicated in the model analysis, resulting in improved orientation and distance to the centre of mass (95–100%). Note the additional benefits of assimilating both SMOS and SSH together for the analysis can be seen for many metrics. A summary statistic (final column in Table 2) confirms that the combined assimilation of SMOS and SSH is the best performing experiment for three of the four months analysed. Improving upon assimilating solely SMOS by 4–7% averaged across all metrics.

While the SMOS plume specific statistics confirm the model can effectively replicate the SMOS plume via the assimilation process, independent observations from Argo floats offer a more robust indication of performance (Table 3).

Table 3. The mean absolute error (MAE) computed for Argo float samples for each experiment (CNTRL, SSH, SMOS, SMOS SSH) assimilation analysis over each month. The Argo float MAE for the SMOS data itself (SMOS DATA) during the analysis period is also shown for comparison. The samples are split into whole domain MAE statistics or inside plume only. The number of Argo float samples over which each MAE is computed is also presented.

	Whole Domain				Inside Plume			
	No. of Argo Float Samples							
	Jan	Apr	Jul	Aug	Jan	Apr	Jul	Aug
	38	17	22	24	0	3	4	3
Analysis Exp. Name	MAE (PSU)				MAE (PSU)			
CNTRL	0.30	0.46	0.39	0.24	—	0.92	1.84	0.79
SSH	0.15	0.60	0.33	0.26	—	0.78	1.14	0.54
SMOS	0.26	0.22	0.28	0.21	—	0.49	0.80	0.33
SMOS SSH	0.21	0.24	0.30	0.21	—	0.51	0.82	0.38
SMOS DATA	0.26	0.28	0.30	0.25	—	0.58	0.80	0.37

For the whole domain, the analysis for the SMOS assimilation experiment produced the smallest Argo MAE for April, July and August, improving over the CNTRL by 13% to 52%. For Argo floats inside the plume as identified by the algorithm as described in Hopkins et al. [4], the SMOS assimilation also produced the smallest Argo MAE comparisons. Curiously, the addition of the SSH, as in the combination of SMOS and SSH often increased errors for Argo float comparisons. Only for January did the additional assimilation of SSH decrease errors. This improvement is clearly because of superior performance of the SSH analysis (without SMOS). Possibly indicating that the assimilated SMOS data may be more inconsistent with the Argo floats, which carries through to the SMOS analysis. This relationship is shown in Figure 5. There is a strong significant correlation (0.95, p-value < 0.01) between the SSH analysis–SMOS data error difference and SSH analysis–SMOS analysis error difference. Dividing the data into each month reveals that most of the January SSH Argo float errors (green circles in Figure 5) are smaller than that of the SMOS data and the SMOS analysis situated in the lower left hand shaded region (<0) within Figure 5.

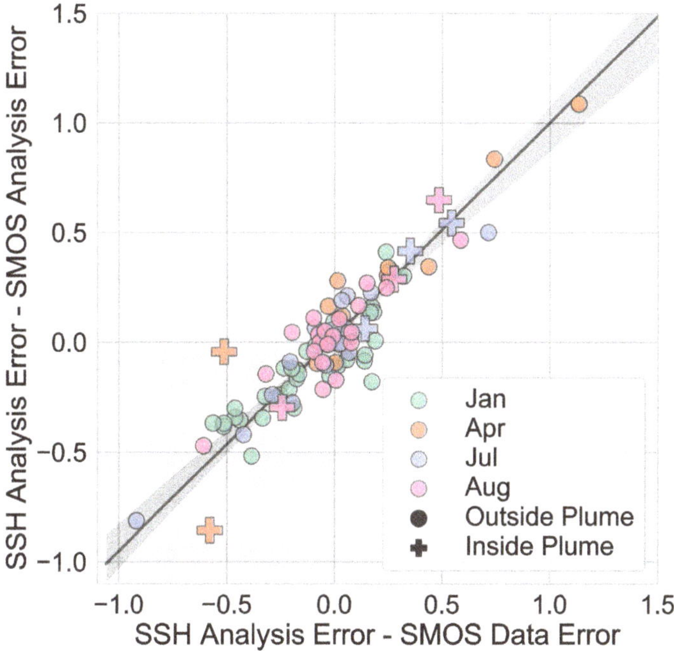

Figure 5. A scatter plot of the Argo float SSH analysis error minus the Argo float SMOS analysis error against the Argo float SSH analysis error minus the Argo float SMOS data error. The data is split into each study month, January (green), April (orange), July (blue) and August (pink). Outside (inside) plume samples are represented as circles (pluses). A linear regression is overlaid with a shaded bootstrap confidence interval. An additional shaded area for values <0 is displayed to highlight Argo float SSH analysis errors that outperform both the SMOS analysis and SMOS data.

The spatial distribution of Argo floats present during the analysis is explored (Figure 6). For January (first row of Figure 6), very few Argo floats are located near the plume, with none independently determined to be inside the plume by the plume-detection algorithm (crosses in Figure 6). Here, the assimilation of SMOS appears to significantly increase errors further south as well as creating negative errors (the model is too saline) further north towards the coast. For April (second row in Figure 6), the CNTRL and SSH assimilation experiments exhibit a cluster of Argo floats with a substantial positive error (too fresh) around the centre of the model. The SMOS assimilation significantly reduces this cluster of errors. For July (third row in Figure 6), the SMOS assimilation slightly decreases large plume-detected Argo float errors in the north. Finally, August (final row in Figure 6) Argo floats exhibit a subtle change in errors from the CNTRL, with the SMOS assimilation reducing errors for the northernmost cluster that compose of the three plume-detected Argo floats.

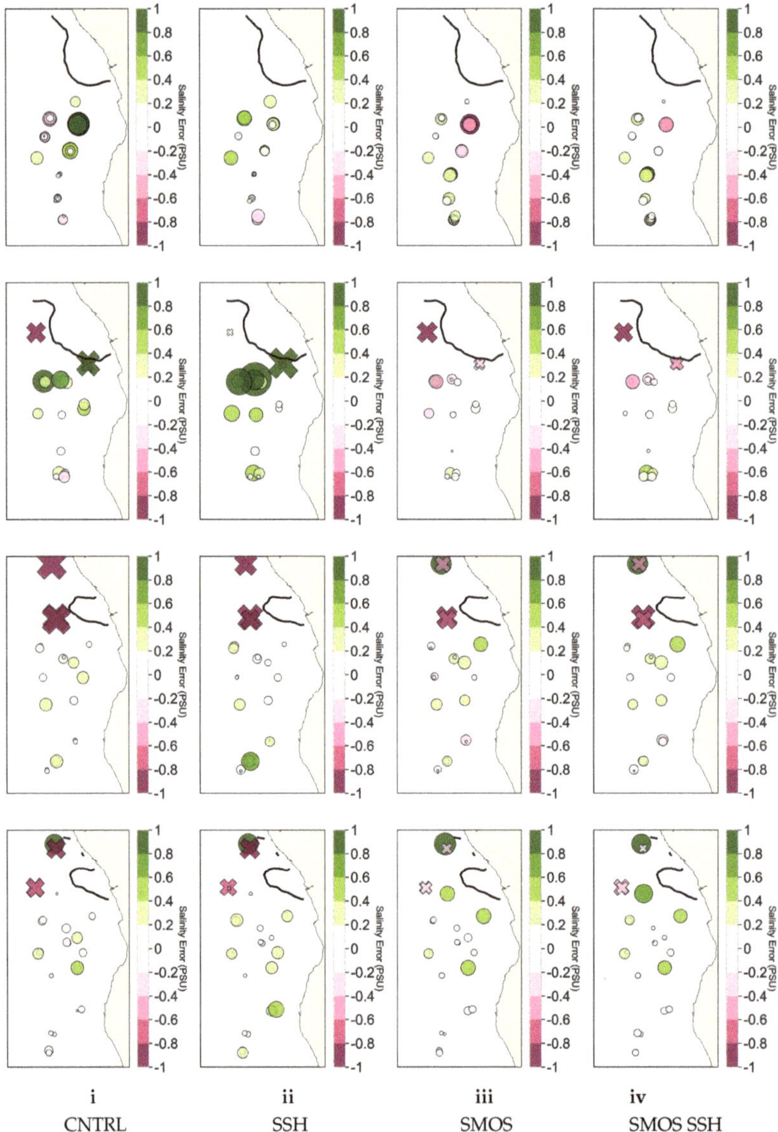

Figure 6. The Argo float surface salinity errors over the January (1st row), April (2nd row), July (3rd row) and August (4th row) analysis for experiments (**i**) CNTRL, (**ii**) SSH, (**iii**) SMOS and (**iv**) SMOS SSH. In addition to the caxis colour scale representing negative (purple) to positive (green) salinity bias (PSU), smaller circle sizes represent smaller biases. Crosses (instead of circles) are alternatively displayed to represent the plume detected samples as seen in Figure 3. The average 34 PSU isohaline computed from the SMOS data for each respective month (analysis period) is overlaid in black.

3.2. Forecast

The 16-day forecasts initialised after the end of the final assimilation analysis cycle for each month also exhibited improvements over the CNTRL for SMOS plume specific statistics (Table 4). For higher discharge months (January and April), all metrics exhibit improvements over the CNTRL of up to 86% in the plume area, 54% in the distance to the centre of mass, 97% in orientation and 41% in salinity within. Conversely, for low discharge months (July and August), only minimal improvements are exhibited over the CNTRL for a handful of metrics, with degradation often present, notably during August. Despite improvements mainly limited to high discharge months, this result shows that the SMOS assimilation can maintain improvements from the analysis up to 16 days.

Table 4. The SMOS plume specific statistics for each experiment (CNTRL, SSH, SMOS, SMOS SSH) assimilation forecast over each study month. An additional forecast denoted SMOS PERSIST is included. Statistics include plume area mean absolute error (MAE), distance to the plume centre of mass from source (COM Dist.) MAE, plume orientation (angle) MAE, inside plume salinity MAE and a metric mean % improvement over the CNTRL.

Exp. Name	Area MAE (km^2)	COM Dist. MAE (km)	Angle MAE (°)	Inside Salinity MAE (PSU)	Mean % Improv.
January					
CNTRL	140,152	33	14.7	2.05	-
SSH	209,978	12	12.1	1.79	11%
SMOS	19,801	16	0.7	1.28	68%
SMOS SSH	5893	8	1.9	1.15	75%
SMOS PERSIST	9174	14	1.1	0.08	85%
April					
CNTRL	264,010	77	9.7	1.01	-
SSH	246,378	70	8.9	1.12	3%
SMOS	43,054	36	3.0	0.60	62%
SMOS SSH	31,586	43	1.2	1.01	55%
SMOS PERSIST	11,433	5	0.4	0.04	95%
July					
CNTRL	9764	86	9.3	2.93	-
SSH	8867	83	18.0	3.28	−23%
SMOS	9685	72	16.4	3.04	−16%
SMOS SSH	8061	65	15.2	3.29	−9%
SMOS PERSIST	2305	3	1.6	0.02	90%
August					
CNTRL	18,121	85	20.2	2.94	-
SSH	5291	52	7.9	3.09	41%
SMOS	7482	109	41.0	3.75	−25%
SMOS SSH	21,754	30	4.4	4.15	21%
SMOS PERSIST	7403	8	1.0	0.09	85%

In combining SMOS with SSH, additional improvements are also exhibited for several plume metrics, more systematically for the forecast (Table 4) than as seen for the analysis period (Table 2). The plume area, for example, improves for an additional 10%, 4% and 17% during the January, April, and July forecast. This improvement is illustrated in Figure 7 with the structure of the plume better replicated for the SMOS SSH.

In addition to the model forecasts, a baseline of persistence was included. Since the SMOS dataset represents a 9-day moving average, good persistence skill in the plume statistics are highly likely, especially in an area with slowly evolving dynamics [23,36], where persistence forecast excels over 16 days [37].

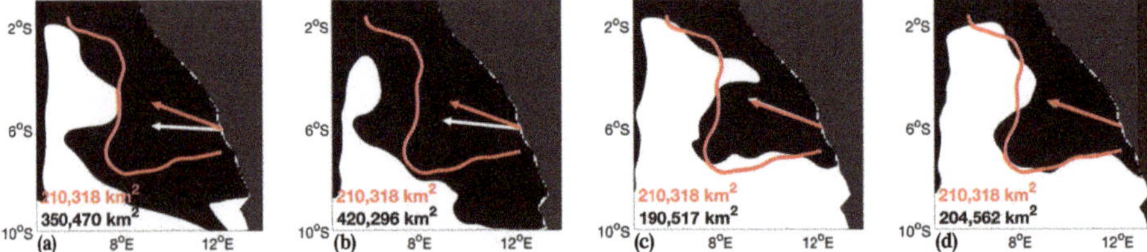

Figure 7. The 34 PSU plume area for (**a**) the control (CNTRL), (**b**) sea surface height (SSH), (**c**) SMOS and (**d**) SMOS SSH assimilation forecast (black filled) and SMOS data (red outlines) averaged for the January forecast. Overlaid are arrows denoting the orientation and distance to centre of mass for each experiment (white) and the SMOS data (red). The bottom left numbers denote the area of the plume for each experiment (black) and the SMOS data (red).

Consequently, for many plume metrics over most months, the SMOS PERSIST forecast did indeed perform the best with the lowest errors and largest average improvement over the CNTRL (85–95%). So while SMOS or SMOS SSH assimilation forecast often outperforms the CNTRL, as seen in Figure 7, SMOS PERSIST is consistently better.

Again it is worth emphasising that the assessment of the SMOS plume specific statistics during the forecast are a measure of whether the model can retain improvements attained during the assimilation, rather than an independent validation of the forecast. This independent validation will instead be examined later for Argo floats present.

Nevertheless, the plume area statistics comes closest to showing some skill in these metrics beyond persistence. Figure 8 illustrates the time series of the plume area error for each experiment, split up into each month. For January and August, the SMOS SSH and SMOS assimilation forecast perform well, with persistence often exhibiting growing errors, contrary to the relatively stable model forecasts. For April and July, persistence is the most beneficial, with SMOS and SMOS SSH forecasts experiencing notable errors in April larger than $-20{,}000$ km^2.

Argo floats present for the forecast period can be used as an independent validation (Table 5). On average, for the whole domain, the Argo float MAE errors for the SMOS or SMOS SSH assimilation experiment were the smallest for three of the four months analysed. This result suggests that persistence has its limitations for independent data, and the SMOS assimilation forecast is valuable.

Table 5. The mean absolute error (MAE) computed for Argo float samples for each experiment (CNTRL, SSH, SMOS, SMOS SSH) assimilation forecast over each month. An additional forecast denoted SMOS PERSIST is included. The Argo float MAE for the SMOS data itself (SMOS DATA) during the forecast period is also shown for comparison. The samples are split into whole domain MAE statistics or inside plume only. The number of Argo float samples over which each MAE is computed is also presented.

	Whole Domain				Inside Plume			
	No. of Float Samples							
	Jan	Apr	Jul	Aug	Jan	Apr	Jul	Aug
	31	21	25	27	5	4	2	1
Forecast Exp. Name	MAE (PSU)				MAE (PSU)			
CNTRL	0.43	0.27	0.28	0.23	0.89	0.67	0.86	0.98
SSH	0.69	0.33	0.19	0.28	2.04	0.74	0.59	0.85
SMOS	0.21	0.33	0.18	0.20	0.36	1.20	0.28	0.89
SMOS SSH	0.19	0.35	0.21	0.21	0.33	1.26	0.48	0.97
SMOS PERSIST	0.22	0.32	0.25	0.22	0.52	0.90	0.31	0.61
SMOS DATA	0.16	0.28	0.17	0.24	0.25	0.86	0.27	0.85

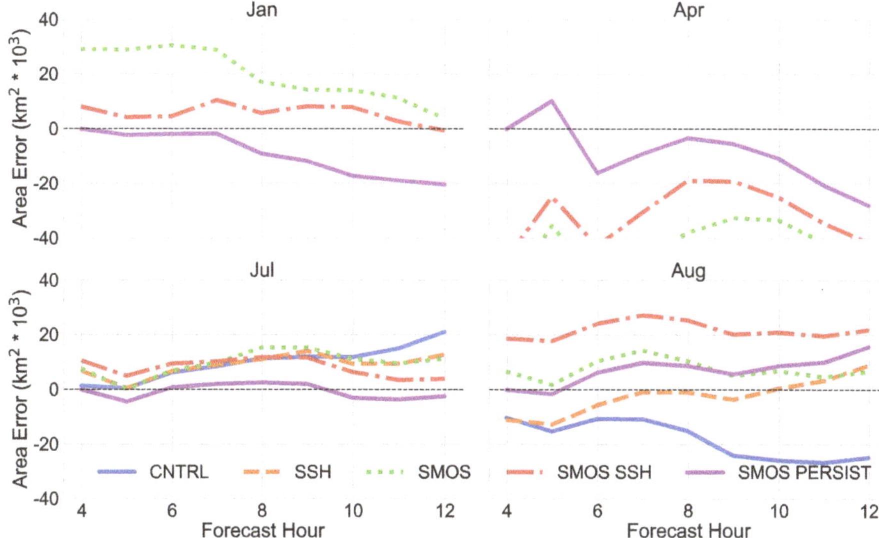

Figure 8. The plume area error as a function of forecast time, split into the four months for each forecast; CNTRL (blue solid line), SSH (yellow dashed line), SMOS (green dotted line), SMOS SSH (red dashed dot line) and SMOS PERSIST (purple solid line). A black dashed line highlights the zero error mark. Note the day four starting date is due to the computation of a 9-day moving average for the model forecast. This average enables an appropriate comparison with the SMOS data set. Forecasts not displayed within a certain month, indicate the errors are beyond the axis limits ($-40{,}000$ to $40{,}000$ km^2), as in January and April for the CNTRL and SSH forecasts.

For January, the SMOS SSH assimilation forecast improves over the CNTRL by 60% and SMOS PERSIST by 14%. For plume-detected Argo floats this improvement is even more pronounced at 63% and 37%. For April, contrary to all previous indications, the CNTRL performs the best. The SMOS data during the April forecast period validates with the Argo floats relatively poorly with an MAE of 0.28 PSU, slightly larger than the CNTRL error at 0.27 PSU. It follows that the CNTRL model is more likely to outperform the SMOS forecast, where a strong correlation ($0.93 < 0.05$) between the CNTRL-SMOS data error difference and CNTRL–SMOS forecast error difference exists (Figure 9). This poor performance is even more apparent for Argo floats inside the plume, where the CNTRL improves over the SMOS forecast by 44%. This degradation in performance is similar to the analysis January period (Table 3), where the SMOS data was found to contain larger errors than the SSH analysis, and thus the resulting SMOS analysis performed inadequately. For July, the SMOS assimilation forecast has the smallest whole domain Argo float MAE, improving over the CNTRL by 36% and over persistence by 28%. The two plume-detected Argo floats also exhibited this trend improving over the CNTRL and persistence by 70% and 10% respectively. Finally, for August, the SMOS assimilation forecast had the smallest Argo float MAE, improving over the control by 13%. Although, the difference between the CNTRL, SMOS forecast, SMOS SSH forecast and SMOS persistence was minimal from 0.01 to 0.04 PSU. A single Argo float was determined to be inside the plume during the August forecast. The location was very close to the edge of the model domain, where boundary errors dominate the model results. Therefore persistence performed well as expected.

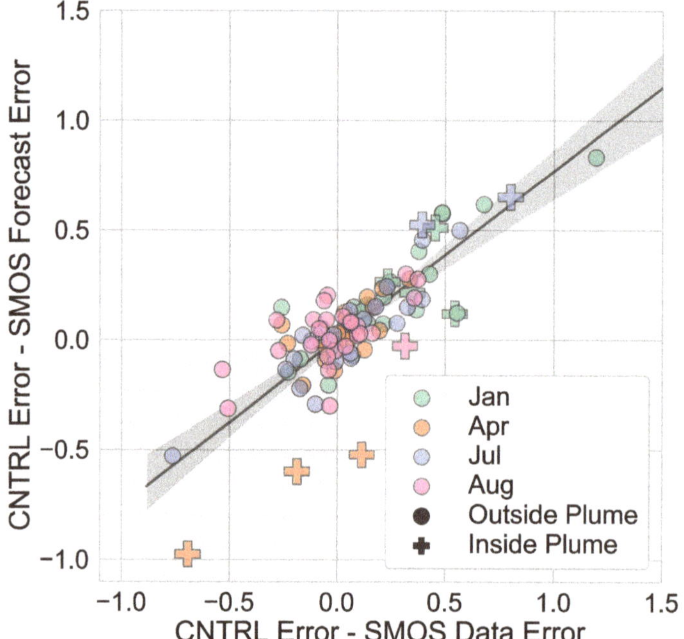

Figure 9. A scatter plot of the Argo float CNTRL error minus the Argo float SMOS forecast error against the Argo float CNTRL error minus the Argo float SMOS data error. The data is split into each study month, January (green), April (orange), July (blue) and August (pink). Outside (inside) plume samples are represented as circles (pluses). A linear regression is overlaid with a shaded bootstrap confidence interval. An additional shaded area for values <0 is displayed to highlight Argo float CNTRL errors that outperform both the SMOS analysis and SMOS data.

We now focus on the spatial distribution of forecast errors for the Argo floats across the domain. Furthermore, we alternatively present the skill score against persistence (Figure 10). Positive values represent improvements over persistence (green triangles), while negative values (purple circles) indicate degradation over persistence. Green pluses and purple crosses are used as an alternative representation for plume-detected Argo floats.

For January (first row of Figure 10), the CNTRL cannot outperform persistence for the majority of the Argo float samples, especially in the north, towards the plume. The SMOS assimilation forecast mostly corrects this, with some Argo floats remaining favoured towards persistence. For April (second row of Figure 10), the CNTRL is the best performing forecast for the majority of Argo floats. The forecast from the assimilation of SMOS does not significantly improve results with both improvements and degradation present. For July (third row of Figure 10), systematic improvements are seen for the assimilation of SSH and SMOS forecast. Interestingly, the improvement of SSH forecast can almost replicate the improvements as seen in the SMOS forecast. Improvements in the circulation are likely contributing to improvements in salinity advection. However, only the SMOS forecast can dramatically improve the plume-detected Argo floats in the north of the domain. Finally, for August (final row of Figure 10), improvements for the SMOS assimilation are mostly exhibited for the central cluster of Argo floats as compared to the CNTRL.

For every Argo float sample available during each month's forecast period (104 in total) the percentage of forecast skill scores above zero (positive values) can be computed. This represents the portion of persistence favoured Argo floats as compared to each experiment forecast. For example, the SMOS and SMOS SSH assimilation forecast contain 55% positive skill scores. This proportion split means these forecasts are 5% more likely to outperform persistence. Much larger than both the CNTRL and SSH experiments at approximately 40%, meaning that persistence is instead 10% more likely to outperform these forecasts.

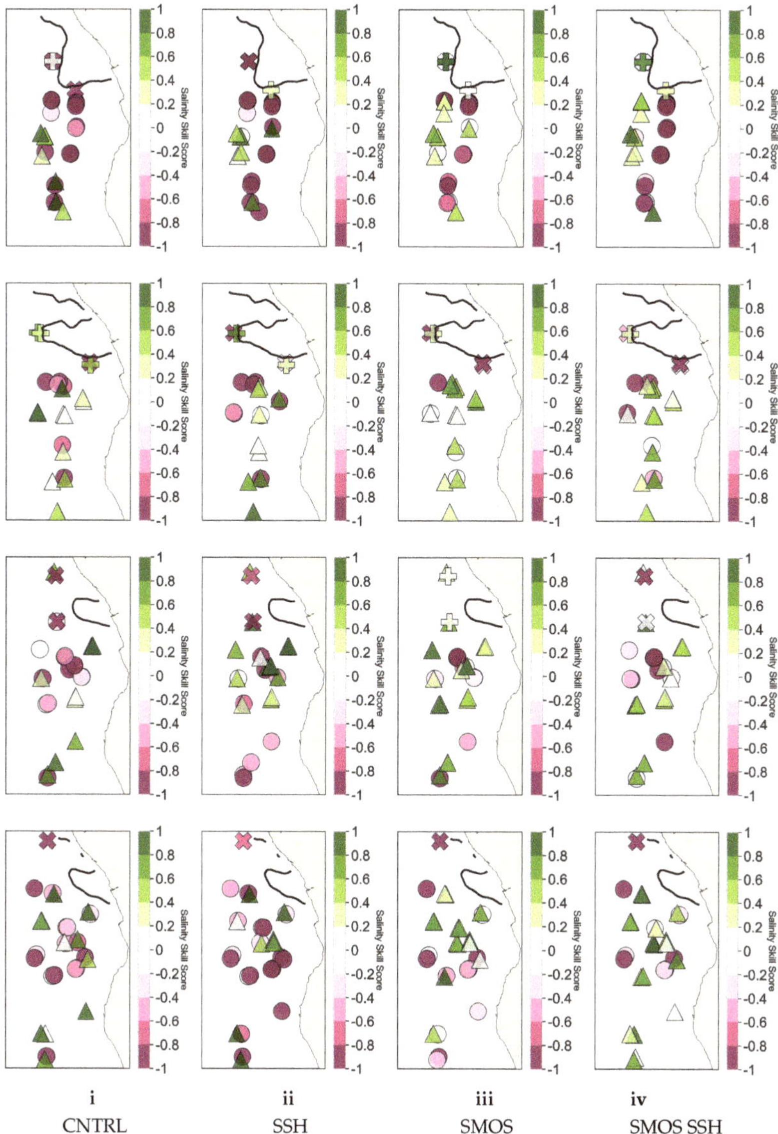

Figure 10. The Argo salinity skill score (persistence baseline) over January (1st row), April (2nd row), July (3rd row) and August (4th row) analysis for experiments (**i**) CNTRL, (**ii**) SSH, (**iii**) SMOS and (**iv**) SMOS SSH. Positive (green triangles) represent model errors smaller than persistence, while negative (purple circles) represent model errors larger than persistence. Purple crosses (instead of triangles) and green pluses (instead of circles) are alternatively displayed to represent the plume detected samples as seen in Figure 3. The average 34 PSU isohaline computed from the SMOS data for each respective month (forecast period) is overlaid in black.

4. Discussion

The CNTRL and SSH assimilation experiments inadequately reproduced observations of the Congo River plume (via SMOS plume statistics and Argo float comparisons). A few hypotheses can be put forth to explain this deficiency. Firstly, the discharge of the plume may be incorrect. The river input is based on 103 years of data which ends in 2005 (Figure 2). The standard deviations for this dataset range from only 10–15% of the mean value. The discharge error is thus unlikely to account for the extreme differences between the observations and the control. Deficiencies in the mixing schemes and resolution could limit the plume simulation, especially near the river mouth. The Angola Basin ROMS model has a resolution of 10 km with two grid points representing the opening of the Congo River channel Recently, Bars et al. [15] employed a multi-scale unstructured mesh model to simulate the Congo river-to-sea continuum, with elements as small as 200 m in the river and estuary and as large 20 km in the deep ocean. They note the river flow is characterised by smaller length scales that require a higher resolution. Therefore, it is clear that the resolution of the Congo within Angola Basin ROMS could be restricting the mixing of freshwater, and thus enabling a very fresh bias to expand into the open ocean where the general ocean circulation could propagate the bias into the basin. Furthermore, extreme gradients in the bathymetry such as those found near the Congo Canyon (originating downstream of the river) are particularly challenging [15]. The surface forcing from ERA-I could also generate errors. For example, local 10 m winds not correctly captured within ERA-I could result in erroneous currents that inaccurately advect the plume. Total precipitation errors will also affect the upper ocean salinity. This changes the evaporation minus precipitation (E-P) value that determines the net flux of freshwater into the ocean. Finally, the boundary conditions from HYCOM could contain an inaccurate salinity field that is propagating into the domain. However, despite these apparent difficulties, the assimilation of SMOS can successfully constrain the plume and adjust any prior bias in the model salinity as seen in the control.

Either the SMOS or SMOS SSH experiments performed the best for the majority of metrics in the analysis and forecast. It is well documented that the assimilation of SSH improves the large-scale ocean circulation [38–42] and this, in turn, could become especially important for the forecast of the plume with the improved currents spreading the adjusted plume (via the SMOS assimilation) more realistically. The free-running model forecast is no longer constrained by the regular daily SMOS assimilation and instead relies on the dynamics of the circulation spreading the far-field plume. While this suggests the assimilation of solely SSH should also improve the plume, deficiencies in CNTRL plume structure are likely already too large to correct from the adjustments in circulation patterns and multivariate correlations in the background error covariance.

Several Argo floats were crucially available throughout the domain for independent verification during each month. The SMOS assimilation analysis and forecasts performed the best most months analysed with the smallest Argo float errors. The addition of the SSH, as in the combined assimilation of SMOS with SSH, often slightly degraded the results. A hypothesis for this degradation is related to the relative weighting of the observations and model within the 4D-Var system. Most Argo float samples were located further away from the plume, where the standard deviation of the model climatology simulation drops below 0.2 PSU (not shown). Therefore with a 1.2 PSU SMOS observation error, the cost function is heavily weighted towards the model, only subtly changing the sea surface salinity field within the model, while also adjusting the salinity field via the multivariate balance options and, the evolution of the tangent linear (TL) and adjoint (AD) models from the SSH assimilation. This is the limitation of using a static covariance matrix within 4D-Var. Advanced hybrid methodologies (Hybrid-4DVar) are likely to improve results. Usually this involves replacing the static background error covariance matrix with an ensemble covariance computed from using an ensemble of non-linear model runs at each assimilation time [43]. Implementing this system into ROMS IS4D-Var is beyond the scope of this study.

While a simple persistence forecast proved to be a good predictor of the majority of plume statistics (computed from the SMOS data), persistence could not provide an adequate forecast for the

Argo floats. However, it was found that when the SMOS data validated poorly against the Argo floats during the analysis (forecast) period, as compared to the CNTRL or SSH analysis (forecast), the SMOS or SMOS SSH analysis (forecast) also performed poorly. While this result is somewhat trivial, it is vital to note that the capability of the model is intrinsically linked to the quality of the data.

The difference in the temporal resolution between the Argo floats, and SMOS data could explain the discrepancy in this comparison. The Argo floats represent the surface salinity at a much shorter time scales than the SMOS data (a 9-day moving average product). Therefore, the plume's extent is highly likely to be underestimated, and details near the plume will not be captured [4]. Still, despite this clear limitation, the SMOS forecast performed well for the majority of months analysed.

Lu et al. [18] showed that for the assimilation of SMOS the root mean squared error (RMSE) of the modelled sea surface salinity field compared with Argo float salinity data within the tropics (20°S–20°N) reduced by approximately 30% as compared to a control. This result held only for the open ocean as Lu et al. [18] used an older version of the BEC data [44]. Nevertheless, an improvement in the model when assimilating SMOS is similarly found in the experiments presented in this study. Köhl et al. [19] assimilated a different SMOS product (processed by the University of Hamburg) globally but found conflicting results to Lu et al. [18] and this study (negative to neutral impacts on the salinity field for the open ocean). Lu et al. [18] noted differences in the observation error covariance, models and data assimilation methods and SMOS datasets as possible reasons for this discrepancy. Finally, Mu et al. [21] recently assimilated a bias-corrected version of the updated SMOS dataset. They found up to 70% improvements with an RMSE less than 0.1 PSU, although it should be noted that their control RMSE was initially 0.3 PSU, smaller than the control for this study (RMSE errors of 0.4-0.7 PSU).

5. Conclusions

A first attempt of assimilating satellite salinity to model a major river plume has been presented. The latest version of SMOS, a satellite salinity product, was recently (2017) processed with an updated technique to reduce errors near the coast. This allows more frequent coastal data, which could be employed in a DA system of the Congo River plume. The ROMS Angola Basin configuration with IS4D-Var assimilated SSH, SMOS or SMOS SSH combined for four months.

The metrics applied to assess the assimilation analysis revealed that the SMOS observations were successfully assimilated during each month and maintained improvements throughout a subsequent forecast. The SMOS structure was well replicated within the model with regards to the plume area, distance to the centre of mass, orientation and average salinity within. Independent Argo float salinity profiles were available during the months analysed. The MAE of the Argo floats during the majority of months analysed reduced as compared to the control and persistence with the assimilation of SMOS. There are limitations to the current model, which are most likely due to a higher horizontal resolution required to resolve the upstream plume mixing. However, it has been shown that the current generation of satellite salinity observations can be successfully used to constrain a large river plume such as that produced by the Congo River. This successful assimilation stage could be the first step in producing operational forecasts of river plumes in the future.

Author Contributions: L.P. and R.T. conceived and designed the experiments. L.P. performed the experiments, analysis and wrote the paper. R.T provided insights and guidance, as well as contributing to the writing-review and editing of the paper.

Funding: This research was funded by the Natural Environment Research Council (NERC) award reference 1660325.

Acknowledgments: This study was supported by the Natural Environment Research Council (NERC). The authors would like to thank the Imperial College High Performance Computing Service (10.14469/hpc/2232) for providing computational resources. The altimeter products were produced by Ssalto/Duacs and distributed by Aviso, with support from Cnes (http://www.aviso.altimetry.fr/duacs/). Argo float data were collected and made freely available by the International Argo Program and the national programs that contribute to it (http://www.argo.ucsd.edu, http://argo.jcommops.org). The Argo Program is part of the Global Ocean

Observing System (http://doi.org/10.17882/42182). We thank two anonymous reviewers whose constructive comments improved the manuscript.

Conflicts of Interest: The authors declare no conflict of interest.

References

1. Dagg, M.; Benner, R.; Lohrenz, S.; Lawrence, D. Transformation of dissolved and particulate materials on continental shelves influenced by large rivers: Plume processes. *Cont. Shelf Res.* **2004**, *24*, 833–858. [CrossRef]
2. Chen, C.T.A.; Zhai, W.; Dai, M. Riverine input and air-sea CO_2 exchanges near the Changjiang (Yangtze River) Estuary: Status quo and implication on possible future changes in metabolic status. *Cont. Shelf Res.* **2008**, *28*, 1476–1482. [CrossRef]
3. Kang, Y.; Pan, D.; Bai, Y.; He, X.; Chen, X.; Chen, C.T.A.; Wang, D. Areas of the global major river plumes. *Acta Oceanol. Sin.* **2013**, *32*, 79–88. [CrossRef]
4. Hopkins, J.; Lucas, M.; Dufau, C.; Sutton, M.; Stum, J.; Lauret, O.; Channelliere, C. Detection and variability of the Congo River plume from satellite derived sea surface temperature, salinity, ocean colour and sea level. *Remote. Sens. Environ.* **2013**, *139*, 365–385. [CrossRef]
5. Braga, E.S.; Andrié, C.; Bourlès, B.; Vangriesheim, A.; Baurand, F.; Chuchla, R. Congo River signature and deep circulation in the eastern Guinea Basin. *Deep-Sea Res. Part I Oceanogr. Res. Pap.* **2004**, *51*, 1057–1073. [CrossRef]
6. Higgins, H.W.; Mackey, D.J.; Clementson, L. Phytoplankton distribution in the Bismarck Sea north of Papua New Guinea: The effect of the Sepik River outflow. *Deep-Sea Res. Part I Oceanogr. Res. Pap.* **2006**, *53*, 1845–1863. [CrossRef]
7. Kouame, K.; Yapo, O.; Mambo, V.; Seka, A.; Tidou, A.; Houenou, P. Physicochemical Characterization of the Waters of the Coastal Rivers and the Lagoonal System of Cote d'Ivoire. *J. Appl. Sci.* **2009**, *9*, 1517–1523. [CrossRef]
8. Yankovsky, A.E.; Chapman, D.C. A simple theory for the fate of buoyant coastal discharges. *J. Phys. Oceanogr.* **1997**, *27*, 1386–1401. [CrossRef]
9. Eisma, D.; van Bennekom, A.J. The Zaire river and estuary and the Zaire outflow in the Atlantic Ocean. *Neth. J. Sea Res.* **1978**, *12*, 255–272. [CrossRef]
10. Denamiel, C.; Budgell, W.P.; Toumi, R. The congo river plume: Impact of the forcing on the far-field and near-field dynamics. *J. Geophys. Res. Ocean.* **2013**, *118*, 964–989. [CrossRef]
11. Signorini, S.R.; Murtugudde, R.G.; McClain, C.R.; Christian, J.R.; Picaut, J.; Busalacchi, A.J. Biological and physical signatures in the tropical and subtropical Atlantic. *J. Geophys. Res. Ocean.* **1999**, *104*, 18367–18382. [CrossRef]
12. Vic, C.; Berger, H.; Tréguier, A.M.; Couvelard, X. Dynamics of an Equatorial River Plume: Theory and Numerical Experiments Applied to the Congo Plume Case. *J. Phys. Oceanogr.* **2014**, *44*, 980–994. [CrossRef]
13. Palma, E.D.; Matano, R.P. An idealized study of near equatorial river plumes. *J. Geophys. Res. Ocean.* **2017**, *122*, 3599–3620. [CrossRef]
14. White, R.H.; Toumi, R. River flow and ocean temperatures: The Congo River. *J. Geophys. Res. Ocean.* **2014**, *119*, 2501–2517. [CrossRef]
15. Bars, Y.L.; Vallaeys, V.; Deleersnijder, É.; Hanert, E.; Carrere, L.; Channelière, C. Unstructured-mesh modeling of the Congo river-to-sea continuum. *Ocean. Dyn.* **2016**, *66*, 589–603. [CrossRef]
16. Liu, Y.; MacCready, P.; Hickey, B.M.; Dever, E.P.; Kosro, P.M.; Banas, N.S. Evaluation of a coastal ocean circulation model for the Columbia River plume in summer 2004. *J. Geophys. Res. Ocean.* **2009**, *114*, C00B04. [CrossRef]
17. Lahoz, W.; Khattatov, B.; Ménard, R. Data assimilation and information. In *Data Assimilation: Making Sense of Observations*, 1st ed.; Lahoz, W., Khattatov, B., Menard, R., Eds.; Springer: Berlin/Heidelberg, Germany, 2010; Chapter 1; pp. 3–12.
18. Lu, Z.; Cheng, L.; Zhu, J.; Lin, R. The complementary role of SMOS sea surface salinity observations for estimating global ocean salinity state. *J. Geophys. Res. Ocean.* **2016**, *121*, 3672–3691. [CrossRef]
19. Köhl, A.; Sena Martins, M.; Stammer, D. Impact of assimilating surface salinity from SMOS on ocean circulation estimates. *J. Geophys. Res. Ocean.* **2014**, *119*, 5449–5464. [CrossRef]

20. Olmedo, E.; Martínez, J.; Turiel, A.; Ballabrera-Poy, J.; Portabella, M. Debiased non-Bayesian retrieval: A novel approach to SMOS Sea Surface Salinity. *Remote. Sens. Environ.* **2017**, *193*, 103–126. [CrossRef]

21. Mu, Z.; Zhang, W.; Wang, P.; Wang, H.; Yang, X. Assimilation of SMOS sea surface salinity in the regional ocean model for South China Sea. *Remote. Sens.* **2019**, *11*, 919. [CrossRef]

22. Moore, A.M.; Arango, H.G.; Broquet, G.; Edwards, C.; Veneziani, M.; Powell, B.; Foley, D.; Doyle, J.D.; Costa, D.; Robinson, P. The Regional Ocean Modeling System (ROMS) 4-dimensional variational data assimilation systems. Part III—Observation impact and observation sensitivity in the California Current System. *Prog. Oceanogr.* **2011**, *91*, 74–94. [CrossRef]

23. Phillipson, L.; Toumi, R. Impact of data assimilation on ocean current forecasts in the Angola Basin. *Ocean. Model.* **2017**, *114*, 45–58. [CrossRef]

24. Chassignet, E.P.; Hurlburt, H.E.; Smedstad, O.M.; Halliwell, G.R.; Hogan, P.J.; Wallcraft, A.J.; Baraille, R.; Bleck, R. The HYCOM (HYbrid Coordinate Ocean Model) data assimilative system. *J. Mar. Syst.* **2007**, *65*, 60–83. [CrossRef]

25. Dee, D.P.; Uppala, S.M.; Simmons, A.J.; Berrisford, P.; Poli, P.; Kobayashi, S.; Andrae, U.; Balmaseda, M.A.; Balsamo, G.; Bauer, P.; et al. The ERA-Interim reanalysis: Configuration and performance of the data assimilation system. *Q. J. R. Meteorol. Soc.* **2011**, *137*, 553–597. [CrossRef]

26. Vorosmarty, C.; Fekete, B.M.; Tucker, B. *Global River Discharge, 1807–1991, V. 1.1 (RivDIS)*; Oak Ridge Natl. Lab. Distrib. Active Arch. Cent.: Oak Ridge, TN, USA, 1998.

27. Meybeck, M.; Ragu, A. *GEMS-GLORI World River Discharge Database*; Lab. de Gologie Appl., Univ. Pierre et Marie Curie: Paris, France, 2012.

28. Alsdorf, D.; Beighley, E.; Laraque, A.; Lee, H.; Tshimanga, R.; O'Loughlin, F.; Mahé, G.; Dinga, B.; Moukandi, G.; Spencer, R.G.M. Opportunities for hydrologic research in the Congo Basin. *Rev. Geophys.* **2016**, *54*, 378–409. [CrossRef]

29. Jackson, B.; Nicholson, S.E.; Klotter, D. Mesoscale Convective Systems over Western Equatorial Africa and Their Relationship to Large-Scale Circulation. *Mon. Weather. Rev.* **2009**, *137*, 1272–1294. [CrossRef]

30. Phillipson, L. Ocean Data Assimilation in the Angola Basin. Ph.D. Thesis, Imperial College London, London, UK, 2018.

31. SMOS-BEC Team. *Quality Report: Validation of Debiased Non-Bayesian BEC advanced Sea Surface Salinity Products*; Technical Report; Barcelona Expert Centre: Barcelona, Spain, 2017.

32. Willis, J.K. Interannual variability in upper ocean heat content, temperature, and thermosteric expansion on global scales. *J. Geophys. Res.* **2004**, *109*, C12036. [CrossRef]

33. Good, S.A.; Martin, M.J.; Rayner, N.A. EN4: Quality controlled ocean temperature and salinity profiles and monthly objective analyses with uncertainty estimates. *J. Geophys. Res. Ocean.* **2013**, *118*, 6704–6716. [CrossRef]

34. Weaver, A.T.; Deltel, C.; Machu, E.; Ricci, S.; Daget, N. A multivariate balance operator for variational ocean data assimilation. *Q. J. R. Meteorol. Soc.* **2006**, *131*, 3605–3625. [CrossRef]

35. Moore, A.M.; Edwards, C.A.; Fiechter, J.; Drake, P.; Neveu, E.; Arango, H.G.; Gürol, S.; Weaver, A.T. A 4D-var analysis system for the california current: A prototype for an operational regional ocean data assimilation system. In *Data Assimilation for Atmospheric, Oceanic and Hydrologic Applications (Vol. II)*; Springer: Berlin/Heidelberg, Germany, 2013; pp. 345–366.

36. Lumpkin, R.; Johnson, G.C. Global ocean surface velocities from drifters: Mean, variance, El Niño–Southern Oscillation response, and seasonal cycle. *J. Geophys. Res. Ocean.* **2013**, *118*, 2992–3006. [CrossRef]

37. Phillipson, L.M.; Toumi, R. The Crossover Time as an Evaluation of Ocean Models Against Persistence. *Geophys. Res. Lett.* **2018**, *45*, 250–257. [CrossRef]

38. Fukumori, I.; Raghunath, R.; Fu, L.L.; Chao, Y. Assimilation of TOPEX/Poseidon altimeter data into a global ocean circulation model: How good are the results? *J. Geophys. Res. Ocean.* **1999**, *104*, 25647–25665. [CrossRef]

39. Dorofeev, V.L.; Korotaev, G.K. Assimilation of the data of satellite altimetry in an eddy-resolving model of circulation of the Black Sea. *Phys. Oceanogr.* **2004**, *14*, 42–56. [CrossRef]

40. Dombrowsky, E.; Bertino, L.; Brassington, G.; Chassignet, E.; Davidson, F.; Hurlburt, H.; Kamachi, M.; Lee, T.; Martin, M.; Mei, S.; et al. GODAE Systems in Operation. *Oceanography.* **2009**, *22*, 80–95. [CrossRef]

41. Moore, A.M.; Arango, H.G.; Broquet, G.; Edwards, C.; Veneziani, M.; Powell, B.; Foley, D.; Doyle, J.D.; Costa, D.; Robinson, P. The Regional Ocean Modeling System (ROMS) 4-dimensional variational data assimilation systems. Part II—Performance and application to the California Current System. *Prog. Oceanogr.* **2011**, *91*, 50–73. [CrossRef]
42. Da Rocha Fragoso, M.; de Carvalho, G.V.; Soares, F.L.M.; Faller, D.G.; de Freitas Assad, L.P.; Toste, R.; Sancho, L.M.B.; Passos, E.N.; Böck, C.S.; Reis, B.; et al. A 4D-variational ocean data assimilation application for Santos Basin, Brazil. *Ocean. Dyn.* **2016**, *66*, 419–434. [CrossRef]
43. Desroziers, G.; Camino, J.T.; Berre, L. 4DEnVar: Link with 4D state formulation of variational assimilation and different possible implementations. *Q. J. R. Meteorol. Soc.* **2014**, *140*, 2097–2110. [CrossRef]
44. Font, J.; Boutin, J.; Reul, N.; Spurgeon, P.; Ballabrera-Poy, J.; Chuprin, A.; Gabarró, C.; Gourrion, J.; Guimbard, S.; Hénocq, C.; et al. SMOS first data analysis for sea surface salinity determination. *Int. J. Remote. Sens.* **2013**, *34*, 3654–3670. [CrossRef]

remote sensing

Article

In Situ and Satellite Sea Surface Salinity in the Algerian Basin Observed through ABACUS Glider Measurements and BEC SMOS Regional Products

Giuseppe Aulicino [1,2,*], Yuri Cotroneo [2], Estrella Olmedo [3], Cinzia Cesarano [1], Giannetta Fusco [2] and Giorgio Budillon [2]

[1]   Department of Life and Environmental Sciences, Università Politecnica delle Marche, 60131 Ancona, Italy; cinziacesarano@gmail.com
[2]   Department of Science and Technology, Università degli Studi di Napoli Parthenope, 80143 Napoli, Italy; cotroneo@uniparthenope.it (Y.C.); giannetta.fusco@uniparthenope.it (G.F.); giorgio.budillon@uniparthenope.it (G.B.)
[3]   Department of Physical Oceanography, Institute of Marine Sciences, CSIC, Barcelona Expert Center, 08003 Barcelona, Spain; olmedo@icm.csic.es
*   Correspondence: g.aulicino@staff.univpm.it

Received: 4 April 2019; Accepted: 4 June 2019; Published: 6 June 2019

Abstract: The Algerian Basin is a key area for the general circulation in the western Mediterranean Sea. The basin has an intense inflow/outflow regime with complex circulation patterns, involving both fresh Atlantic water and more saline Mediterranean water. Several studies have demonstrated the advantages of the combined use of autonomous underwater vehicles, such as gliders, with remotely sensed products (e.g., altimetry, MUR SST) to observe meso- and submesoscale structures and their properties. An important contribution could come from a new generation of enhanced satellite sea surface salinity (SSS) products, e.g., those provided by the Soil Moisture and Ocean Salinity (SMOS) mission. In this paper, we assess the advantages of using Barcelona Expert Center (BEC) SMOS SSS products, obtained through a combination of debiased non-Bayesian retrieval, DINEOF (data interpolating empirical orthogonal functions) and multifractal fusion with high resolution sea surface temperature (OSTIA SST) maps. Such an aim was reached by comparing SMOS Level-3 (L3) and Level-4 (L4) SSS products with in situ high resolution glider measurements collected in the framework of the Algerian Basin Circulation Unmanned Survey (ABACUS) observational program conducted in the Algerian Basin during falls 2014–2016. Results show that different levels of confidence between in situ and satellite measurements can be achieved according to the spatial scales of variability. Although SMOS values slightly underestimate in situ observations (mean difference is −0.14 (−0.11)), with a standard deviation of 0.25 (0.26) for L3 (L4) products), at basin scale, the enhanced SMOS products well represent the salinity patterns described by the ABACUS data.

Keywords: sea surface salinity; BEC SMOS products; Mediterranean Sea; Algerian Basin; ABACUS gliders

---

1. Introduction

The Algerian Basin (hereafter AB) is a wide and deep transit region located in the western Mediterranean Sea, of which it constitutes the southern part. This basin is about 750 km wide at 38.5°N between the coast of Spain and the Sardinia Channel, while the latitudinal extension between the Algerian Coast and the Balearic Islands (at 2.5°E) is about 250 km. The AB is characterized by the presence at surface levels of both fairly fresh water coming from the Atlantic (Atlantic water, hereafter AW) and more saline waters, which typically reside in the Mediterranean region (Mediterranean water, MW). According to different stages of mixing, geographical position, and residence time in the Mediterranean Sea, AW properties vary across this basin, ranging between 14–27 °C and 36.5–38.0, in terms of potential temperature and salinity respectively. The Levantine intermediate water (LIW), typified by subsurface temperature and salinity maxima, is also present at intermediate levels (400–900 m depth), with a typical potential temperature lower than 13.5 °C and a salinity of about 38.5. The western Mediterranean deep water (WMDW) occupies deeper layers, being characterized by temperatures of about 13 °C and salinity values ranging between 38 and 38.9 [1–3].

The general circulation of these water masses is strongly influenced by both the intense inflow/outflow regime [2,4] and the complex circulation patterns [5–7], which act in the AB at different spatial and temporal scales [8]. In particular, AW carried by the Algerian Current (AC) generates several fresh-core coastal eddies that propagate downstream [9,10] and promote water mass mixing, thus affecting the spatial distribution of salinity, and, consequently, the Mediterranean Sea surface circulation. These mesoscale energetic structures also have marked repercussions on nutrient injection (removal) into (out of) the euphotic layer. In fact, a large anticorrelation between sea level anomalies and phytoplankton biomass has been previously observed in the AB [11], thus suggesting a clear response of biological activity to the shoaling/deepening of isopycnals [12]. As part of the southwestern Mediterranean region, this basin is also known to be particularly responsive to climate change [13,14], and the Mediterranean waters flowing in this area have already shown significant trends at different depths in both temperature and salinity [15–18].

Moreover, the convection processes occurring in the northwestern Mediterranean feed the formation of western Mediterranean deep water (hereafter WMDW), which contributes to the Mediterranean thermohaline circulation. These deep waters have become saltier and warmer for at least the past 40 years at rates of about 0.015 and 0.04 °C per decade [19], but the budget of salt involved in the formation of WMDW is still under investigation [20–22]. WMDW convection is also associated with the vertical mixing that enriches the surface layer with critical nutrients, thus contributing to spring bloom and primary production rates [23]. These phenomena have been intensively studied in recent years, but a complete knowledge of the mechanisms acting in the AB is still pending. Thus, both accurate observations of surface salinity and reliable estimates of salt budgets are essential to support the scientific efforts to model past and future evolution of Mediterranean climate and give an answer to these open questions.

Although the Mediterranean region is strongly affected by radio frequency interference (RFI) and systematic biases due to the coast contamination, European Space Agency (ESA) soil moisture and ocean salinity (SMOS) [24,25] sea surface salinity (SSS) products have been already used to detect mesoscale structures and reconstruct coherent currents in the AB [26]. Nevertheless, evident biases were found when analyzing SSS values in comparison with Argo floats. To overcome these problems, a new set of SMOS SSS enhanced products [27] has been obtained at the Barcelona Expert Center (BEC) through a combination of debiased non-Bayesian retrieval [28], deletion of time-dependent residual biases by means of DINEOF (data interpolating empirical orthogonal functions) [29], and multifractal fusion with high resolution sea surface temperature (OSTIA SST) maps [30].

Several studies have assessed the advantages of multiplatform ocean monitoring over different scales, including mesoscale, and at different latitudes [31–37]. Recent experiments demonstrated that combining new technologies for in situ data collection, mainly autonomous underwater vehicles (AUVs), and reliable satellite data in the AB (e.g., [12,38,39]), provides useful contributions to properly address state-of-the-art scientific challenges [40–42]. AUVs allow the collection of high resolution physical and biochemical data along the water column from surface to 975 m depth [17], while their use in combination with remote sensing observations provides a better understanding of basin-scale processes, such as those influencing the southwestern Mediterranean Sea dynamics [38,43].

Furthermore, information collected through AUVs in the AB has already confirmed their usefulness in the validation of new satellite altimetry products, e.g., those coming from the SARAL/AltiKa [7] and the new Sentinel-3 mission [44]; AUV activities in the AB are also fully involved in the framework of upcoming satellite missions, such as the Surface Water and Ocean Topography (SWOT) wide-swath radar interferometer [45,46].

In such a context, this study compares ABACUS (Algerian Basin Circulation Unmanned Survey) in situ high resolution glider measurements collected in the AB during fall 2014–2016 with co-located SMOS enhanced SSS L3 and L4 products, provided by BEC, in order to confirm that retrieving reliable SMOS SSS in the Mediterranean region is indeed possible.

ABACUS glider cruises were developed along a repeated monitoring line across the AB, and allowed the collection of a huge dataset of physical and biological ocean parameters in the first 975 m of the water column. Even though the present study is restricted to a particular area, i.e., the AB, we strongly believe that the scientific community could use our results to enlarge the knowledge of the western Mediterranean Sea and to refine the use of glider missions to validate SMOS SSS products in other regional-scale studies (e.g., those focusing on the Agulhas Leakage, the Malvinas Confluence, the Gulf Stream). BEC SMOS SSS L3 and L4 products, as well as glider measurements and mission strategies, are described in Section 2. Results and discussion are presented in Section 3. Finally, Section 4 reports prominent conclusions.

2. Data and Methods

2.1. Glider In Situ Observations

In 2014, a new repeated monitoring line was created in the western Mediterranean, in the framework of the ABACUS project. Since then, five glider missions have been carried out in the AB from 2014 to 2019, permitting a significant overlap with the satellite SMOS mission and products. In the present work, the glider data collected during the autumns from 2014 to 2016 were used (Figure 1). Additional ABACUS campaigns were developed during fall 2017 and spring 2018 and 2019, but the collected measurements are still under quality control analyses and have not been released yet. Conversely, the ABACUS 2014–2016 dataset was fully quality controlled and is available through a public repository at https://dx.doi.org/10.25704/b200-3vf5. This dataset includes a total of eight glider transects realized between the Island of Mallorca and the Algerian Coast to collect temperature, salinity, turbidity, oxygen, and chlorophyll concentration measurements between surface and 975 m depth.

Each mission had an average duration of about 40 days and was always performed in the same season, i.e., between September and December. The timing of the missions was accurately planned in order to provide synoptic in situ observations with respect to the satellite SARAL/AltiKa and Sentinel-3A passages. In 2014 and 2015, after the realization of the defined transects, the glider was deviated from the monitoring line in order to sample specific mesoscale structures identified through near real time satellite altimetry and SST maps (Figure 1).

ABACUS data were collected through Slocum G2 deep gliders diving with an angle of 26° with the sea surface, at an average vertical speed of 0.18 ± 0.02 m/s. The resulting net horizontal velocity is about 0.36 m/s.

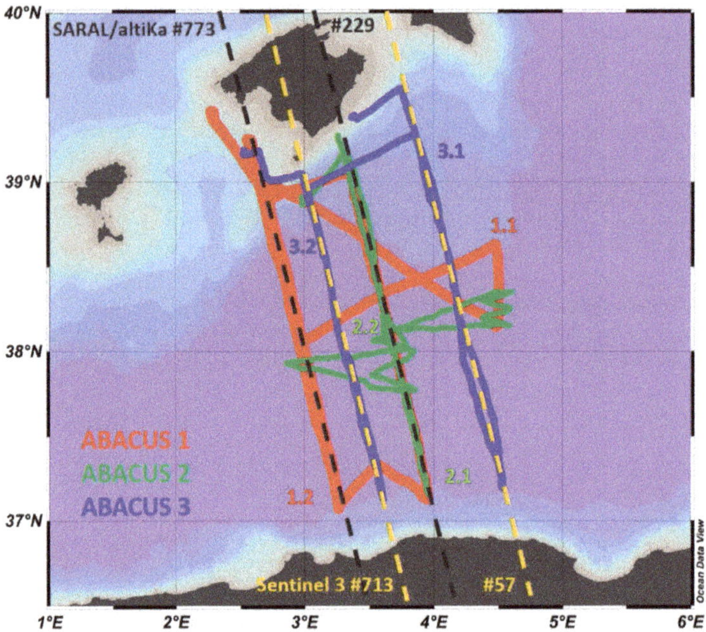

Figure 1. Glider tracks during the missions: ABACUS (Algerian Basin Circulation Unmanned Survey) 1 (red dots), ABACUS 2 (green dots), and ABACUS 3 (blue dots); sub-transects are identified through numbers as in Section 3. The groundtracks of the SARAL/AltiKa (grey dashed lines) and Sentinel 3 (yellow dashed lines) satellites over the study area are also shown. Adapted from Reference [17].

The two autonomous platforms used during the ABACUS surveys were equipped with the same instrumentation: a glider-customized CTD by Seabird measuring temperature, salinity (derived from conductivity), and depth; a two-channel combo fluorometer and turbidity sensor by WetLabs; and an oxygen optode to measure absolute oxygen concentration and saturation by AADI. Temperature and salinity were sampled to full diving depth (0–975 m depth). Oxygen data were collected to the same depth, while the acquisition of the other optical parameters ceased at 300 m depth. Physical parameters (temperature and salinity) were sampled at 1/2 Hz, resulting in a vertical resolution of 0.4 m along the water column. CTD details are listed in Table 1. Furthermore, it is important to remark that glider measurements were not precisely at the sea surface; however, they were closer to the ocean–atmosphere interface than Argo can usually achieve. In fact, the shallowest glider observations range between 0.22–0.99 m depth during ABACUS surveys, while Argo floats usually stop pumping water around 2–5 m depth.

Table 1. Sampling rate and vertical resolution of ABACUS gliders' conductivity-temperature-depth (CTD) sensors (* full scale range).

Parameter	Instrument	Sampling Rate (Hz)	Vertical Resolution (m)	Depth Range (m)	Accuracy	Resolution
Conductivity (C)	Seabird GPCTD				C: ±0.0003 S/m	C: 0.00001 S/m
Temperature (T)	glider payload	1/2	0.4	0 to −975	T: ±0.002 °C	T: 0.001 °C
Depth (D)	pumped CTD				D: ±0.1% fsr *	D: 0.002% fsr *

Temperature and salinity sensors were regularly calibrated after every glider mission in order to guarantee the quality of the measurements. After each mission, data were transferred from the internal glider memory to the SOCIB Data Center, where data processing was carried out and production of delayed time NetCDF files (i.e., level 1 and level 2) occurred before web dissemination of the data.

In the present study, we make use of the level 2 salinity dataset that includes regularly sampled vertical profiles obtained by interpolation of level 1 data from each up or downcast. Data processing included thermal lag correction, which was applied following a specific procedure developed for gliders [47], filtering, and 1 m bin vertical averaging [48]. Nevertheless, an additional quality control procedure was developed and regularly performed at University of Naples "Parthenope" to identify and discard bad data and artifacts that were still present after the level 2 processing. This procedure included a single-point spike control, an interpolation of single missing data along the profiles, and a five-point running mean along the depth; finally, an iterative comparison between adjacent profiles was performed [17]. As for salinity measurements, the detailed results previously reported [17] confirm that ABACUS glider uncertainties are smaller than the typical values that characterize the AB natural variability at surface layer, and thus are accurate enough to capture and correctly describe the main thermohaline properties of the AB water masses and their variability.

Finally, it is important to remark that during the 2014 surveys, only the downcast samplings were collected, breaking the surface at every 8 km; both downcasts and upcasts were gathered in the mission conducted in 2015 with no changes in surface breaking; in 2016, the glider was programmed to break surface after every profile (4 km), acquiring both upcast and downcast data. These changes in the sampling and navigation strategy improved the data collection in the very surface layer (depth <20 m), and provided a more suitable dataset for the chased comparison with satellite data. Indeed, spatial resolution of in situ measurements was improved from 4 km to 2 km, and the number of samples collected in the 0–10 m depth layer increased significantly (i.e., from 25 in 2014 to 150 in 2016, for a regular 12 km interval).

2.2. SMOS L3 and L4 Sea Surface Salinity Products

ABACUS in situ observations were compared to co-located L3 and L4 maps of SMOS SSS, recently obtained at the BEC through the methodology developed by Reference [27]. This methodology takes advantage of a combination of the new retrieval debiased non-Bayesian algorithm, DINEOF, and multifractal fusion to retrieve enhanced SSS fields over the North Atlantic Ocean and the Mediterranean Sea.

The debiased non-Bayesian aims to mitigate the systematic biases associated with the SMOS acquisitions. These are the biases that are expected to be constant under the same acquisition conditions (i.e., those associated to the same location, satellite overpass direction, and antenna coordinates). This methodology has been demonstrated to mitigate a large part of the errors produced by land–sea contamination and errors associated to permanent sources of RFI. However, some residual errors associated with intermittent sources of RFI and residual land–sea contamination were still present in the resulting retrievals. Parts of the temporal dependent residual errors have been characterized and corrected using DINEOF decomposition. The empirical orthogonal functions correlated with the temporal dependent residual errors were removed from the SMOS salinity maps. At the end, the resulting L3 maps provided root mean square differences with respect to an Argo salinity of 0.3 (see Table 2 in Reference [27]). After applying both corrections, the spatial distribution of the errors is pretty uniform in the western Mediterranean (see Figure 5 in [27], first plot of the second row). This suggests that the errors associated with land–sea contamination and RFI in the western Mediterranean have been mitigated. Conversely, systematic negative biases still appear in the eastern part of the basin, of which regions appear more degraded due to RFI contamination.

Furthermore, multifractal fusions were applied to BEC SMOS L3 maps in order to increase the spatial and temporal resolution of the SMOS salinity maps. This was done using OSTIA SST maps as a template [49]. The resulting L4 maps showed coherent (and better resolved) spatial structures in some specific regions, as the Alboran Sea and the Gulf of Lion. Due to limitations of the fusion methodology and of the effective spatial resolution of the template, the effect of the fusion method sometimes produces an additional smoothing to the SSS maps, preventing an actual improvement of the effective resolution of the L3 maps. This phenomenon has also been observed during the present study; although several structures are better resolved by L4 products, we did not observed an evident improvement of the effective resolution for all the analyzed SSS maps of the AB.

Nonetheless, these products present an improved accuracy in the Mediterranean basin with respect to the objectively analyzed maps and the L4 products previously available, i.e., a reduction of both the seasonal bias and the standard deviation of the error [27].

In this study, we used L3 nine-day binned maps at 0.25° and L4 daily maps at 0.05°, which are freely available at http://bec.icm.csic.es/ocean-experimental-dataset-mediterranean/. L3 maps are generated by using 9 days of SMOS acquisitions and by applying a scheme of objective analysis. They are generated daily on a grid of 0.25° × 0.25°. L4 maps are generated by fusing the nine-day L3 maps with daily OSTIA SST maps, so that the day of the SST map corresponds to the central day of the SMOS nine-day period. The L4 SMOS SSS maps inherit the resolution of the OSTIA SST maps (0.05° × 0.05°).

The co-location of SMOS and glider SSS was performed comparing the uppermost (0–10 m) SSS measurements provided by the ABACUS gliders at the instant t0, with the SMOS SSS field given by the nine-day map (t0 at 4 days). ABACUS SSS deeper than 10 m were not considered in this study.

2.3. Altimetry and Sea Surface Temperature Maps

In this study, we combined altimetry observations and SST information in order to improve the description of the main features of the western Mediterranean Sea during the analyzed glider surveys. The final goal was to evaluate the presence of filaments and mesoscale structures that could have caused the observed difference between SMOS and glider SSS values.

As for altimetry, we used Mediterranean Sea gridded L4 daily absolute dynamic topography (ADT) and geostrophic currents computed with respect to a twenty year 2012 mean by Optimal Interpolation, merging the measurement from all altimeter missions (i.e., Jason-3, Sentinel-3A, HY-2A, SARAL/AltiKa, Cryosat-2, Jason-2, Jason-1, T/P, ENVISAT, GFO, ERS1/2) through the DUACS multimission altimeter data processing system [50]. To produce reprocessed maps of ADT in delayed-time, this system uses the along-track altimeter missions from products called SEALEVEL*_PHY_L3_REP_OBSERVATIONS_008_*. The datasets are available from the CMEMS web portal (http://marine.copernicus.eu/services-portfolio/access-to-products/).

L4 SST foundation data were provided by the NASA Jet Propulsion Laboratory Physical Oceanography Distributed Active Archive Center (PODAAC). We opted for MUR-SST foundation products (full name JPL-L4_GHRSSTSSTfnd-MUR-GLOB-v02.0-fv04.1) that spanned from June 2002 to the present [51]. The dataset is available at ftp://podaac-ftp.jpl.nasa.gov/allData/ghrsst/data/L4/ GLOB/JPL/MUR as a netCDF file containing daily global SST data at a spatial resolution of 0.01° in longitude–latitude coordinates (about 1 km intervals). These maps are made mostly from satellite SST measurements, with help from surface observations from ships and buoys. Ultra-high resolution is achieved for the MUR-SST foundation dataset through the application of a multi-resolution variational analysis. In general, this dataset compares well in magnitude and phase with the lower resolution products (e.g., OSTIA); it also allows identification of interesting finer scale structures that are most likely due to meso- and submesoscale eddies.

3. Results and Discussion

We compared SMOS SSS, which represents few centimeters of depth, and the corresponding integrated value of the pixel area, with the mean salinity value provided by the first 10m depth of ABACUS observations. This subset was chosen according to the analysis of SSS values measured by ABACUS glider. To this aim, observations collected in the first 10 m depth were divided into three bands, i.e., 0–3 m, 3–6 m, and 6–10 m depth. The example reported in Figure 2 shows that SSS did not present a significant variability with depth in this very surface layer during the ABACUS 1 glider surveys; this result reflects the thermocline location at depths greater than 10 m during these cruises [17]. Similar results were obtained for the ABACUS 2 and ABACUS 3 observations.

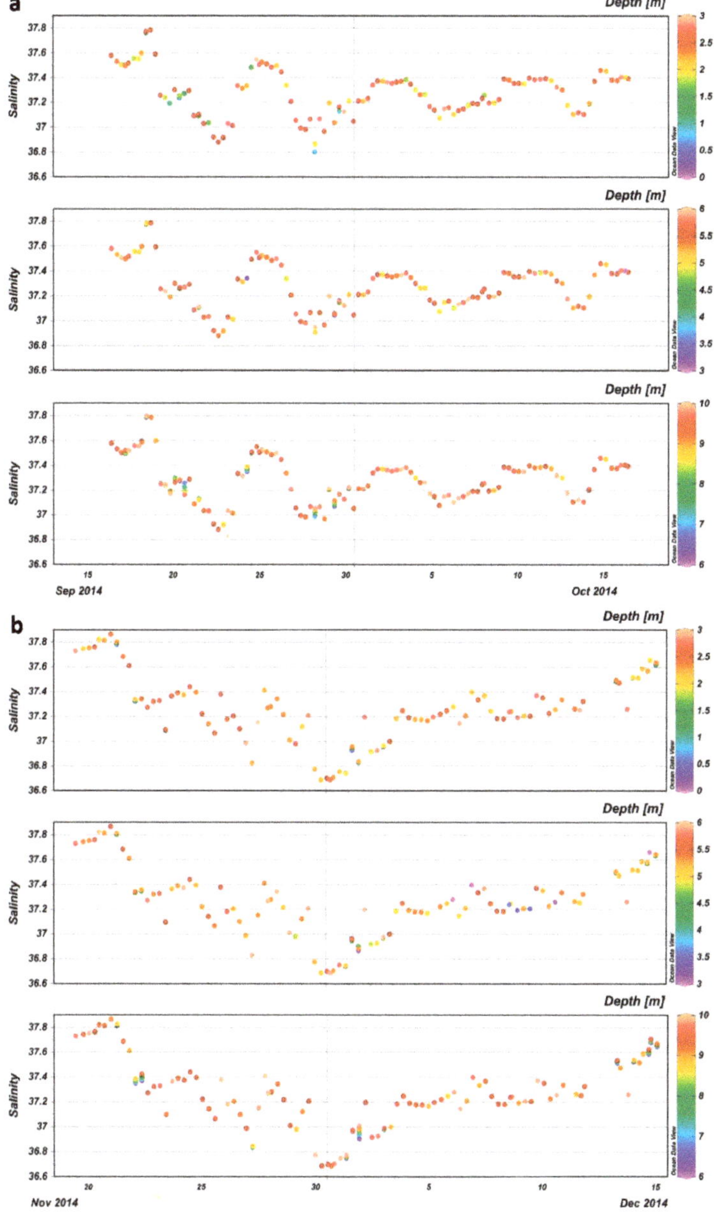

Figure 2. Glider sea surface salinity (SSS) collected in the first 10 m depth during the ABACUS missions 1.1 (**a**) and 1.2 (**b**). For each survey, observations are split into three bands, i.e., 0–3 m, 3–6 m, and 6–10 m depth.

We then spatially averaged the ABACUS salinity of the 0–3 m layer in the same pixel of SMOS. The resulting averaged value was compared with SMOS L3 and L4 maps available for different dates (Figure 3).

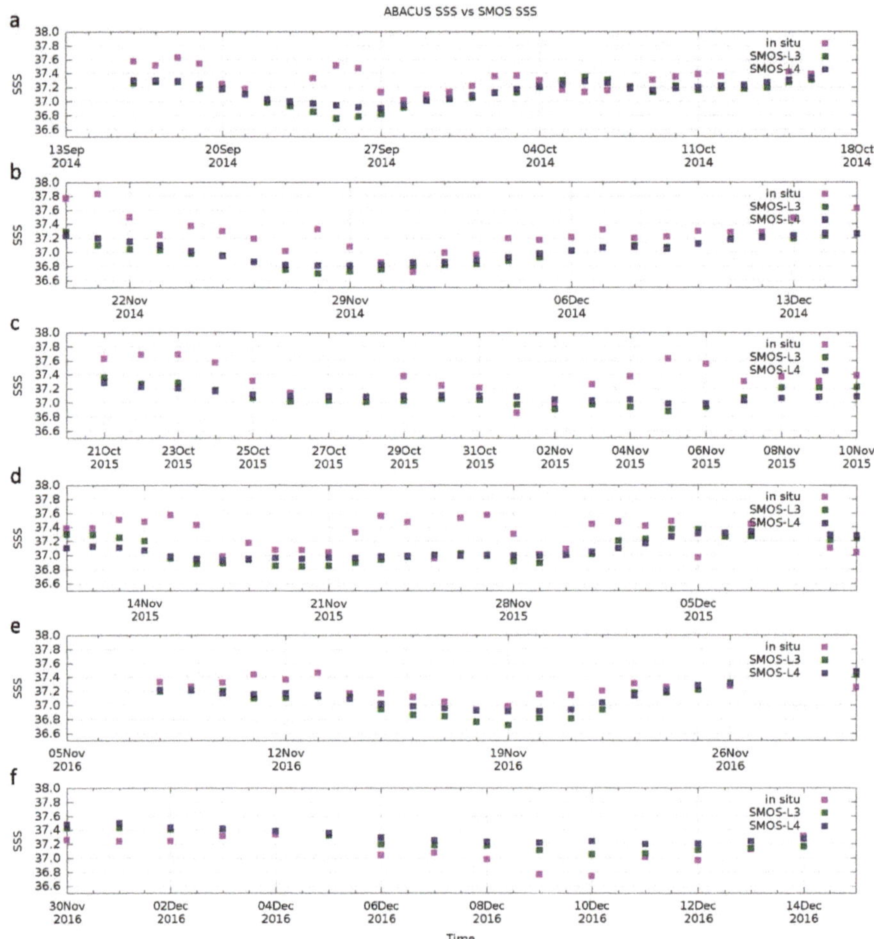

Figure 3. Comparison between averaged in situ glider salinities (purple dots) and SMOS L3 (green dots) and L4 (blue dots) SSS during the missions (**a**) ABACUS 1.1 (Sept–Oct 2014), (**b**) ABACUS 1.2 (Nov–Dec 2014), (**c**) ABACUS 2.1 (Oct–Nov 2015), (**d**) ABACUS 2.2 (Nov–Dec 2015), (**e**) ABACUS 3.1 (Nov 2016), and (**f**) ABACUS 3.2 (Dec 2016).

3.1. Overall Results

In spite of the limitation associated with the differences in the spatial and temporal resolutions, the enhanced SMOS products represent well the salinity patterns described by in situ SSS, although slightly underestimating ABACUS observations. Over 6799 comparisons, a mean difference of −0.14 was estimated between SMOS L3 and ABACUS SSS, the standard deviation of difference (std) being 0.25. The achieved results are similar to those previously estimated by Argo floats (−0.16, with a std of 0.34) [27], and fall within the range of variability that typically characterizes AB surface salinities during the September–December season [17]. The very small uncertainties associated with the glider measurements do not seem to appreciably condition these statistics.

Figure 3 shows that SMOS L3 retrievals represent well the qualitative evolution of surface salinity along the glider transects. Nevertheless, SMOS estimations were still far to be coincident with in situ high resolution measurements, and their smoothed behavior prevented them reproducing local variability, possibly due to submesoscale structures, lenses, and other small scale phenomena intercepted by gliders and recorded in the in situ dataset. In this sense, the main statistics of the analyzed ABACUS surveys, reported in Table 2, suggest that SMOS L3 retrieval capability can vary largely across different glider experiments, depending on local dynamics and ocean structures.

Similar results were achieved from the comparison with SMOS L4 products that returned a mean difference (SMOS-ABACUS) of −0.11 and a std of 0.26; linear correlation among satellite and in situ datasets (R-values in Table 2) was not improved significantly by L4 products, except during the ABACUS 1 survey. This suggests that specific features captured by in situ observations are not resolved by SMOS products, even the L4 one.

However, the advantages of the downscaling algorithm used for producing L4 products emerge when looking at the latitudinal comparisons between ABACUS and SMOS products. Figures 4 and 5 show that L4 SSS are generally saltier than L3, thus being typically closer to in situ SSS measurements. This is more evident at lower latitudes, i.e., between 37.1 and 37.5° N, where SMOS L4 SSS are saltier than L3 by a factor of up to 0.2. On the other hand, L4 estimations can be smoother than L3, thus a lower correspondence to glider observation was found in other sub-regions of the study area, i.e., between 38.2 and 38.6° N, where L3 products manage to capture part of the finer scale phenomena that L4 products are not able to describe.

In general, we can assess that, for the most of the co-locations, SMOS measured fresher SSS than the in situ, with larger discrepancies at the edges than in the middle of the transects. Nevertheless, some positive differences (SMOS saltier than in situ) were observed locally (i.e., in proximity of the AC during the December 2016 cruise and in short segments of the other transects), although usually lower in absolute value than those typically observed in case of fresher SMOS SSS anomalies.

The largest differences were located at the northern and southern edges of the transects during the four glider surveys carried out in September/October 2014 (Figures 4a and 5a), November/December 2014 (Figures 4b and 5b), October/November 2015 (Figures 4c and 5c), and December 2016 (Figures 4e and 5e). This could suggest that differences are due to some kind of residual land–sea contamination or non-permanent RFI in the satellite data. However, during November 2016 (Figures 4d and 5d), the maximum differences were not in the edges but in the latitude range of 38.0–38.2° N, that is, in the middle of the AB, and too far from the coast.

Although we were conscious that several studies (e.g., References [26,52]) previously demonstrated that interpolated SMOS SSS fields are able to resolve intense mesoscale structures that generally coincide with those identified by gliders and described by altimetry and MODIS images [7,12], we tried to correlate the ABACUS-SMOS differences that came to light with the presence of oceanographic mesoscale events that the fine spatial–temporal resolution of the glider is able to capture but the satellite probably cannot. This was intended to identify a possible origin of the increase of the error in the observed sub-regions. To this aim, we analyzed satellite SST and altimetry maps over the study area in order to evaluate: (i) the presence of interesting mesoscale structures; and (ii) the eventual interactions with peculiar water masses, in particular at the very active northern and southern edges of the ABACUS transects which are characterized by the proximity of the Mallorca Channel and the AC, respectively.

As supposed, our results show that lower correlation values coincide with transects (i.e., ABACUS 1.1, ABACUS 2.1, and ABACUS 2.2) characterized by zone of sharp salinity changes due to intense mesoscale activity and AC meandering. Hence, we conclude that BEC SMOS products capture well the broad resolution of the salinity pattern, but cannot properly resolve the finer scale variability that needs higher resolution to be detected. Several case studies from the analyzed ABACUS surveys are discussed in detail in the following subsections.

Table 2. Main statistics (mean; standard deviation; linear correlation) of daily differences computed between BEC SMOS L3/L4 SSS and ABACUS in situ SSS along the analyzed glider transects.

Mission	Start	End	SMOS L3-ABACUS	std	R-Value	SMOS L4-ABACUS	std	R-Value
ABACUS 1.1	20140913	20141018	−0.17	0.21	0.23	−0.13	0.17	0.39
ABACUS 1.2	20141120	20141215	−0.28	0.17	0.76	−0.24	0.16	0.80
ABACUS 2.1	20151020	20151110	−0.25	0.20	0.52	−0.23	0.22	0.37
ABACUS 2.2	20151111	20151211	−0.22	0.24	0.22	−0.22	0.24	0.25
ABACUS 3.1	20161105	20161130	−0.17	0.14	0.72	−0.11	0.12	0.65
ABACUS 3.2	20161130	20161215	0.13	0.12	0.78	0.19	0.13	0.73

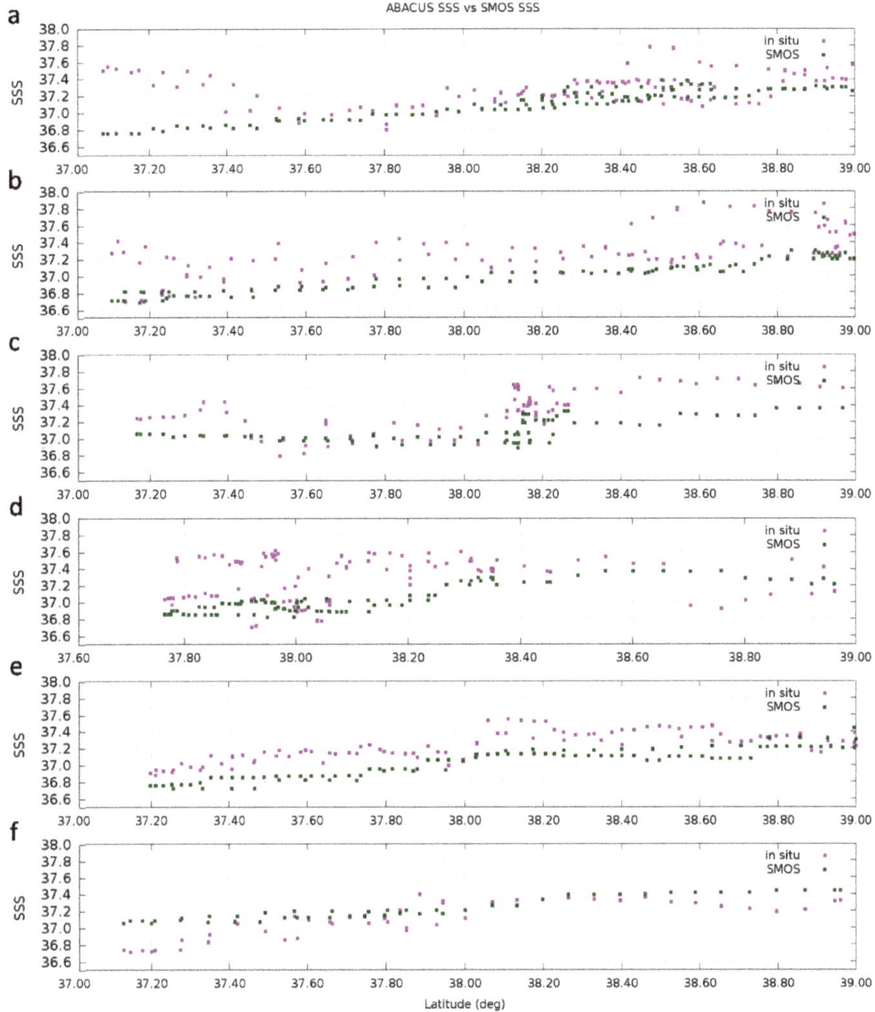

Figure 4. Comparison between averaged in situ glider salinities (purple dots) and SMOS L3 SSS (green dots) along latitude, during the missions (**a**) ABACUS 1.1 (Sept–Oct 2014), (**b**) ABACUS 1.2 (Nov–Dec 2014), (**c**) ABACUS 2.1 (Oct–Nov 2015), (**d**) ABACUS 2.2 (Nov–Dec 2015), (**e**) ABACUS 3.1 (Nov 2016), and (**f**) ABACUS 3.2 (Dec 2016).

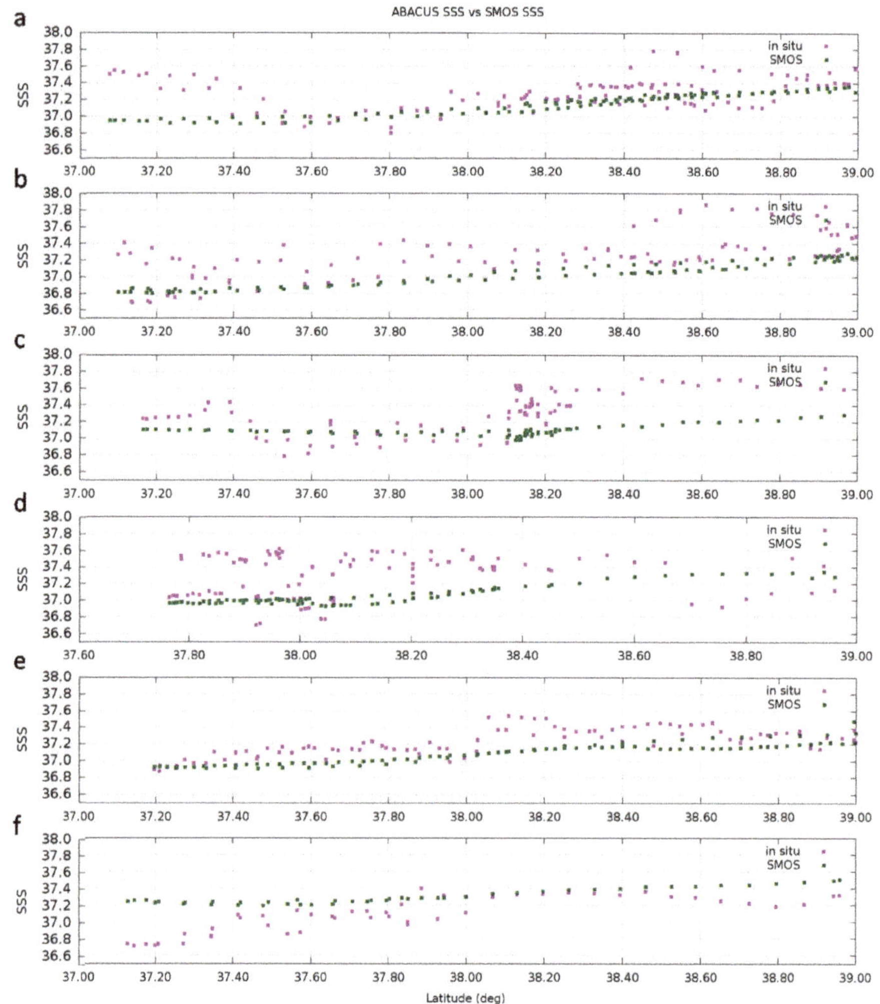

Figure 5. Comparison between averaged in situ glider salinities (purple dots) and SMOS L4 SSS (green dots) along latitude, during the missions (**a**) ABACUS 1.1 (Sept–Oct 201), (**b**) ABACUS 1.2 (Nov–Dec 201), (**c**) ABACUS 2.1 (Oct–Nov 2015), (**d**) ABACUS 2.2 (Nov–Dec 2015), (**e**) ABACUS 3.1 (Nov 2016), and (**f**) ABACUS 3.2 (Dec 2016).

3.2. ABACUS 1 Surveys

During the ABACUS experiment carried out in September–October 2014, the most significant difference between in situ and SMOS SSS was detected at the southern edge of the glider transect, i.e., between 37.1 and 37.5°N (Figures 4a and 5a). The glider crossed this region between 24 and 27 September (Figure 3a); the transect's southernmost waypoint was hit on 25 Sept (at 23:47 UTC). This means that the largest differences were estimated when the glider was navigating across the border of the AC influence zone. This area is typically characterized by large instabilities in the surface layer, with meandering activity that allows AW transported by AC to intrude into the more homogenous resident surface water of the AB. At the time of SMOS-glider co-location on 26 September (Figure 6), for example, an intrusion of fresher AW coming from the AC was present in the 0–200m depth section of the water column (see Figure 10a in Reference [7]), but could not reach the surface due to the presence of a local cap of saltier water (up to 37.55). Since SMOS products are strongly influenced by the presence of the fresher AC, which characterizes the area at large scale, they seem unable to capture the local variability associated with the meandering instability. The downscaling provided in SMOS L4 maps improves the capability to identify finer scale features, but cannot completely resolve them.

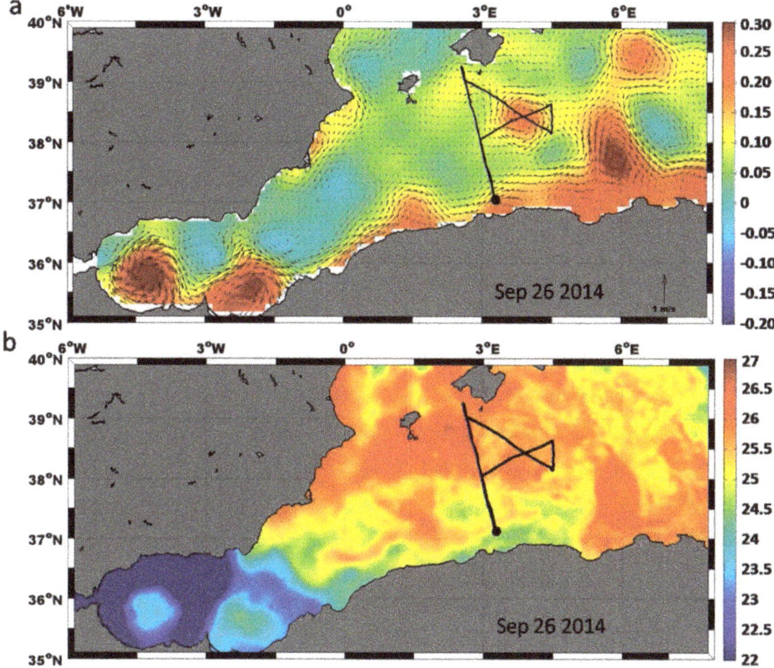

Figure 6. (**a**) All-sat-merged AVISO Absolute Dynamic Topography (ADT) and (**b**) ultra-high-resolution Sea Surface Temperature (SST) maps of the Alboran Sea and the Algerian Basin on 26 September 2014; geostrophic currents are represented through black arrows. Glider track in black. Large black dots represent glider location on each date. ADT in meters, SST in °C.

Another significant difference can be found at the beginning of the ABACUS 1.1 survey, when the glider was crossing the area included between 38.4 and 38.6°N on its way south from Mallorca to the AC. In situ SSS showed a rapid and intense increase (up to 0.5 in very few kilometers), with single salinity values up to 37.79 on 19 September, that SMOS L3 and L4 products could not identify (Figures 3–5, subplots a). The glider was enough far from the coast to prevent any impact of land contamination, so the observed sudden increase of salinity could indeed be due to the presence of the anticyclonic eddy centered at 2.21°E 38.61°N crossed by the ABACUS glider along its eastern edge (Figure 7).

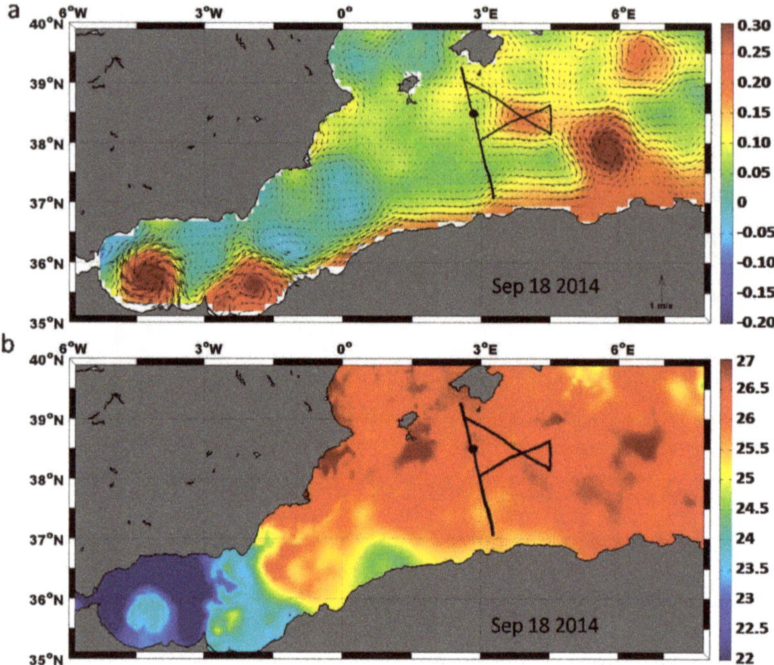

Figure 7. (**a**) All-sat-merged AVISO ADT and (**b**) ultra-high-resolution SST maps of the Alboran Sea and the Algerian Basin on 18 September 2014; geostrophic currents are represented through black arrows. Glider track in black. Large black dots represent glider location on each date. ADT in meters, SST in °C.

The region included between 38.4 and 38.8°N is typically characterized by a meandering salinity front in the upper layers which dominates the density distribution, often associated with water salinity asymmetry at the northeastern (saltier) and the southeastern (fresher) edges of the region [53]. A previous study [12] focusing on the mesoscale structure located on the eastern side of the glider transect, and centered at 38.34°N 3.83°E on 18 September (Figures 6 and 7), showed that this eddy was typically characterized by shoaling of isolines on its borders for all the physical parameters, as well as for chlorophyll concentration. They also computed vertical velocities across the eddy to demonstrate the presence of positive (upward) velocities on its southeastern border, thus suggesting that this mechanism may upwell sub-surface and intermediate water (and associated nutrients) to the photic layer.

Thus, we suggest that the spotty saltier in situ values (37.47–37.79) collected on 16–19 September could represent the signature of upwelled saltier sub-surface waters which were intercepted by the ABACUS glider when navigating across the eastern border of the small eddy centered at 2.21°E 38.61°N.

Although SMOS L3 and L4 products retrieve slightly higher SSS in this region in comparison to the rest of the central part of the Mallorca–AC transect, they cannot identify the maximum values observed by the glider, confirming their limits in describing the small scale variability which characterizes the areas of intense local activity. SMOS L4 SSS maps of the Alboran Sea and the AB on 18 September and 26 September 2014 (Figure 8) provide an illuminating example of their scale limitations.

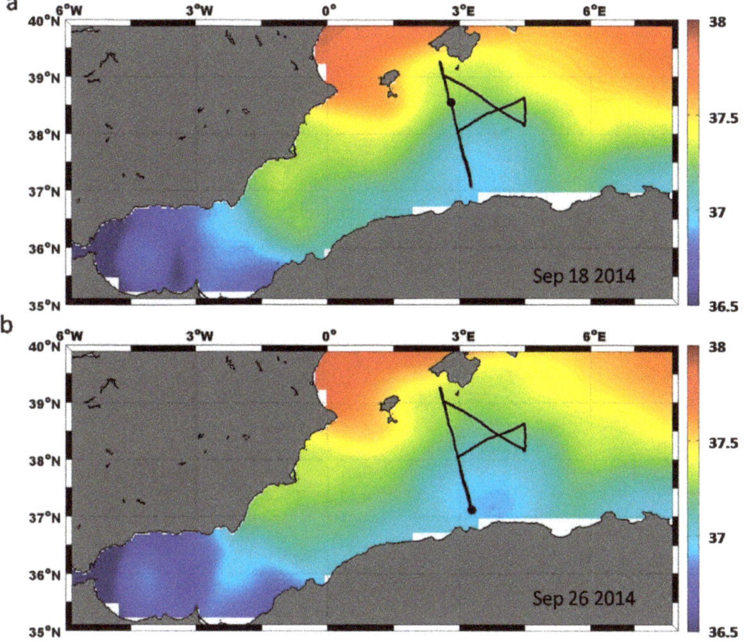

Figure 8. SMOS L4 Sea Surface Salinity (SSS) maps of the Alboran Sea and the Algerian Basin on 18 September and 26 September 2014—ABACUS 1.1 survey. Glider track in black.

On the other hand, SMOS SSS retrievals correspond well with in situ measurements during the "butterfly shaped" monitoring (2–12October) of the mesoscale anticyclonic eddy located at the eastern side of the ABACUS main transect, which detached from the AC and was centered at 38.34°N 3.83°E on 18 September (Figures 6 and 7). As usual, SMOS SSS were generally slightly fresher than those measured in situ, but they were saltier when the glider was crossing the northeastern region of the monitored eddy (4–7 October).

Similar results were achieved for the ABACUS 1.2 experiment, with significant differences between in situ and SMOS SSS identified at the very beginning of the survey (i.e., between 38.4 and 39.0°N), when the glider was diving southward from Mallorca to the AC (Figures 4b and 5b), and close to its southern edge (i.e., between 37.1 and 37.5°N), when the glider entered the region influenced by the proximity of the AC. Smaller but significant differences were also found in the central part of the glider transect (i.e., between 37.7 and 38.0°N).

The SST maps of the AB at the beginning/end of this survey (Figure 9) present the peculiar features of this sub-region during the fall season, pointing out the presence of the colder AC flow. Over its area of influence, the SMOS SSS values remain in the middle of those collected by the two passes of the glider, which registered fresher and saltier salinities during its southward and northward passages, respectively (Figures 4b and 5b). SMOS captured the general salinity pattern, but missed the finer scale variability observed in situ. On the other hand, a very good agreement between glider and satellite SSS can be observed during the second part of this experiment (1–12 December) when the glider navigated northward along the easternmost SARAL-AltiKa #229 groundtrack.

These results are consistent with the observed dynamics of the AB (Figure 10), showing that the glider crossed (i) an area of very intense ocean instability, characterized by the presence of two small cyclonic structures, on its way south; (ii) a much more homogeneous region, delimited and controlled by an intense long-living anticyclonic eddy, on its way north.

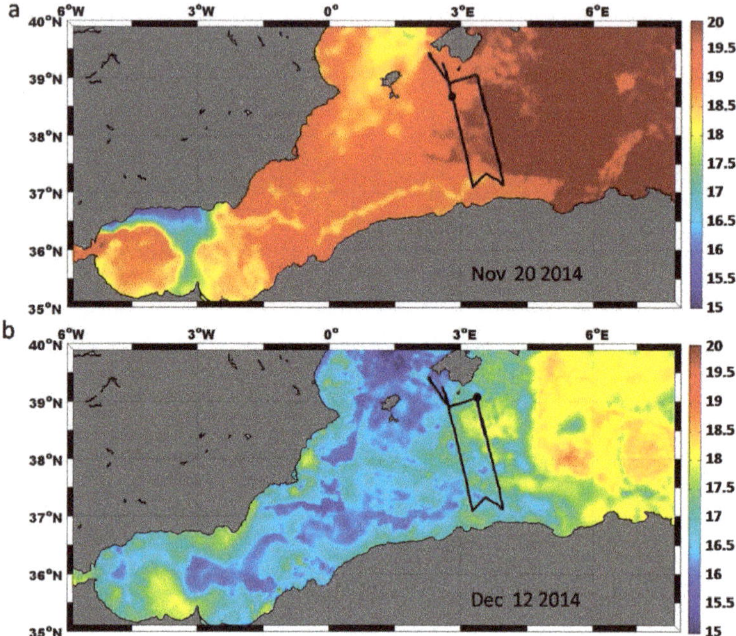

Figure 9. Ultra-high-resolution SST maps of the Alboran Sea and the Algerian Basin on (**a**) 20 November 2014 and (**b**) 12 December 2014. Glider track in black. Large black dots represent glider location on each date. SST in °C.

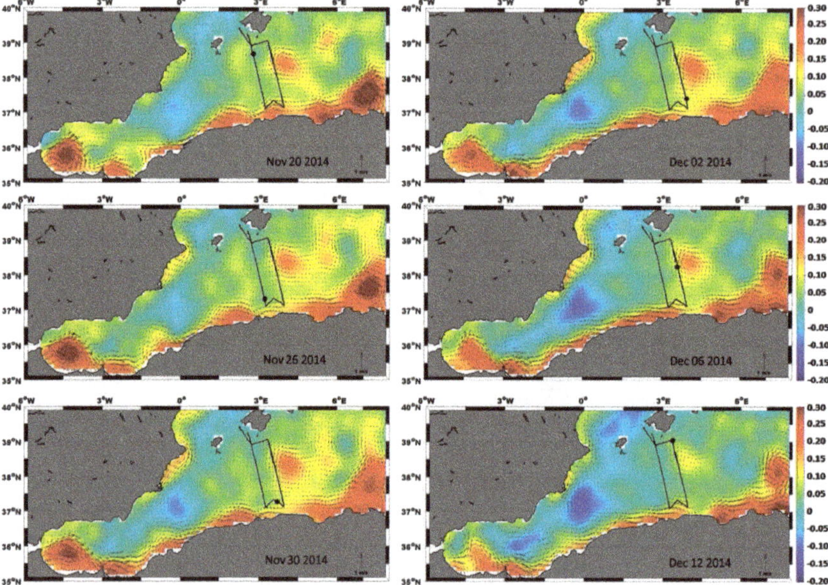

Figure 10. All-sat-merged AVISO ADT maps of the Alboran Sea and the Algerian Basin on six representative dates of the ABACUS 1.2 survey. Glider track in black. Large black dots represent glider location on each date. Geostrophic currents are represented through black arrows. ADT in meters.

The local variability measured when hitting the mesoscale structures in Figure 10 was always associated with a lower correspondence of SMOS SSS retrievals; this happened on 20 and 26 November, when the glider was crossing two small cyclonic eddies, and on 6 December, when it was lining the western border of a large anticyclonic eddy. SMOS limitations to retrieve finer scale reliable SSS are evident in Figure 11.

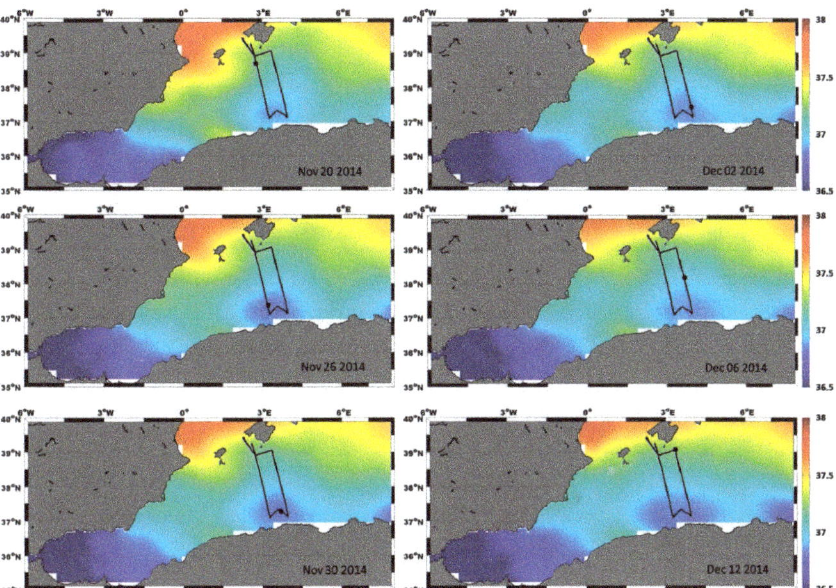

Figure 11. SMOS L4 SSS maps of the Alboran Sea and the Algerian Basin on six representative dates of the ABACUS 1.2 survey. Glider track in black.

3.3. ABACUS 2 Surveys

As in ABACUS 1, an evident fresh SMOS anomaly was found in the northern part of the ABACUS 2 transect, i.e., between 38.1 and 39.0°N (Figure 4c), during the glider southward survey from 21 to 23 October 2015 (Figure 3c). This impressive anomaly was not present during the second glider passage at the end of the ABACUS 2 mission, when satellite and in situ measurements correspond much more. Nevertheless, it is interesting to observe that on 5–6 December 2015 (Figure 3d) SMOS SSS looked saltier than those measured in situ in the northern sector of the transect, i.e., between 38.7 and 39.0°N (Figure 4d). As usual, these differences were attenuated in L4 products (Figure 3c,d and Figure 5c,d).

SST conditions (Figure 12) and ocean circulation features (Figure 13) over the AB during the southward and northward transects suggest that the observed differences between SMOS and in situ salinity values could be ascribed to the presence of an intense mesoscale circulation south of the Mallorca Island, which was passed through by the ABACUS glider during its southward transect along the SARAL-AltiKa groundtrack #229 (Figure 1). In fact, during the first week of this survey, the glider crossed the western border of a small but intense anticyclonic eddy moving westward (Figure 13). This eddy was not present anymore at the end of this survey in late November–early December, so that satellite and in situ observations agree better.

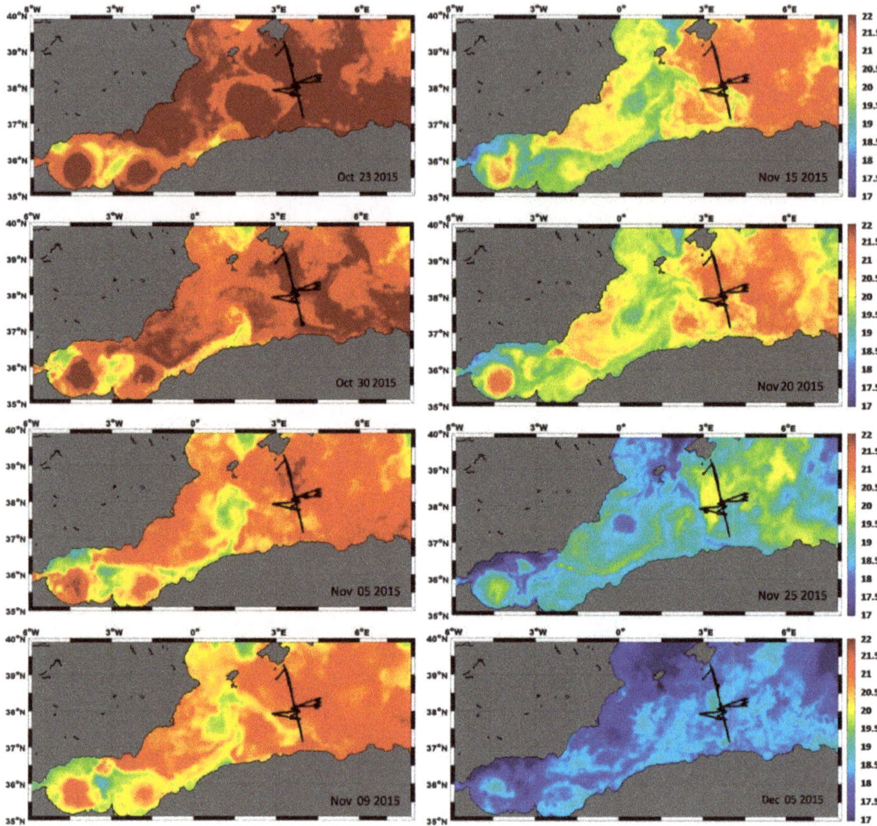

Figure 12. Ultra-high-resolution SST maps of the Alboran Sea and the Algerian Basin on eight representative dates of the ABACUS 2 survey. Glider track in black. Large black dots represent glider location on each date. ADT in meters. A different colorscale is used for October maps in order to highlight the observed ocean surface features.

As for the salty SMOS anomaly on 5–6 December, it was completely due to an abrupt decrease in the in situ salinities. The available SST and SSH images do not provide any clear explanation of this observation, except the fact that, on this date, the glider met a filament of warmer sea water that patched the area south of Mallorca during late autumn 2015 (Figure 12). This filament would have been out of the spatial scales resolved by SMOS satellite (Figure 14).

SMOS L3 SSS values were generally fresher than in situ ones also in the latitude range 37.1–37.5°N, i.e., in proximity of the area of influence of the AC border (Figure 4c). As in 2014, this anomaly was reduced but still present in L4 products (Figure 5c). The glider went through this area between 28 and 31 October 2015, and hit waypoint south on 30 October. Figure 12 highlights the presence of a colder filament characterized by a SST of about 20°C in this area, which was crossed by the ABACUS glider at the latitude of approximately 37.4°N; its presence was revealed by in situ measurements, but not by SMOS retrievals (Figures 4c and 5c).

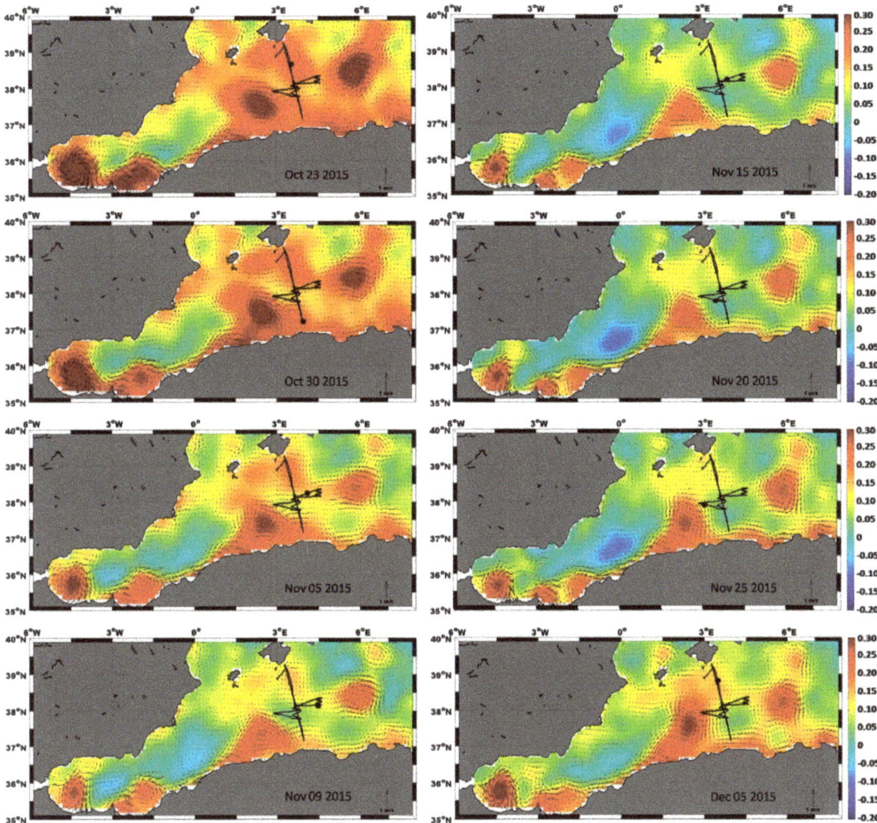

Figure 13. All-sat-merged AVISO ADT maps of the Alboran Sea and the Algerian Basin on eight representative dates of the ABACUS 2 survey. Glider track in black. Large black dots represent glider location on each date. Geostrophic currents are represented through black arrows. ADT in meters.

Figure 5c points out also the issue of different temporal scales between satellite and in situ observations. The ABACUS glider covered the transect section in the latitude range 37.7–38.1°N twice at a short temporal distance. During its way southward (24–27 October), it measured fresher SSS, while saltier values were observed on its way back northward (15 November). Despite the short temporal distance, salinity can differ up to 0.4. SMOS SSS retrievals over this latitude sector presented intermediate values between the two passages; again, the general feature was well described, but local short-time variability was not captured.

Significant SMOS fresh anomalies were also recognizable in the center of the AB during the second half of November, when the glider carried out a sort of butterfly monitoring activity on the two sides of the main SARAL-AltiKa groundtrack. On the eastern side, the glider reached and monitored (9–12 November) the border of an anticyclonic eddy moving away eastwards (Figure 13); SMOS maps seem to capture well part of the local variability as observed in situ (Figure 3d). On the western side, a cyclonic eddy was entirely crossed by the ABACUS glider (22–30 November) up to its border region, with another anticyclonic structure (hit on 26 November) detaching from the AC (Figure 13). Although the general salinity pattern was captured (Figure 14), in this case, both L3 and L4 SMOS SSS largely underestimated glider measurements (Figure 3d).

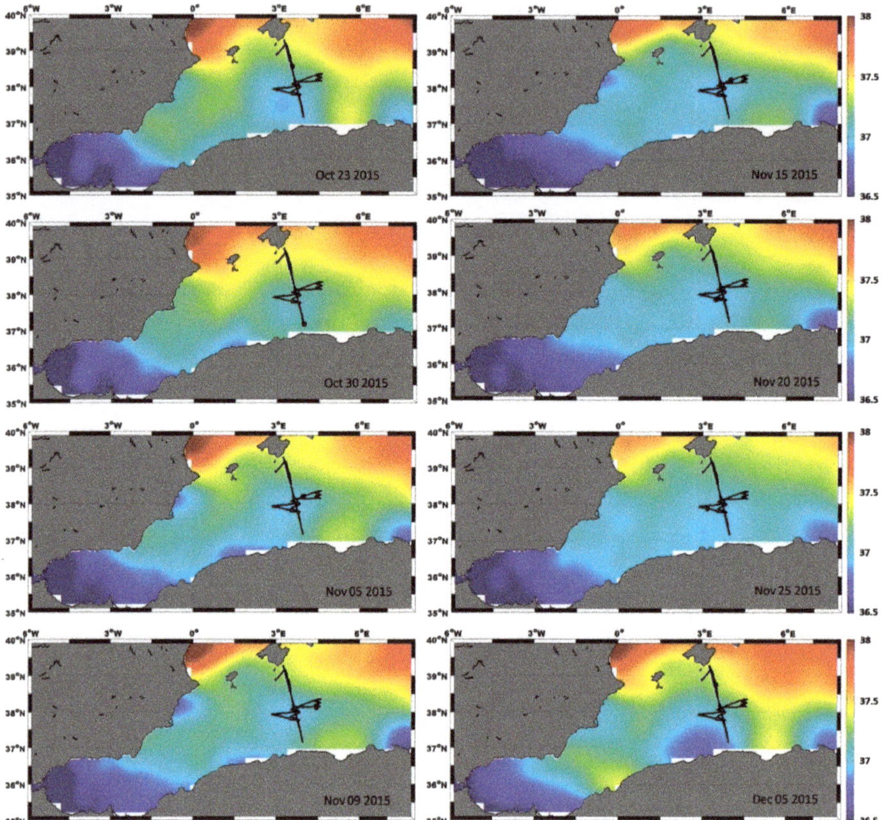

Figure 14. SMOS L4 SSS maps of the Alboran Sea and the Algerian Basin on eight representative dates of the ABACUS 2 survey. Glider track in black.

3.4. ABACUS 3 Surveys

During fall 2016, both L3 and L4 SMOS retrievals were slightly fresher, but in very good agreement with in situ measurements (Figure 3e,f). This could also have been favored by the improved resolution in glider surface acquisitions during this survey.

Nonetheless, the latitudinal comparison pointed out some fresh SMOS anomalies between 38.0 and 38.2°N, during the glider dives along the Sentinel 3 satellite groundtrack #57 (hereafter ABACUS 3.1 cruise). The glider covered this sector twice during ABACUS 3.1, first on 11–13 November (southward leg), and then on 21–23 November (northward leg). These fresh SMOS anomalies were smaller than those observed in the previous campaigns, and disappeared during the following glider passages at the same latitude (December 2016), along the Sentinel groundtrack#713.

The AVISO SSH maps in Figure 15 show that a similar circulation pattern persisted over the AB during both the southward and the northward legs along the Sentinel groundtrack #57 (November 2016). This eastern transect was completely dominated by the presence of an intense anticyclonic eddy that the glider went through, along its north–south axis, during both passages. This mesoscale structure was much more intense during the southward leg; this could explain the moderate fresh anomalies observed in SMOS SSS on 11–13 November.

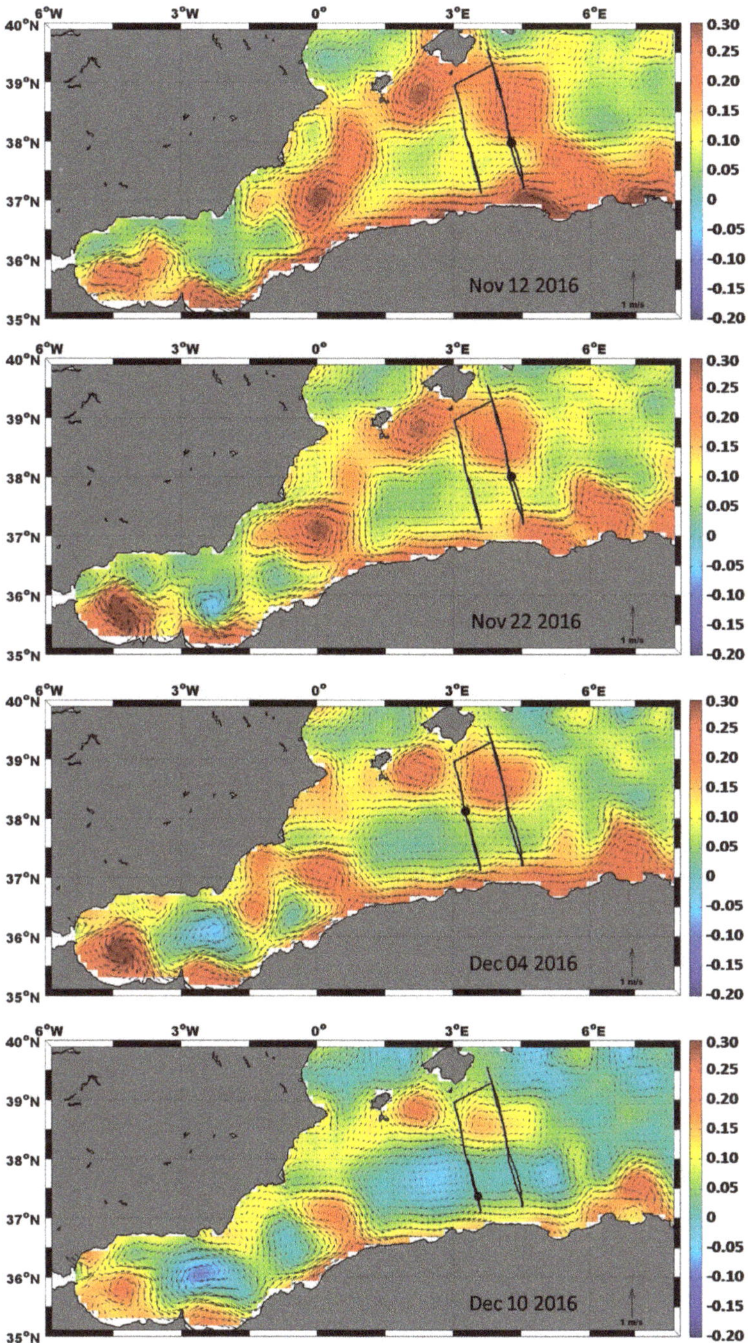

Figure 15. All-sat-merged AVISO ADT maps of the Alboran Sea and the Algerian Basin on four representative dates of the ABACUS 3 survey. Glider track in black. Large black dots represent glider location on each date. Geostrophic currents are represented through black arrows. ADT in meters.

Conversely, SMOS retrievals could not capture the evident SSS decrease measured by the glider along its western transect (hereafter ABACUS 3.2). In particular, salty SMOS anomalies, unexpectedly accentuated in L4 products, were identified along most of its northward leg, i.e., between 6 and 12 December 2016 (Figure 3f). As for the rest of ABACUS 3.2, satellite and in situ measurements agree very well. Figure 15 suggests that the observed differences could be due to the presence of an intense cyclonic mesoscale structure developing in the southern sector of this transect, the western boundary of which was crossed by the glider from 6 to 12 December 2016. On these dates, SMOS L4 SSS maps did not capture any evident salinity decrease associated with the presence of this mesoscale structure (Figure 16).

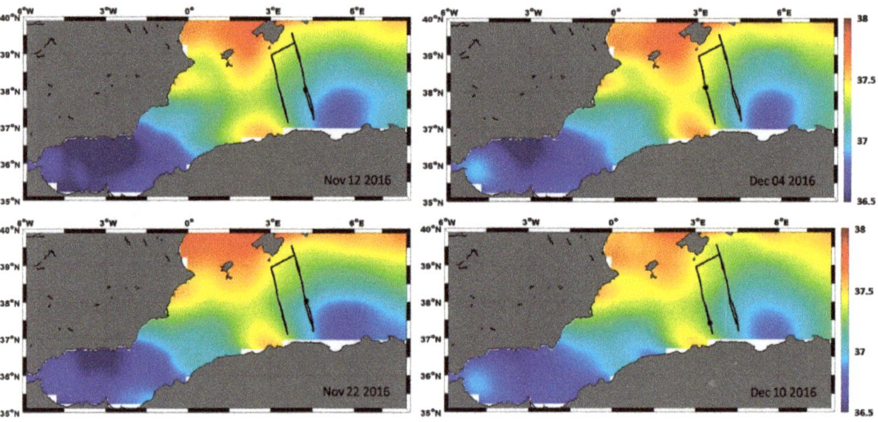

Figure 16. SMOS L4 SSS maps of the Alboran Sea and the Algerian Basin on four representative dates of the ABACUS 3 survey. Glider track in black.

4. Conclusions

The comparison with the ABACUS in situ measurements shows that, in spite of the limitation of the differences associated with the spatial and temporal resolutions, the analyzed BEC enhanced SMOS SSS products provide a coherent description of the salinity patterns in the AB. This supports the results previously reported [27], which demonstrated that the statistical filtering criteria, the EOF decomposition, and the debiased non-Bayesian retrieval-DINEOF-fusion approach behind the L3 and L4 products improve the salinity retrievals with respect to official SMOS Level 2 Ocean Salinity products.

SMOS retrieved L3 and L4 SSS values were generally slightly fresher (approximately −0.11, −0.14) than those collected in situ, which also captured more salinity variations, including those usually associated with local variability. Larger fresh SMOS anomalies emerged in proximity of dynamic mesoscale structures, i.e., in the area of influence of the AC and/or in presence of intense cyclonic and anticyclonic eddies. Only a few cases showed salty SMOS anomalies, e.g., when observing an area characterized by the presence of an intense cyclonic eddy during the ABACUS 3.2 survey.

As expected, the BEC L4 corrected products generally improved satellite retrievals, reducing differences with in situ measurements when a fresh SMOS anomaly was observed. However, the effective resolution of the L4 SMOS products still cannot capture all the small scale structures that in situ data do.

These results suggest that (i) SMOS retrievals are able to describe well the large scale salinity patterns in the AB; (ii) the existing differences are associated with neither systematic biases nor land contamination, but depend on the actual satellite capability of resolving local variability due to intense and rapidly evolving mesoscale dynamics.

On one hand, although the native resolution of the satellite is about 45 km, the effective resolution of the L3 product is driven by the correlation radii used in its generation (175, 125, and 75 km). These radii are too large for describing parts of the mesoscale dynamics of this region. On the other hand, the results obtained from the L4 products indicate that the simplified scheme of multifractal fusion (scalar approach [30]) is not enough to fully describe the fast, small-scale dynamics present in this region, and that more complex schemes are needed (such as the vector approach [30] that was used in Reference [26] for capturing eddies in AB). Therefore, since the large scale salinity patterns are well described, future releases of these products should now be focused on using different interpolation/fusion schemes to improve the effective spatial and temporal resolutions of the SMOS SSS products. Since systematic negative biases still appear in the eastern Mediterranean Sea, a better mitigation of RFI contamination will then be required in these regions, for example by improving the quality of the brightness temperature with methodologies [54,55] that have already been proven to improve salinity retrieval in coastal areas [56]. Furthermore, all the ABACUS high resolution in situ data used in this study come from three glider surveys carried out in the AB during fall season (i.e., September to December). Additional efforts should set up BEC SMOS L3 and L4 products for evaluation during the rest of the year. To this aim, since 2018, new research efforts have been invested by present authors to realize additional ABACUS glider experiments in the AB during springtime (e.g., ABACUS 4.2 was carried out in May 2018). An analysis at the seasonal scale might also improve the comparison in this region and extend the application of a similar approach in other Mediterranean regions, also favoring the investigation and test of the future releases of the SMOS and the other missions (e.g., NASA Soil Moisture Active Passive—SMAP) SSS products.

Author Contributions: Conceptualization, G.A., Y.C. and E.O.; Data curation, G.A., Y.C., E.O., C.C. and G.F.; Formal analysis, G.A., Y.C., E.O. and C.C.; Funding acquisition, Y.C., G.F. and G.B.; Investigation, G.A. and E.O.; Methodology, G.A. and Y.C.; Supervision, G.F. and G.B.; Validation, G.A. and Y.C.; Writing—original draft, G.A., Y.C. and E.O.; Writing—review & editing, G.A., Y.C., E.O., C.C., G.F. and G.B.

Funding: The ABACUS 1 missions (2014) were supported by the Joint European Research Infrastructure network for Coastal Observatories (JERICO) TransNational Access (TNA) third call (grant agreement no. 262584). The research leading to ABACUS 3 (2016) was supported by the European Union's H2020 Framework Programme (h2020-INFRAIA-2014-2015) (JERICO-NEXT grant agreement no. 654410). The activities described in this paper were developed in the framework of the Italian Flagship Project RITMARE. This work was supported by the Spanish R+D plan under projects L-BAND (ESP2017-89463-C3-1-R) and PROMISES (ESP2015-67549-C3-2) and from European Space Agency by means of the contracts SMOS ESL L2OS, CCI+SSS.

Acknowledgments: The authors would like to thank the referees and the academic editor for the helpful and careful comments. The authors are particularly grateful to the SOCIB glider facility team and the Data Centre and Engineering and Technology Deployment staff for their efficient cooperation. The SOCIB Data Server hosts ABACUS data which are available at https://doi.org/10.25704/b200-3vf5. BEC L3 and L4 maps of SMOS SSS are freely available at http://bec.icm.csic.es/ocean-experimental-dataset-mediterranean/. The altimeter products were produced by Ssalto/Duacs and freely distributed by CMEMS (www. http://marine.copernicus.eu/). The GHRSST Level 4 MUR Global Foundation Sea Surface Temperature Analysis (Ver. 2) provided by JPL MUR MEaSUREs Project 2010, PO.DAAC, CA, USA, have been accessed at https://doi.org/10.5067/GHGMR-4FJ01.

Conflicts of Interest: The authors declare no conflict of interest.

References

1. Millot, C. Some features of the Algerian Current. *J. Geophys. Res.* **1985**, *90*, 7169. [CrossRef]
2. Millot, C. Circulation in the Western Mediterranean Sea. *J. Mar. Syst.* **1999**, *20*, 423–442. [CrossRef]
3. Millot, C.; Candela, J.; Fuda, J.-L.; Tber, Y. Large warming and salinification of the Mediterranean outflow due to changes in its composition. *Deep Sea Res. Part I Oceanogr. Res. Pap.* **2006**, *53*, 656–666. [CrossRef]
4. Puillat, I.; Taupier-Letage, I.; Millot, C. Algerian Eddies lifetime can near 3 years. *J. Mar. Syst.* **2002**, *31*, 245–259. [CrossRef]
5. Testor, P.; Send, U.; Gascard, J.-C.; Millot, C.; Béranger, K.; Taupier-Letage, I.; Taupier-Letage, I. The mean circulation of the southwestern Mediterranean Sea: Algerian Gyres. *J. Geophys. Res.* **2005**, *110*, 11017. [CrossRef]

6. Pascual, A.; Bouffard, J.; Ruiz, S.; Nardelli, B.B.; Vidal-Vijande, E.; Escudier, R.; Sayol, J.M.; Orfila, A. Recent improvements in mesoscale characterization of the western Mediterranean Sea: Synergy between satellite altimetry and other observational approaches. *Sci. Mar.* **2013**, *77*, 19–36. [CrossRef]

7. Aulicino, G.; Cotroneo, Y.; Ruiz, S.; Román, A.S.; Pascual, A.; Fusco, G.; Tintoré, J.; Budillon, G. Monitoring the Algerian Basin through glider observations, satellite altimetry and numerical simulations along a SARAL/AltiKa track. *J. Mar. Syst.* **2018**, *179*, 55–71. [CrossRef]

8. Vidal-Vijande, E.; Pascual, A.; Barnier, B.; Molines, J.M.; Tintoré, J. Analysis of a 44-year hindcast for the Mediterranean Sea: Comparison with altimetry and in situ observations. *Sci. Mar.* **2011**, *75*, 71–86. [CrossRef]

9. Taupier-Letage, I.; Puillat, I.; Millot, C.; Raimbault, P. Biological response to mesoscale eddies in the Algerian Basin. *J. Geophys. Res.* **2003**, *108*, 3245–3267. [CrossRef]

10. Pessini, F.; Olita, A.; Cotroneo, Y.; Perilli, A. Mesoscale Eddies in the Algerian Basin: Do they differ as a function of their formation site? *Ocean Sci. Discuss.* **2018**, *14*, 1–26. [CrossRef]

11. Olita, A.; Ribotti, A.; Sorgente, R.; Fazioli, L.; Perilli, A. SLA-chlorophyll-a variability and covariability in the Algero-Provençal Basin (1997–2007) through combined use of EOF and wavelet analysis of satellite data. *Ocean Dyn.* **2011**, *61*, 89–102. [CrossRef]

12. Cotroneo, Y.; Aulicino, G.; Ruiz, S.; Pascual, A.; Budillon, G.; Fusco, G.; Tintoré, J. Glider and satellite high resolution monitoring of a mesoscale eddy in the algerian basin: Effects on the mixed layer depth and biochemistry. *J. Mar. Syst.* **2016**, *162*, 73–88. [CrossRef]

13. Gualdi, S.; Artale, V.; Adani, M.; Dell'Aquila, A.; Elizalde, A.; Harzallah, A.; L'Hévéder, B.; May, W.; Oddo, P.; Ruti, P.; et al. The CIRCE Simulations: Regional Climate Change Projections with Realistic Representation of the Mediterranean Sea. *Bull. Am. Meteorol. Soc.* **2013**, *94*, 65–81. [CrossRef]

14. Schroeder, K.; Chiggiato, J.; Josey, S.A.; Borghini, M.; Aracri, S.; Sparnocchia, S. Rapid response to climate change in a marginal sea. *Sci. Rep.* **2017**, *7*, 4065. [CrossRef] [PubMed]

15. Fusco, G.; Artale, V.; Cotroneo, Y.; Sannino, G. Thermohaline variability of Mediterranean Water in the Gulf of Cadiz, 1948–1999. *Deep Sea Res. Part I Oceanogr. Res. Pap.* **2008**, *55*, 1624–1638. [CrossRef]

16. Budillon, G.; Cotroneo, Y.; Fusco, G.; Rivaro, P. Variability of the Mediterranean Deep and Bottom Waters: Some Recent Evidences in the Western Basin. *In CIESM Workshop Monographs* **2009**, *38*, 132.

17. Cotroneo, Y.; Aulicino, G.; Ruiz, S.; Román, A.S.; Tomàs, M.T.; Pascual, A.; Fusco, G.; Heslop, E.; Tintoré, J.; Budillon, G. Glider data collected during the Algerian Basin Circulation Unmanned Survey. *Earth Syst. Sci. Data* **2019**, *11*, 147–161. [CrossRef]

18. Durante, S.; Schroeder, K.; Mazzei, L.; Pierini, S.; Borghini, M.; Sparnocchia, S. Permanent Thermohaline Staircases in the Tyrrhenian Sea. *Geophys. Res. Lett.* **2019**, *46*, 1562–1570. [CrossRef]

19. Borghini, M.; Bryden, H.; Schroeder, K.; Sparnocchia, S.; Vetrano, A. The Mediterranean is becoming saltier. *Ocean Sci.* **2014**, *10*, 693–700. [CrossRef]

20. Drobinski, P.; Ducrocq, V.; Allen, J.; Alpert, P.; Anagnostou, E.; Béranger, K.; Borga, M.; Braud, I.; Chanzy, A.; Davolio, S.; et al. HyMeX, a 10-year multidisciplinary project on the Mediterranean water 1 cycle. *Bull. Am. Meteorol. Soc.* **2014**, *95*, 1063–1082. [CrossRef]

21. Estournel, C.; Testor, P.; Damien, P.; D'Ortenzio, F.; Marsaleix, P.; Conan, P.; Kessouri, F.; De Madron, X.D.; Coppola, L.; Lellouche, J.-M.; et al. High resolution modeling of dense water formation in the north-western Mediterranean during winter 2012–2013: Processes and budget. *J. Geophys. Res. Oceans* **2016**, *121*, 5367–5392. [CrossRef]

22. Waldman, R.; Somot, S.; Herrmann, M.; Testor, P.; Estournel, C.; Sevault, F.; Prieur, L.; Mortier, L.; Coppola, L.; Taillandier, V.; et al. Estimating dense water volume and its evolution for the year 2012-2013 in the Northwestern Mediterranean Sea: An observing system simulation experiment approach. *J. Geophys. Res. Oceans* **2016**, *121*, 6696–6716. [CrossRef]

23. Herrmann, M.; Estournel, C.; Adloff, F.; Diaz, F. Impact of climate change on the northwestern Mediterranean Sea pelagic planktonic ecosystem and associated carbon cycle. *J. Geophys. Res. Oceans* **2014**, *119*, 5815–5836. [CrossRef]

24. Borges, A.; Hahne, A.; Mecklenburg, S.; Font, J.; Camps, A.; Martín-Neira, M.; Boutin, J.; Reul, N.; Kerr, Y.H. SMOS: The Challenging Sea Surface Salinity Measurement From Space. *Proc. IEEE* **2010**, *98*, 649–665.

25. Mecklenburg, S.; Drusch, M.; Kerr, Y.H.; Font, J.; Martin-Neira, M.; Delwart, S.; Buanadicha, G.; Reul, N.; Daganzo-Eurebio, E.; Oliva, R.; et al. ESA's Soil Moisture and Ocean Salinity Mission: Mission Performance and Operations. *IEEE Trans. Geosci. Remote Sens.* **2012**, *50*, 1354–1366. [CrossRef]

26. Isern-Fontanet, J.; Olmedo, E.; Turiel, A.; Ballabrera-Poy, J.; García-Ladona, E. Retrieval of eddy dynamics from SMOS sea surface salinity measurements in the Algerian Basin (Mediterranean Sea). *Geophys. Res. Lett.* **2016**, *43*, 6427–6434. [CrossRef]

27. Olmedo, E.; Taupier-Letage, I.; Turiel, A.; Alvera-Azcaràte, A. Improving SMOS Sea Surface Salinity in the Western Mediterranean Sea through Multivariate and Multifractal Analysis. *Remote. Sens.* **2018**, *10*, 485. [CrossRef]

28. Olmedo, E.; Martínez, J.; Turiel, A.; Ballabrera-Poy, J.; Portabella, M. Debiased non-Bayesian retrieval: A novel approach to SMOS Sea Surface Salinity. *Remote. Sens. Environ.* **2017**, *193*, 103–126. [CrossRef]

29. Alvera-Azcarate, A.; Barth, A.; Parard, G.; Beckers, J.-M. Analysis of SMOS sea surface salinity data using DINEOF. *Remote. Sens. Environ.* **2016**, *180*, 137–145. [CrossRef]

30. Olmedo, E.; Martínez, J.; Umbert, M.; Hoareau, N.; Portabella, M.; Ballabrera-Poy, J.; Turiel, A. Improving time and space resolution of SMOS salinity maps using multifractal fusion. *Remote. Sens. Environ.* **2016**, *180*, 246–263. [CrossRef]

31. Tintore, J.; Vizoso, G.; Casas, B.; Heslop, E.; Pascual, A.; Orfila, A.; Ruiz, S.; Martínez-Ledesma, M.; Torner, M.; Cusí, S.; et al. SOCIB: The Balearic Islands Coastal Ocean Observing and Forecasting System Responding to Science, Technology and Society Needs. *Mar. Technol. Soc. J.* **2013**, *47*, 101–117. [CrossRef]

32. Mangoni, O.; Saggiomo, V.; Bolinesi, F.; Margiotta, F.; Budillon, G.; Cotroneo, Y.; Misic, C.; Rivaro, P.; Saggiomo, M. Phytoplankton blooms during austral summer in the Ross Sea, Antarctica: Driving factors and trophic implications. *PLoS ONE* **2017**, *12*, e0176033. [CrossRef] [PubMed]

33. Cipollini, P.; Calafat, F.M.; Jevrejeva, S.; Melet, A.; Prandi, P. Monitoring Sea Level in the Coastal Zone with Satellite Altimetry and Tide Gauges. In *Integrative Study of the Mean Sea Level and Its Components*; Springer: Cham, Switzerland, 2017; pp. 35–59.

34. Misic, C.; Harriague, A.C.; Mangoni, O.; Aulicino, G.; Castagno, P.; Cotroneo, Y. Effects of physical constraints on the lability of POM during summer in the Ross Sea. *J. Mar. Syst.* **2017**, *166*, 132–143. [CrossRef]

35. Rivaro, P.; Ianni, C.; Langone, L.; Ori, C.; Aulicino, G.; Cotroneo, Y.; Saggiomo, M.; Mangoni, O. Physical and biological forcing of mesoscale variability in the carbonate system of the Ross Sea (Antarctica) during summer 2014. *J. Mar. Syst.* **2017**, *166*, 144–158. [CrossRef]

36. Rivaro, P.; Ardini, F.; Grotti, M.; Aulicino, G.; Cotroneo, Y.; Fusco, G.; Mangoni, O.; Bolinesi, F.; Saggiomo, M.; Celussi, M. Mesoscale variability related to iron speciation in a coastal Ross Sea area (Antarctica) during summer 2014. *Chem. Ecol.* **2019**, *35*, 1–19. [CrossRef]

37. Wadhams, P.; Aulicino, G.; Parmiggiani, F.; Persson, P.O.G.; Holt, B. Pancake Ice Thickness Mapping in the Beaufort Sea From Wave Dispersion Observed in SAR Imagery. *J. Geophys. Res. Oceans* **2018**, *123*, 2213–2237. [CrossRef]

38. Aulicino, G.; Cotroneo, Y.; Lacava, T.; Sileo, G.; Fusco, G.; Carlon, R.; Satriano, V.; Pergola, N.; Tramutoli, V.; Budillon, G. Results of the first Wave Glider experiment in the southern Tyrrhenian Sea. *Adv. Oceanogr. Limnol.* **2016**, *7*, 16–35. [CrossRef]

39. Pascual, A.; Ruiz, S.; Olita, A.; Troupin, C.; Claret, M.; Casas, B.; Mourre, B.; Poulain, P.-M.; Tovar-Sanchez, A.; Capet, A.; et al. A Multiplatform Experiment to Unravel Meso- and Submesoscale Processes in an Intense Front (AlborEx). *Front. Mar. Sci.* **2017**, *4*, 263. [CrossRef]

40. Bouffard, J.; Pascual, A.; Ruiz, S.; Faugère, Y.; Tintore, J. Coastal and mesoscale dynamics characterization using altimetry and gliders: A case study in the Balearic Sea. *J. Geophys. Res.* **2010**, *115*, 10029. [CrossRef]

41. Bosse, A.; Testor, P.; Mortier, L.; Prieur, L.; Taillandier, V.; D'Ortenzio, F.; Coppola, L. Spreading of Levantine Intermediate Waters by submesoscale coherent vortices in the northwestern Mediterranean Sea as observed with gliders. *J. Geophys. Res. Oceans* **2015**, *120*, 1599–1622. [CrossRef]

42. Rudnick, D.L. Ocean Research Enabled by Underwater Gliders. *Annu. Rev. Mar. Sci.* **2016**, *8*, 519–541. [CrossRef] [PubMed]

43. Ruiz, S.; Pascual, A.; Garau, B.; Pujol, I.; Tintore, J. Vertical motion in the upper ocean from glider and altimetry data. *Geophys. Res. Lett.* **2009**, *36*, L14607. [CrossRef]

44. Heslop, E.E.; Sánchez-Román, A.; Pascual, A.; Rodríguez, D.; Reeve, K.A.; Faugère, Y.; Raynal, M.; Sánchez-Román, A.; Sánchez-Román, A. Sentinel-3A Views Ocean Variability More Accurately at Finer Resolution. *Geophys. Res. Lett.* **2017**, *44*, 12367–12374. [CrossRef]

45. Barceló-Llull, B.; Pascual, A.; Mason, E.; Mulet, S. Comparing a Multivariate Global Ocean State Estimate With High-Resolution in Situ Data: An Anticyclonic Intrathermocline Eddy Near the Canary Islands. *Front. Mar. Sci.* **2018**, *5*, 66. [CrossRef]

46. Barceló-Llull, B.; Pascual, A.; Ruiz, S.; Escudier, R.; Torner, M.; Tintoré, J. Temporal and Spatial Hydrodynamic Variability in the Mallorca Channel (Western Mediterranean Sea) From 8 Years of Underwater Glider Data. *J. Geophys. Res. Oceans* **2019**, *124*, 2769–2786. [CrossRef]

47. Garau, B.; Ruiz, S.; Zhang, W.G.; Pascual, A.; Heslop, E.; Kerfoot, J.; Tintore, J. Thermal Lag Correction on Slocum CTD Glider Data. *J. Atmospheric Ocean. Technol.* **2011**, *28*, 1065–1071. [CrossRef]

48. Troupin, C.; Beltran, J.P.; Heslop, E.; Torner, M.; Garau, B.; Allen, J.; Ruiz, S.; Tintoré, J. A toolbox for glider data processing and management. *Meth. Oceanogr.* **2016**, *13–14*, 13–23. [CrossRef]

49. Donlon, C.J.; Martin, M.; Stark, J.; Roberts-Jones, J.; Fiedler, E.; Wimmer, W. The Operational Sea Surface Temperature and Sea Ice Analysis (OSTIA) system. *Remote. Sens. Environ.* **2012**, *116*, 140–158. [CrossRef]

50. Taburet, G.; Sanchez-Roman, A.; Ballarotta, M.; Pujol, M.-I.; Legeais, J.-F.; Fournier, F.; Faugere, Y.; Dibarboure, G. DUACS DT-2018: 25 years of reprocessed sea level altimeter products. *Ocean Sci. Discuss.* **2019**. [CrossRef]

51. Chin, T.M.; Vazquez, J.; Armstrong, E. A multi-scale high-resolution analysis of global sea surface temperature. *Remote Sens. Environ.* **2017**, *200*, 154–169. [CrossRef]

52. Buongiorno Nardelli, B.; Droghei, R.; Santoleri, R. Multi-dimensional interpolation of SMOS sea surface salinity with surface temperature and in situ salinity data. *Remote Sens. Environ.* **2016**, *180*, 392–402. [CrossRef]

53. Barceló-Llull, B.; Pascual, A.; Díaz Barroso, L.; Sánchez-Román, A.; Casas, B.; Muñoz, C.; Torner, M.; Alou, E.; Cutolo, E.; Mourre, B.; et al. *PRE-SWOT Cruise Report. Mesoscale and Sub-Mesoscale Vertical Exchanges from Multi-Platform Experiments and Supporting Modeling Simulations: Anticipating SWOT Launch (CTM2016-78607-P)*; Technical report V5.0; CSIC-UIB—Instituto Mediterráneo de EstudiosAvanzados (IMEDEA): Esporles, Spain, 2018; p. 138. [CrossRef]

54. González-Gambau, V.; Turiel, A.; Olmedo, E.; Martínez, J.; Corbella, I.; Camps, A.; Sampling, N. A new image reconstruction algorithm for SMOS. *IEEE Geosci. Remote Sens. Lett.* **2016**, *54*, 2314–2328. [CrossRef]

55. González-Gambau, V.; Olmedo, E.; Turiel, A.; Martínez, J.; Ballabrera-Poy, J.; Portabella, M.; Piles, M. Enhancing SMOS brightness temperaturas over the ocean using nodal sampling image reconstruction technique. *Remote Sens. Environ.* **2016**, *180*, 205–220. [CrossRef]

56. Turiel, A.; Gonzalez-Gambau, V.; Olmedo, E.; Martinez, J.; Duran, I. Improvements on Calibration and Image Reconstruction of SMOS for Salinity Retrievals in Coastal Regions. *IEEE J. Sel. Top. Appl. Earth Obs. Remote. Sens.* **2017**, *10*, 3064–3078.

 *remote sensing*

Article

Real-time Reconstruction of Surface Velocities from Satellite Observations in the Alboran Sea

Jordi Isern-Fontanet [1,2,*], Emilio García-Ladona [1], José Antonio Jiménez-Madrid [1,3], Estrella Olmedo [1,2], Marcos García-Sotillo [4], Alejandro Orfila [5] and Antonio Turiel [1,2]

[1] Institut de Ciències del Mar (CSIC), 08003 Barcelona, Spain; emilio@icm.csic.es (E.G.-L.); jmadrid@icm.csic.es (J.A.J.-M.); olmedo@icm.csic.es (E.O.); turiel@icm.csic.es (A.T.)
[2] Barcelona Expert Centre in Remote Sensing, 08003 Barcelona, Spain
[3] Departament de Teoria del Senyal i Comunicacions, Universitat Politécnica de Catalunya, 08034 Barcelona, Spain
[4] Puertos del Estado, 28042 Madrid, Spain; marcos@puertos.es
[5] Institut Mediterrani d'Estudis Avançats (CSIC), 07190 Esporles, Spain; aorfila@imedea.uib-csic.es
* Correspondence: jisern@icm.csic.es

Received: 23 December 2019; Accepted: 20 February 2020; Published: 22 February 2020

Abstract: Surface currents in the Alboran Sea are characterized by a very fast evolution that is not well captured by altimetric maps due to sampling limitations. On the contrary, satellite infrared measurements provide high resolution synoptic images of the ocean at high temporal rate, allowing to capture the evolution of the flow. The capability of Surface Quasi-Geostrophic (SQG) dynamics to retrieve surface currents from thermal images was evaluated by comparing resulting velocities with in situ observations provided by surface drifters. A difficulty encountered comes from the lack of information about ocean salinity. We propose to exploit the strong relationship between salinity and temperature to identify water masses with distinctive salinity in satellite images and use this information to correct buoyancy. Once corrected, our results show that the SQG approach can retrieve ocean currents slightly better to that of near-real-time currents derived from altimetry in general, but much better in areas badly sampled by altimeters such as the area to the east of the Strait of Gibraltar. Although this area is far from the geostrophic equilibrium, the results show that the good sampling of infrared radiometers allows at least retrieving the direction of ocean currents in this area. The proposed approach can be used in other areas of the ocean for which water masses with distinctive salinity can be identified from satellite observations.

Keywords: sea surface temperature; altimetry; surface quasi-geostrophic equations; surface currents

1. Introduction

The Strait of Gibraltar is the natural connection of the Mediterranean Sea with the world ocean, being a hot spot area in many senses. There, diverse and separated hydrodynamic phenomena occur over a wide range of spatial and temporal scales (from the small scales and submesoscale phenomena to inter-annual and climate variability). The complex hydrodynamic conditions influence the rich marine ecosystem in the adjacent coastal areas, e.g., [1–3]. Furthermore, the region is of great socioeconomic relevance as a key passage for marine trade. Almost 1/6 of global sea traffic and 1/5 of global oil traffic transits through the Mediterranean basin, still being the shortest route between Europe and Asia [4]. Then, it is not surprising that the Strait of Gibraltar and the adjacent Alboran Sea have been traditionally a focus area for many research and monitoring efforts since a long time ago, e.g., [5–7]. Although the main characteristics of the prominent processes are reasonably known and modeling efforts capture most of them, an adequate operational forecast of the hydrodynamics conditions in this region remains a challenging task [8]. Roughly, the water mass structure is close to a two-layer system characterized by an upper surface layer of Atlantic waters entering into the basin and denser Mediterranean waters outflowing below [9]. The inflow of Atlantic water and the outflow of Mediterranean water are constrained by hydraulic control in the channel where the bottom relief, stratification, tidal, and wind regimes determine the variability of water exchanges through the strait, which are therefore linked to basin scale variability, e.g., [10,11].

The jet of Atlantic water forms and configures a quasi-permanent vortex or gyre in the western part (western Alboran vortex, WAG) of the Alboran Sea, which progresses further into the second half after Cape Tres Forcas (around 3°W) where it forms a second gyre (eastern Alboran gyre, EAG) and continues further to the east attached to the African coast as the Algerian current, e.g., [12,13] (see Figure 1). This is the dominant general pattern of the surface circulation in the Alboran Sea, particularly in summer, emerging from the analysis of time series of sea level maps and model reanalysis [14,15] and often observed by in situ cruises, e.g., [13]. Besides, the gyres may collapse or migrate until they are restored in the spring and early summer. These events appear to be induced by high frequency processes (tides and atmospheric fluctuations) requiring high resolution in space and time to be adequately analyzed and studied [8].

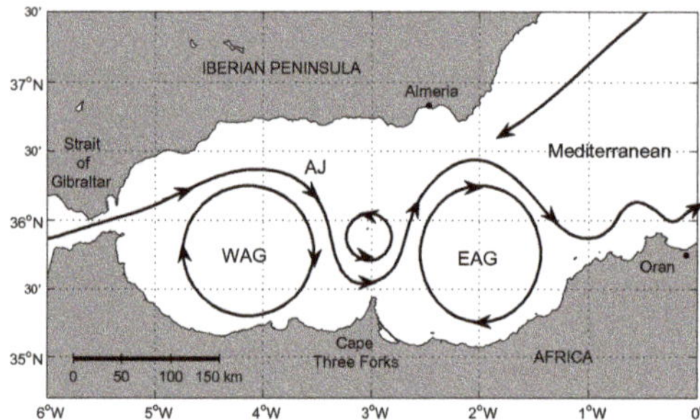

Figure 1. Scheme of the surface circulation in the Alboran Sea showing the western and eastern Alboran gyres (WAG and EAG) as well as the jet of Atlantic waters (AJ), from Sánchez-Garrido et al. [8].

Attempts to forecast the oceanographic conditions during a recent oceanographic experiment, the MEDESS-GIB experiment [16], have shown that models reasonably reproduce the main pattern but fail to reproduce the variability of short scales and the details of the evolution of the Atlantic inflow around WAG, e.g., [17]. Once models have the necessary spatial resolution and are able to reproduce the observed physical processes, as was the case of the MEDESS-GIB experiment, a way to improve the subsequent forecast is to have initial conditions and analyzed fields as close as possible to the truth. Furthermore, in the worst case, when and where operational systems are not working well and depending on the involved scales, an empirical approach using only real time field observations and assuming some kind of persistence may provide a reasonable first guess. Sea surface temperature (SST) is quite satisfactorily retrieved in real time and with enough resolution to reach fields at submesoscale. For the ocean velocity, altimetry offers the possibility to build maps of velocity fields interpolating along-track information. In this case, the resolution attained can reach the ocean mesoscale, although with limitations in terms of accuracy and reliability, e.g., [18] and real time is not possible.

The operational estimation of ocean velocities from satellite observations remains a major problem in satellite oceanography. At present, multiple methods to estimate ocean currents from SST have been proposed with a wide range of performances, see [19] for a review on this subject. In this study, we analyzed the possibility to retrieve real time high resolution fields making use of the surface quasi-geostrophic theory (SQG) [20,21]. SQG offers the theoretical body to derive high resolution surface velocity fields from a single infrared SST image [21–26]. This capability is of key importance for operational applications because it extends its usability in comparison with other techniques such as maximum cross correlation or optical flow that need a sequence of cloud-free images [19]. Two conditions are necessary to apply the SQG framework to SST images: surface density fluctuations have to be strong enough to capture a significant amount of the near-surface dynamics [22,27] and SST has to be a proxy of density anomalies at the base of the mixed layer [28]. In a pioneering work, LaCasce and Mahadevan [29] demonstrated the applicability of the SQG framework in the Alboran Sea and showed that it was possible to retrieve the full 3D structure of ocean density and velocity fields from SST fields. Their reconstructed fields were quite similar to those observed in a CTD cruise; however, the work was flawed by the assumption that three-day cruises could be considered synoptic. Consequently, they had errors in both horizontal and vertical velocities.

In the present study, we went further by validating the reliability of using a time series of ocean velocity fields from SST covering the full area between the Strait of Gibraltar and the Alboran Sea. In particular, we analyzed the performance of the SQG approach when applied to infrared satellite measurements and compared to velocities derived from surface drifters. Moreover, we explored new approaches to overcome the limitations imposed by the lack of observations of ocean salinity.

This paper is organized as follows. We first briefly present the dataset used in Section 2. We develop in detail the methodology to derive velocity fields applying a SQG-based methodology in Section 3. In Section 4, we present and validate the results comparing with field data from the MEDESS-GIB experiment. We finally discuss and conclude the major outcomes.

2. Data

On the frame of the MEDESS-4MS project (EU MED Program), an intensive Lagrangian experiment was organized in the Strait of Gibraltar to validate and test the operational systems running in this area [16]. The experiment consisted od a quasi-synoptic deployment network of surface drifters distributed along the Strait of Gibraltar (Figure 2). The experiment started on 9 September and lasted around three months from September to December 2014. The drifters used in the experiment were mostly of CODE type [30] dragged at 1.5 m from the surface and a few units of oil-spills tracking drifters not used in this study, all set up with a sampling rate of 30 min. The dataset is available at PANGAEA (Data Publisher for Earth and Environmental Science) repository and all the quality control procedures and first view of the trajectories were described by Sotillo et al. [17].

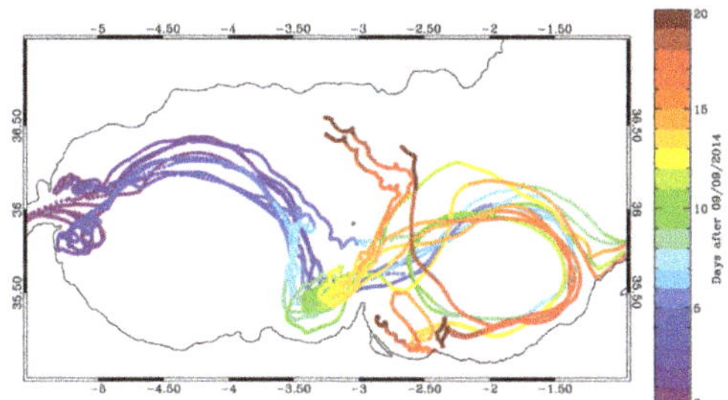

Figure 2. Trajectories of surface drifters for the MEDGIB-GIB experiment. Color scale corresponds to time with respect to 9 September 2014.

Geostrophic velocities used were derived from Near Real Time Absolute Dynamic Topography Maps (NRT-MADT) for the Mediterranean Sea generated by AVISO altimetry and distributed by the GlobCurrent project. Velocities were estimated from NRT-MADT using a nine-point stencil length, as described by Arbic et al. [31]. During the period, most drifters remained in the Alboran Sea (from 9 September to 29 September, Figure 2), but only 10 images with favorable cloud coverage were available. These images correspond to the AVHRR instrument from NOAA 19 platform for the period from 9 September at 03:08 GMT to 14 September at 02:12 GMT. Consequently, only the WAG could be simultaneously sampled by infrared instruments and drifters.

The used SSS corresponds to the Mediterranean and North Atlantic SMOS SSS maps V2.0 product from the Barcelona Expert Centre in Remote Sensing (BEC). These maps were derived from L1B Microwave Brightness Temperatures (MBT) products measured by SMOS and provided by ESA. Then, SMOS SSS daily L3 maps at $1/4° \times 1/4°$ resolution were produced by means of a successive corrections analysis applied over time periods of nine days using influence radii adapted to the Mediterranean Sea, see [32] and reference therein.

3. Reconstruction of Velocities from Thermal Images

3.1. Theoretical Background

Under the geostrophic approximation, surface velocities $\vec{v}(\vec{x})$ are non-divergent and are determined from a stream function $\psi(\vec{x})$

$$\vec{v}(\vec{x}) = \vec{e}_z \times \nabla \psi(\vec{x}), \tag{1}$$

which is proportional to sea surface height ($\eta(\vec{x})$, SSH), i.e.

$$\psi(\vec{x}) = \frac{g}{f_0} \eta(\vec{x}). \tag{2}$$

Here, $\vec{e}_z$ is the vertical unit vector, $\vec{x} = (x, y)$, f_0 is the local Coriolis parameter, and g is the gravity constant.

When there is a strong correlation between sea surface buoyancy ($b_s(\vec{x})$, SSB) and potential vorticity (PV), it is possible to invert the quasi-geostrophic PV equations and derive the stream-function from SSB. For a semi-infinite ocean with constant stratification $N(z) = n_0 f_0$, the stream function at any depth z is given by

$$\hat{\psi}(\vec{k}, z) = \frac{1}{n_e f_0 k} \hat{b}_s(\vec{k}) \exp(n_0 k z), \tag{3}$$

where $\hat{}$ stands for the Fourier transform, n_e is an effective Prandtl ratio that takes into account the stratification and the interior PV, $\vec{k} = (k_x, k_y)$ is the wavevector, and $k = \|\vec{k}\|$ [22]. This is the so-called effective surface wuasi-geostrophic (eSQG) model.

Temperature and salinity are related to buoyancy through

$$b(\vec{x}) = -\frac{g}{\rho_0} \left[\alpha(T(\vec{x}) - T_0) + \beta(S(\vec{x}) - S_0)\right], \tag{4}$$

where $\alpha < 0$ is the thermal expansion coefficient, $-\beta > 0$ is the haline contraction coefficient, and T_0 and S_0 are reference temperature and salinity, respectively. Then, assuming that temperature and salinity anomalies are related, Equation (4) can be approximated as

$$b(\vec{x}) \approx -\frac{g}{\rho_0} \left[\alpha'(T(\vec{x}) - T_0)\right], \tag{5}$$

with α' being an effective thermal expansion coefficient that takes into account the partial compensation between salinity and temperature [28]. Introducing this approximation to Equation (3), it is possible to derive surface currents from satellite observations of sea surface temperatures ($T_s(\vec{x})$, SST) [28]:

$$\hat{\psi}(\vec{k}, z) = -\frac{\alpha'}{\rho_0 n_e f_0 k} \hat{T}_s(\vec{k}) \exp(n_0 k z). \tag{6}$$

Notice that this approach requires settin up the energy level of these velocities because of the presence of a free parameter $\alpha' n_e^{-1}$ to take into account both interior PV and the partial compensation of salinity and temperature [28]. Interestingly, Equation (3) can also be used to derive buoyancy from SSH [28]

$$\hat{b}(\vec{k}, z) = n_e g k \hat{\eta}(\vec{k}) \exp(n_0 k z). \tag{7}$$

Equations (2), (6), and (7) relate key dynamical fields with satellite observations of SST and SSH.

3.2. Implementation

The diagnosis of surface currents from thermal images is straightforward using the SQG approach, if spatial resolutions below 5–10 km are not necessary. Following Isern-Fontanet and Hascoët [33], the surface stream function ($z = 0$) is computed by applying Equation (6) to channel 4 brightness temperatures (BT) instead of SST due to their lower levels of noise. Before this, data gaps in BT images are filled using the approach of Isern-Fontanet et al. [28] and they are interpolated onto a doubly periodic regular grid. Then, geostrophic velocities are obtained using centered finite-differences and the free parameter $\alpha' n_e^{-1}$ in Equation (6) is fixed imposing that the resulting velocity field has the same energy level as the velocity field derived from altimetry [28], i.e.,

$$\langle E_T \rangle \equiv \langle E_\eta \rangle, \tag{8}$$

where $E_T(\vec{x})$ is the kinetic energy of the velocity field derived from thermal images, $E_\eta(\vec{x})$ is the kinetic energy of the velocity field derived from altimetry, and $\langle \cdot \rangle$ stands for the spatial average. Notice, however, that the velocity field derived from thermal images has a much higher spatial resolution than the field derived from altimetry. Therefore, $E_T(\vec{x})$ is first low-pass filtered with a cut-off wavelength of 60 km in order to determine the constant $\alpha' n_e^{-1}$. It is worth mentioning that it is not necessary to have simultaneous observations of BT and SSH to calibrate the velocity field. Moreover, it is possible to use any other independent measurement of the velocities such as those obtained from HF radars, current-meters, or drifting buoys.

Figure 3 shows two examples of the SST representative of the period under study. These images unveil the presence of at least three different water masses: the warm waters associated to the WAG and EAG, a very warm water filling the northeast part of the Alboran Sea as well as the nearby Algerian basin, and a tongue of cold water newly entered from the Strait of Gibraltar that separates the previous ones. For this period, the EAG was smaller than the average situation and the front separating fresher waters from the saltier Mediterranean ones, usually located in the line Almeria–Oran, was displaced allowing Mediterranean waters to penetrate the Alboran basin, see [8] and references therein. The velocity field derived from thermal images correctly reproduce the currents associated to the WAG and partially the currents associated to the EAG. However, it also shows an intense current flowing from the southeast in the middle of the basin, i.e., along the limits of warm waters filling in the northeast of the image, which is not coherent with drifter trajectories (compare Figures 2 and 3). These waters are, very likely, saline enough to dominate density and make this water mass denser than the waters of the Alboran vortices, rather than lighter as it would be expected from observed temperatures alone. Several arguments support this. First, previous in situ measurements showed that resident Mediterranean waters are saltier than the waters that form the Alboran vortices with typical salinities $S > 38$ psu, e.g., [34]. Second, SSS maps derived from SMOS show the penetration of salty waters into the Alboran basing from the east (Figure 4). Third, buoyancy derived from altimetric maps using Equation (7) shows that this water mass, which is represented by the lined area in Figure 4, has less buoyancy and, consequently, it has to be more saline. Consequently, velocities have the opposite sign compared to velocities observed from drifters.

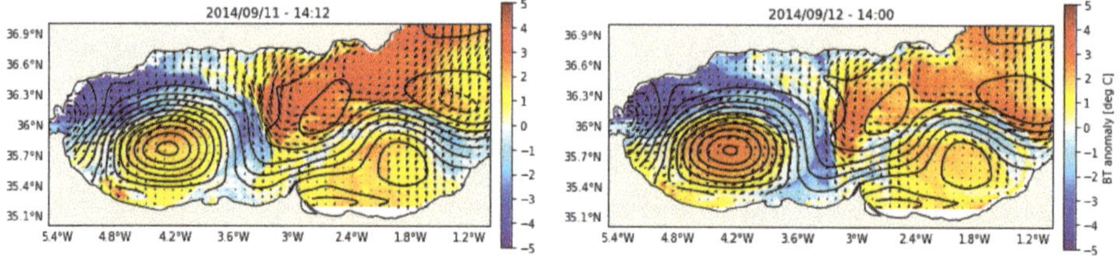

Figure 3. Brightness temperature anomaly derived from AVHRR's channel 4 measurements corresponding to: 11 September at 14:12 GMT (**left**); and 12 September at 14:00 GMT (**right**). Arrows show the velocity field derived from the images.

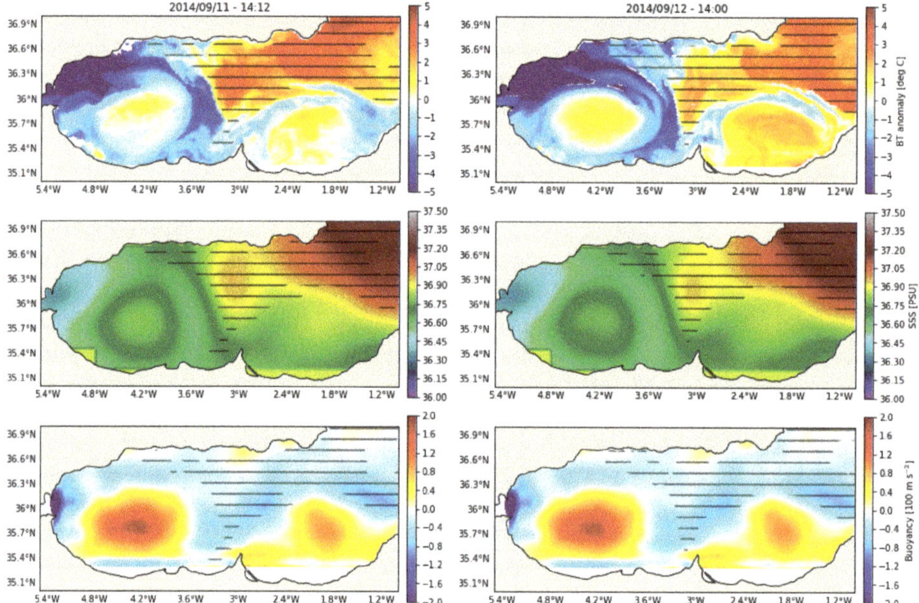

Figure 4. Brightness temperature anomaly derived from satellite measurements (**top**); sea surface salinity derived from SMOS (**middle**); and surface buoyancy derived from altimetric maps (**bottom**) corresponding to: 11 September at 14:12 GMT (**left**); and 12 September (**right**) at 14:00 GMT. Lined area identifies resident Mediterranean waters obtained using the method described in Section 3.3.

3.3. Phase Correction

The first approach to compute surface currents is to assume that temperature is a good proxy of density and, thus, the approach given by Equation (5) is valid. While this is a reasonable approach for the Alboran vortices and surrounding waters [29], results (Section 3.2, Figure 3) show that this is not valid for the warmer waters of the northeast. Consequently, the direct application of the eSQG model to these images produces velocities in the wrong sense in some parts of the image during this period of the year. Moreover, SSS from SMOS is not yet able to provide the necessary spatial resolution to correct full resolution infrared SST data, is not available in real time, and still slightly underestimates the dynamical range of SSS, see [26], for a more detailed discussion.

High resolution SST can still be used to derived currents if the dynamics in the area under study is dominated by the presence of water masses with distinctive salinity. Indeed, suppose there are M water masses with constant salinities S_m, where $m = 1, \ldots, M$. Then, Equation (4) becomes

$$b(\vec{x}) = -\frac{g}{\rho_0}\left[\alpha(T(\vec{x}) - T_0) + \beta \sum_{m=1}^{M} (S_m - S_0)\theta_m(\vec{x})\right],$$ (9)

where $\theta_m(\vec{x})$ is a function such that

$$\theta_m(\vec{x}) = \begin{cases} 0 & \text{if } S(\vec{x}) \neq S_m \\ 1 & \text{if } S(\vec{x}) = S_m \end{cases}$$ (10)

with the additional property that

$$\sum_{m=1}^{N} \theta_m(\vec{x}) = 1.$$ (11)

This function segments the domain into non-overlapping areas with different constant salinities. Notice that these functions are also a function of time. If, as in the Alboran Sea, only two water masses are present, the above buoyancy can be written as

$$b(\vec{x}) = -\frac{g}{\rho_0}\left[\alpha(T(\vec{x}) - T_0) + \beta\Delta S(1 - 2\theta_{med}(\vec{x}))\right],$$ (12)

where the salinity of Atlantic waters is $S_{atl} = S_0 - \Delta S$, the salinity of the Mediterranean waters is $S_{med} = S_0 + \Delta S$, and $\theta_{med}(\vec{x})$ gives the extension of Mediterranean waters. The major practical difficulty is to determine the extension of Mediterranean waters, i.e to determine the function $\theta_{med}(\vec{x})$. Here, we compute $\theta_{med}(\vec{x})$ by computing the temperature anomaly and extracting the different water masses as the connected areas with the same sign of temperature anomaly. This idea is implemented by decomposing thermal images applying the *à trous* wavelets algorithm with B_3-splines [33] and then removing the largest and the shortest scales. Finally, the wide, warm, and saltier mass located in the northeast can be identified as the largest connected area with positive temperature anomaly (lined area in Figure 4). Once the resident Mediterranean water mass has been identified, the effect of salinity can be corrected. To prove the concept and avoid numerical issues associated with the sharp transition of SSS, Equation (12) is further simplified to

$$b(\vec{x}) \approx -\frac{g}{\rho_0}\left[\alpha'(T(\vec{x}) - T_0)(1 - 2\theta_{med}(\vec{x}))\right].$$ (13)

This approach implies that gradients of temperature and salinity tend to be aligned but with a change of sign in the different parts of the area under study (see Figure 2 in Isern-Fontanet et al. [24] and discussions in Isern-Fontanet et al. [24], Isern-Fontanet et al. [26]).

The application of Equation (6) to the corrected temperature anomalies given by Equation (13) provides more realistic results (Figure 5). Indeed, Atlantic waters now flow from west to east, as seen in previous studies [35] and in agreement with the trajectories of the drifters released during the MEDESS-GIB experiment (Figure 2). It is important to take into account that the transfer function between the stream function and temperature (or buoyancy) scales as $\sim k^{-1}$, which implies certain dominance of larger scales over smaller scales. This has as a consequence that the incorrect determination of the buoyancy sign for the Mediterranean waters can have a non-local impact. Therefore, ocean velocities around the WAG are wrongly diagnosed if this is not taken into account. Finally, its worth mentioning that the use of this crude approach (Equation (13)) instead of the approximation given by Equation (12) may affect the sign of the small scales in the Mediterranean waters. Nevertheless, we could not investigate this limitation due to the lack of data.

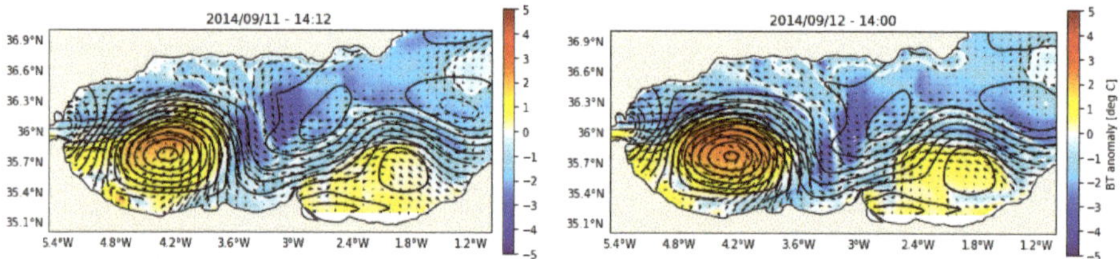

Figure 5. Corrected brightness temperature anomaly corresponding to: 11 September at 14:12 GMT (**left**); and 12 September (**right**) at 14:00 GMT. Arrows show the velocity field derived from the images.

4. Comparison between Velocity Estimations

The eSQG model, together with the phase correction approach described in Section 3.3, were used to derive velocities from all the available thermal images (Figure 6 shows six of them). The resulting fields exhibit more temporal variability of the WAG than the velocity derived from SSH (Figures 6 and 7). Indeed, thermal images shows that the WAG changes its shape significantly at temporal scales of the order of 12 h, but this variability is not observed in altimetric maps, which show an almost stationary vortex. A second major difference in the western part of the basin is the circulation east of the Strait of Gibraltar, the area between the strait and the vortex approximately limited by the meridian located at 4.8°W. There, altimetric maps show a northwards current while thermal images generate an eastwards current, indicating entrance of Atlantic waters into the Mediterranean Sea. The comparison between velocities derived from temperature and from sea level also show differences in the time evolution of the EAG. While the first shows an evolving vortex, altimetry shows an almost stationary bipolar structure. The differences between altimetric and thermal images are also evident comparing BT and buoyancy derived from SSH (Figure 4). It is worth mentioning that, contrary to SST, buoyancy derived from SSH tends to be more representative of the patterns below the mixed layer rather than the ocean surface [28] and, consequently, fields may not be fully comparable. Nevertheless, the temporal evolution of this vortex and its characteristics points to sampling limitations of the altimetric measurements rather than the difference between density anomalies in the Mixed Layer and below it. Finally, both fields seem to capture the separation of the west–east flow from the WAG located around longitude 3.55°W. The exception is for the AVHRR image of 10 September at 14:24, which shows currents that do not agree with other satellite images or altimetric maps. The reason for this disagreement relies on the failure of the phase correction approach for this particular image. Indeed, the separation between the EAG and the large extension of warm waters is not good enough to correct for their contribution to the flow on this side of the basin.

The velocity fields derived from altimetry and thermal images were compared to the in situ measurements provided by the drifters of the MEDESS-GIB experiment. A first comparison comes from the juxtaposition of trajectories with the instantaneous velocity fields (Figures 6 and 7), which shows a strong tendency of drifter trajectories to move tangent to the velocity fields. There are, however, some exceptions, particularly in the area east of the Strait of Gibraltar where some drifters have swirling trajectories that do not agree with any of the velocity fields derived from satellite measurements. Although the temporal length of available thermal images is not long enough to analyze the penetration of drifters in the EAG, it is long enough to see the detachment of drifters from the WAG and their path towards the east, which is in agreement with the velocities seen from satellites. Notice that, in this area, swirling trajectories are also present and in disagreement with both velocity fields.

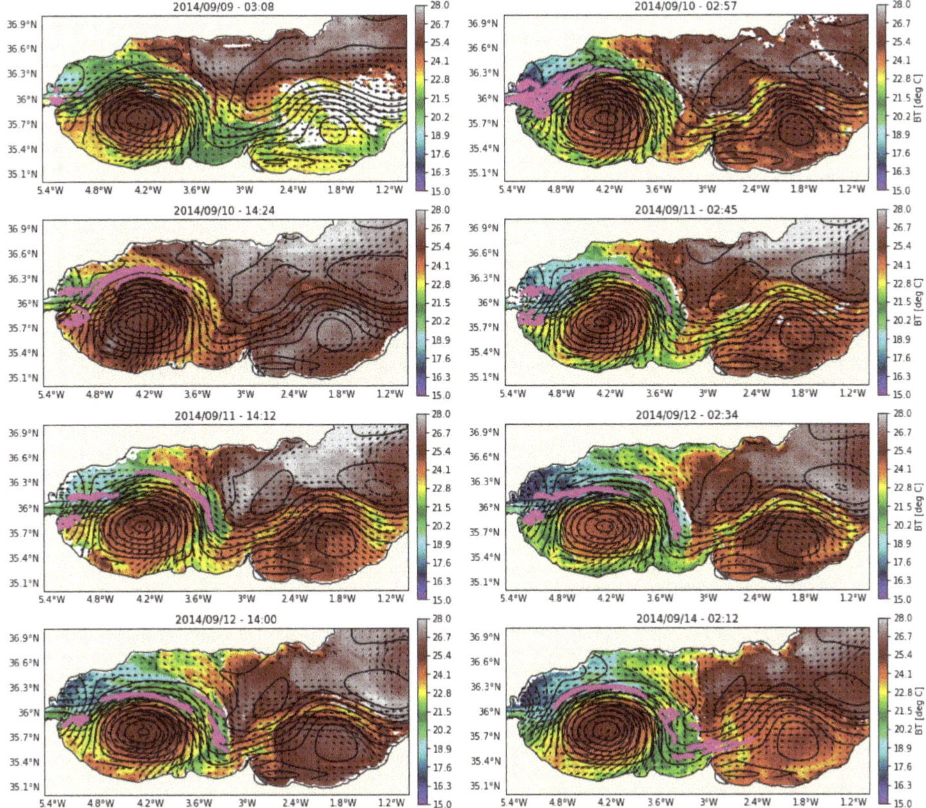

Figure 6. Sequence of corrected brightness temperature with the associated geostrophic velocities. Magenta dots correspond to the position of drifters within the the period of ±0.5 days around the map date.

The velocities derived from surface drifters were compared to surface velocities derived from satellite observations by interpolating the latter onto the position and time of drifters (see Figure 8). Since no correction for the presence of inertial oscillations was applied, only those drifter segments that did not show small loops were taken from the whole set of drifters depicted in Figure 2. The global comparison between velocity components suggests a slightly better agreement between the velocities derived from SST than those derived from SSH (Figure 9). This can be seen in the global correlations between reconstructed velocities and drifter velocities, which are $r_u = 0.63$ and $r_v = 0.74$ for the velocities derived from SST and $r_u = 0.48$ and $r_v = 0.66$ for the velocities derived from SSH. Since some ageostrophic corrections modify the speed but do not modify the direction of the geostrophic current, e.g., the cyclostrophic flow, it is interesting to compare velocity directions. Here, the correlation between the velocity directions diagnosed from satellite and those derived from drifter trajectories are quite similar: $r_\theta = 0.74$ for SST and $r_\theta = 0.71$ for SSH. Notice, however, that all correlations are well below 0.9. Another difference between the two types of velocities derived from satellite observations is their kinetic energy. Although velocities derived from SST are calibrated with altimetric data, the lack of small scale signals in altimetric maps implies that higher resolution velocities derived from SST have to be low-pass filtered for calibration (Equation (8)) but the resulting field has higher kinetic energies closer to the kinetic energy of surface drifters (Figure 9).

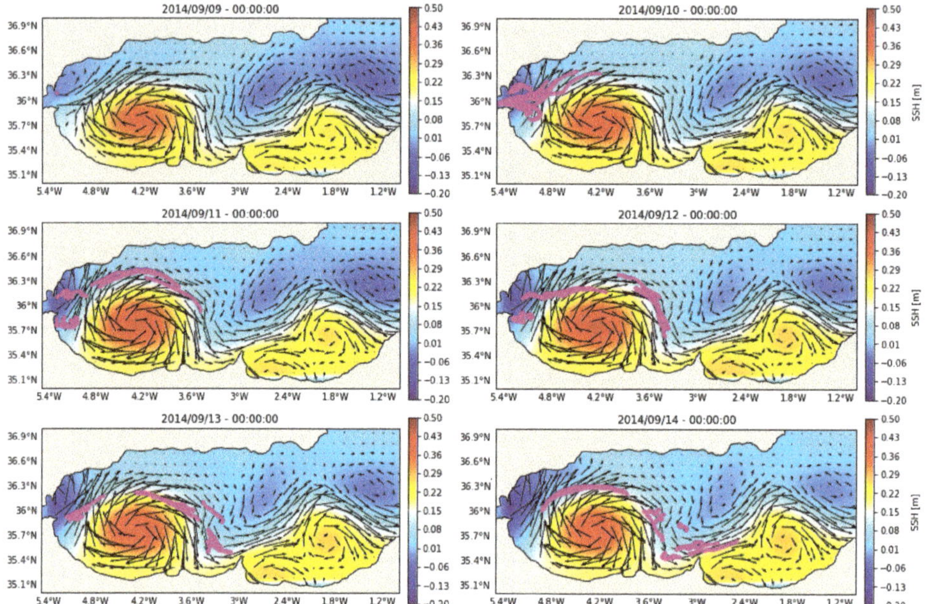

Figure 7. Absolute dynamic topography and associated geostrophic velocities. Magenta dots correspond to the position of drifters within the the period of ±0.5 days around the map date.

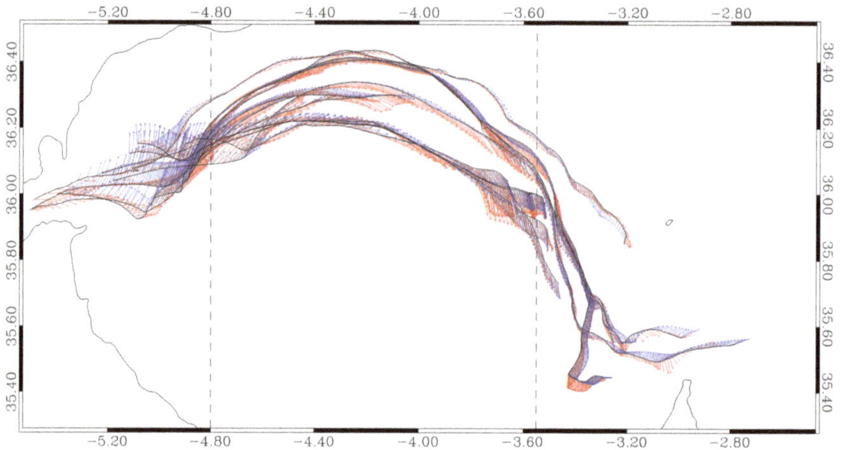

Figure 8. Drifter trajectories for the period under analysis and velocities derived from the IR radiometer (red) and the Radar Altimeter (blue) interpolated onto drifter position and time. The vertical line indicates the longitude used to separate the area east of Gibraltar from the WAG. Black lines correspond to the same trajectories shown in the upper panel.

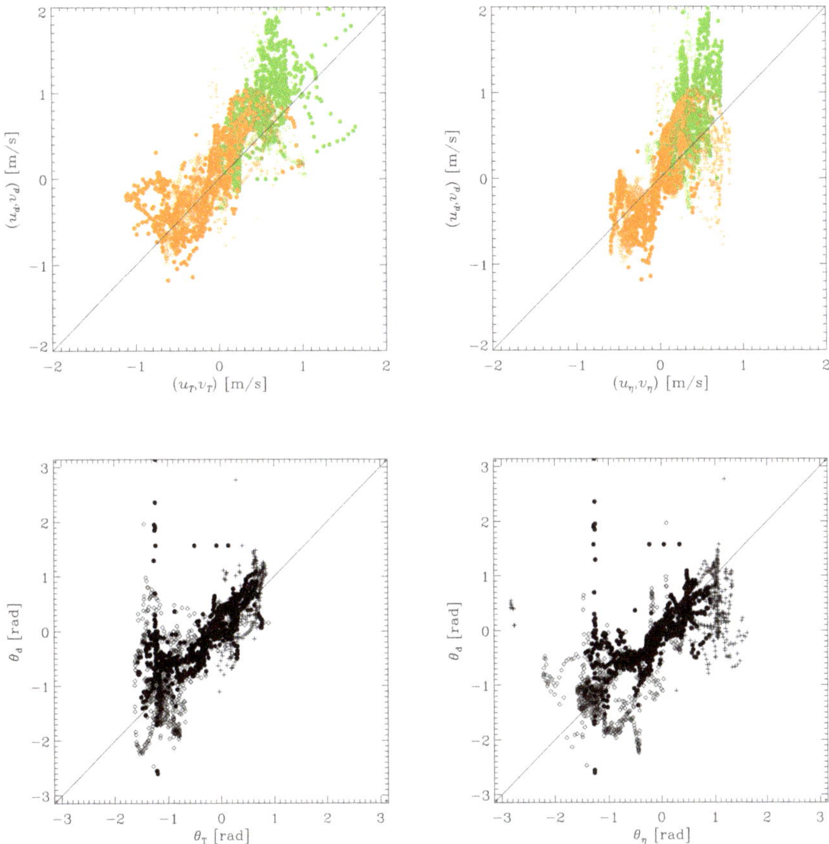

Figure 9. Scatter plot between the drifter velocities (ordinate) and the velocities derived from thermal images (abscissa, **left**) and altimetry (abscissa, **right**). Green corresponds to zonal velocity, orange to meridional velocity, and black to the velocity angle with respect to the east–west direction. Diamonds correspond to the area east of the Strait of Gibraltar, dots to the central part of the WAG, and crosses to the eastern area of the vortex (see Figure 8).

Figure 8 shows a relatively good agreement in the direction of currents, i.e., small angles between velocities and drifter trajectories, in the WAG for both the velocities derived from altimetry and velocities derived from thermal images. However, there is an important difference in the velocity angles observed in the velocities derived from SST with respect to drifter velocities for values around -1.5 rad (Figure 9). This discrepancy is mainly due to the velocities in the eastern edge of the WAG. Moreover, between the Strait of Gibraltar and the WAG, altimetric velocities have angles that can be up to 90° with respect to the trajectories of drifters while, velocities derived from thermal images show significantly smaller angles. The above observations and the patterns seen in the velocity fields (Figures 6 and 7) suggest that the agreement between satellite derived and in situ velocities may not be homogeneous and point to define three different areas: the area east of Gibraltar (the area between the Strait of Gibraltar and meridian located at 4.8°W), the WAG (between the meridians located at 4.8°W and 3.45°W), and the bifurcation area (east of the meridian located at 3.45°W), where part of the flow is trapped by the WAG and part of it flows eastwards towards the EAG (see Figure 8).

These areas were used to better quantify the agreement between in situ and satellite velocities. Taylor diagrams confirm that the ability to reconstruct velocities is better for WAG than for any other area (Figure 10), although velocities derived from temperature have slightly higher correlations for the direction. In the area east of the bifurcation point, on the contrary, the performance is better for altimetry, although correlations are quite low for the orientation of the current. The opposite situation is found in the area east of Gibraltar, in which altimetry shows very poor performance. Interestingly, the meridional velocity is better reconstructed than the zonal velocity for both SSH- and SST-derived velocities.

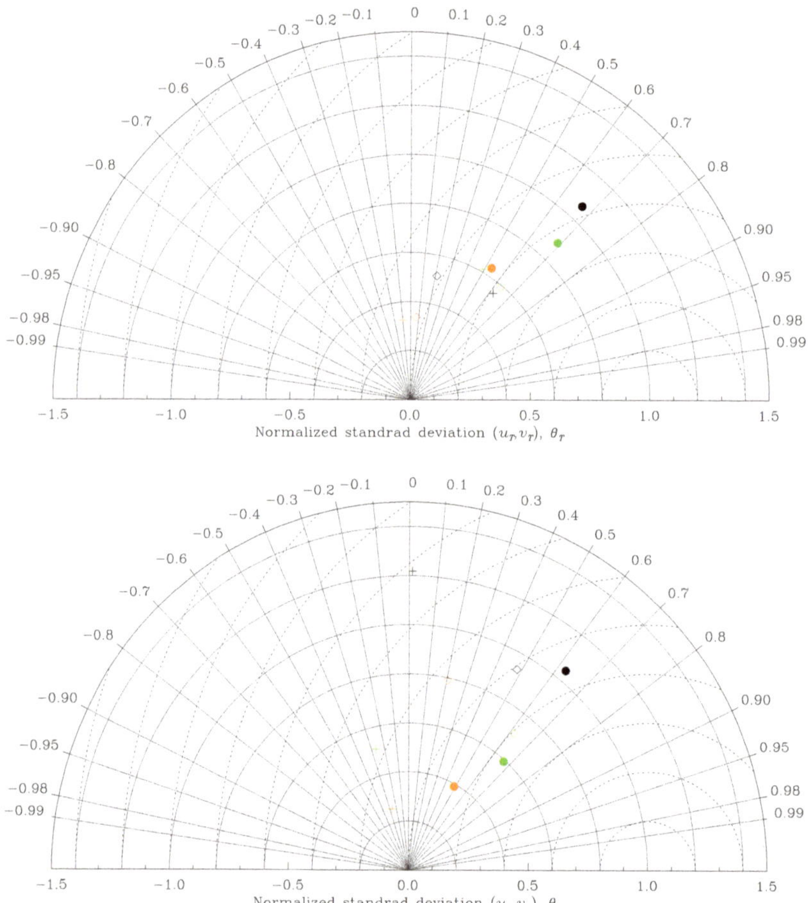

Figure 10. Taylor diagram for zonal (orange) and meridional (green) velocity components and velocity direction (black) for the area east of Gibraltar (cross), WAG (circle) and stagnation point (diamond): (**top**) velocities derived from thermal images; and (**bottom**) velocities derived from altimetry. To facilitate the comparison, standard deviations were normalized by the standard deviations of drifters given in Table 1.

Table 1. Standard deviation of zonal velocity (σ_u), meridional velocity (σ_v), and velocity direction (σ_u) derived from the selected drifter trajectories shown in Figure 8.

Area	σ_u [ms^{-1}]	σ_v [ms^{-1}]	σ_θ [deg]
Western vortex	0.42	0.45	32.5
east of Gibraltar	0.52	0.41	30.7
West of bifurcation point	0.21	0.26	37.8

Finally, the topology of the two velocity fields derived from satellite observations was unveiled by using the methodology presented by Jiménez Madrid and Mancho [36], which has been already applied on altimeter datasets over the area of the Kuroshio Current [37]. The idea was to produce a two-dimensional field covering the region, which computes at each point the arc length of the trajectory passing for the point. More precisely, by using synthetic trajectories obtained integrating forwards and backwards for a determined amount of time and getting the distance travelled, the arc lengths of the trajectory are computed for the central instant of the time series to have the maximum possible time period to integrate, i.e., 2.5 days (see Appendix A for technical details). Figure 11 depicts the arc length plots for both velocity fields. One can see both velocity fields are able to capture the two-vortex patterns typical of the Alboran Sea but with significant differences. First, altimetric field is more stable, i.e., vortex cores are clearly delimited and the flow of Atlantic water is very well defined by large arc length values. This indicates, as already stated, that the temporal evolution of the altimetric field is slow. On the contrary, thermal field shows the effect of rapidly evolving WAG producing a vaguer structure but it is still possible to see the vortex core and its contour. It is worth mentioning that the detection of EAG is challenging due to the similar temperature it has with respect to the Mediterranean waters. It is only seen thanks to the flow of fresh waters that clearly separates both structures. Notice the strong differences between them. Second, the thermal field is able to capture the entrance of Atlantic waters while the altimetric field is not. Indeed, the shadowed area in Figure 11 is due to the particles that escape from the domain; consequently, it is not possible to track their evolution for the entire time period considered. It is worth remarking that, in the plot corresponding to the altimetric field, there are big regions of very low arc length values (purple color in the figure), showing the limitation of the altimetric fields when one approaches the coast in comparison to the velocities derived from SST.

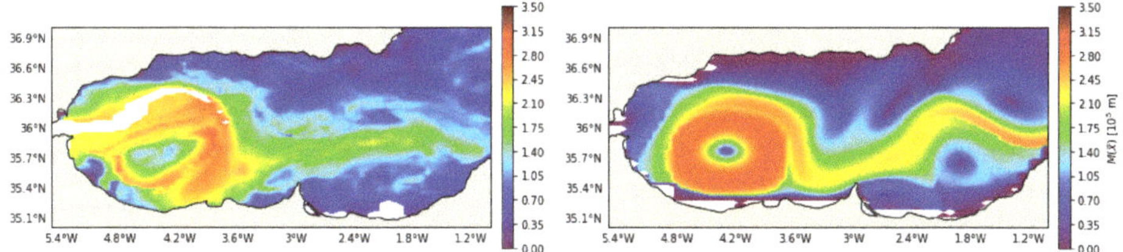

Figure 11. Arc length of the trajectory for the central time of the period analyzed for the velocity fields derived from: thermal images (**top**); and altimetry (**bottom**). The white area west of the Strait of Gibraltar corresponds to the point at which particles escape through the Strait of Gibraltar when integrating backwards.

5. Discussion

The circulation in the Alboran Sea is usually dominated by the presence of two large vortices that dominate the upper 100–300 m of the ocean [34,38]. These conditions are favorable to the exploitation of the eSQG approximation to retrieve surface current, as shown by LaCasce and Mahadevan [29] and confirmed by the results shown in Section 4. This opens the door to use SST to retrieve surface currents in this area, which allows resolving both the high frequency evolution of these vortices and the smaller scales structures that surround them [12]. On the contrary, Altimeters have spatial and temporal sampling limitations. Alboran vortices, however, are large enough to be properly sampled by altimeters. Consequently, both velocities derived from infrared radiometers and velocities derived from radar altimeters show similar capabilities when compared with the trajectories of drifting buoys. Nevertheless, there are some relevant exceptions: the area east of Gibraltar. This area is not properly sampled by altimeters and, therefore, velocities diagnosed in this zone are significantly different from those observed by drifting buoys. On the contrary, velocities derived from thermal images give velocities with directions relatively close to the observed by drifters ($\langle|\theta_\eta - \theta_d|\rangle = 40.2°$ while $\langle|\theta_T - \theta_d|\rangle = 19.3°$). The comparison between drifters and the geostrophic velocities derived from SST or SSH, however, has an important drawback: drifter trajectories have ageostrophic contributions not taken into account in the derivation of surface currents from SST nor SSH. In particular, drifters are affected by wind [39] and inertial oscillations as well as ageostrophic contributions such as the cyclostrophic flow associated to the curvature of the Alboran vortex. Furthermore, drifter speed in the area east of Gibraltar showed some temporal behavior that suggest a relevant contribution from tides.

Although thermal images and altimeter maps have similar performances when the flow is dominated by structures large enough to be sampled by altimeters, both approaches have different latencies, with that of currents derived from SST being of the order of few hours (typically less than 4 h) (This is the time passed between satellite measurement and the delivery of surface currents derived from them. Notice that the time needed to estimate currents from SST is of the order of the time needed to compute the FFT). Of course, velocities have to be calibrated and independent measurements such as the ones provided by altimeters are necessary; however, when only the energy level has to be fixed, simultaneous data are not needed. On the contrary, it requires some time to get accurate altimetric measurements and near-real time maps can only use past data, which further limits the capability to resolve two-dimensional structures if not enough altimeters are available [18]. Thermal images have, however, two main limitations. First, they can only be used under cloud-free situations, which restricts their applicability (particularly for those methods requiring consecutive cloud free images such optical flow and MCC). It is worth mentioning that the Mediterranean Sea is a quite favorable situation, with a relatively large fraction of cloud-free images ($\sim$40%). For moving clouds, gaps can be filled combining several images. One possible approach would be to compute vorticity from SST for several images, calibrate them, combine vorticities for a short enough period of time, and, finally, recompute the stream function inverting the vorticity.

A second limitation, as discussed above, is the need to correct for the impact of SSS. Presently, SSS measurements are not available in real time, although important advances have been done and it is already possible to detect the signature of the Atlantic waters in the Algerian basin [26] and the key patterns in the Alboran Sea with low resolutions. This implies that other approaches are needed. Here, we explored a simple method that consists of detecting Mediterranean waters from thermal images and modifing y buoyancy to take into account the effect of salinity. Although this approach is very crude, it shows that it is able to correct the role of salinity and provide results similar to those provided by altimeters. This raises a question: To what extent can the approach here proposed be used in other areas of the ocean? As shown in Section 3.3, there are two necessary conditions: the existence of water masses with salinities homogeneous enough to be considered constant and a way to identify such water masses from existing satellite observations. Multiple strategies are possible to identify water masses. In areas dominated by the presence of vortices of distinctive salinity, e.g., the Algerian basin [26], vortex identification techniques can be used [40,41]. In areas where SST univocally identifies water masses, temperature can be used to identify the extension of water masses. In other cases, image processing techniques combined with oceanographic knowledge can be applied to satellite observations of temperature or chlorophyll to segment the image. Future improvements could include the use of new SSS products (if they are available in real time), climatological SSS, or the exploitation of the buoyancy derived from altimetry to give a better estimation of density anomalies. However, these approaches require the implementation of accurate numerical methods that are beyond the scope of this study.

6. Conclusions

This study provides further evidence that the SQG approach is able to reconstruct velocities in the Alboran Sea. Moreover, the results show that altimetric measurements do s not capture the temporal evolution of the WAG while infrared measurements do, which can be used to derive surface velocities in this area, although the role of salinity can limit this approach. Nevertheless, the strong relationship between salinity and temperature allows identifying the water masses and correct for the contribution of salinity using image processing techniques. Once corrected, the SQG approach can retrieve ocean currents slightly better to that of near-real-time currents in general but much better in areas badly sampled by altimeters such as the area east of the Strait of Gibraltar. Although this area is far from the geostrophic equilibrium, the above results show that the good sampling capabilities of infrared radiometers allows at least retrieving the direction of ocean currents in this area. These results can be extended to other areas of the ocean, provided that water masses with relatively homogeneous salinity can be identified from remote sensing measurements.

Author Contributions: Conceptualization, J.I.-F. and E.G.-L.; Data curation, A.O. and M.G.-S.; Investigation, J.I.-F., E.G.-L., J.A.J.-M., A.O., E.O., M.G.-S. and A.T.; Methodology, J.I.-F., J.A.J.-M. and E.O.; Software, J.I.-F.; Supervision, J.I.-F.; Writing—original draft, J.I.-F.; Writing—review & editing, J.I.-F., E.G.-L. and J.A.J.-M. All authors have read and agreed to the published version of the manuscript.

Funding: This research was funded by European Space Agency grant number 4000109513/13/I-LG and by Spanish Ministry of Economy and Competitiveness and the European Regional Development Fund grants numbers ESP2015-67549-C3-1-R and CTM2016-79474-R. JIF was funded by Fundación General CSIC through Programa ComFuturo.

Acknowledgments: This work was funded by the European Space Agency through the GlobCurrent Data User Element project (4000109513/13/I-LG) and by the Spanish Ministry of Economy and Competitiveness, through the projects PROMISES (ESP2015-67549-C3-1-R), COSMO (CTM2016-79474-R) and the ERDF (European Regional Development Fund). Financial support by Fundación General CSIC (Programa ComFuturo) is also acknowledged. In situ measurements are available through the PANGAEA (Data Publisher for Earth and Environmental Science) repository, with the following doi:10.1594/PANGAEA.853701. Altimetric data was produced by the GlobCurrent Data User Element project and is available through its web site: http://www.globcurrent.org. Infrared images are available through the Insitut de Ciències del Mar (CSIC) at https://coo.icm.csic.es/site-page/satellite-data as well as Sea Surface Salinity Maps. We would like to thank the anonymous reviewers who help to improve the present manuscript.

Appendix A. Arc Length Of Trajectories

In this part, we explain the technical details to obtain the figures for the arc length of trajectories. We define the function M, which associates to each initial position $\vec{x}^*$ in the region the arc length of the trajectory that passes through $\vec{x}^*$ at the selected time t^* and then the trajectory is advected forward and backward during an integration time τ. More precisely, adapting the definition from [36] to the notation used here, we have the dynamical system

$$\frac{d\vec{x}}{dt} = \vec{v}. \tag{A1}$$

Let $\vec{j}(\vec{x}^*, t^*, t)$ denote a trajectory of the dynamical system in Equation (A1) which at time t^* passes through the point $\vec{x}^*$. Then, for any initial condition $\vec{x}^*$ in an open set included in the area of study, we define the function $M(\vec{x}^*)_{t^*,\tau}$ as

$$M(\vec{x}^*)_{t^*,\tau} = \int_{t^*-\tau}^{t^*+\tau} \left[\left(\frac{dj_u}{dt}\right)^2 + \left(\frac{dj_v}{dt}\right)^2 \right]^{\frac{1}{2}} dt, \tag{A2}$$

where $j_u(\vec{x}^*, t^*, t)$ and $j_v(\vec{x}^*, t^*, t)$ are the zonal and meridional components of $\vec{j}(\vec{x}^*, t^*, t)$, respectively. $M(\vec{x}^*)_{t^*, \tau}$ is a function that associates to each initial condition $\vec{x}^*$ the arc length of the trajectory that at time t^* passes through $\vec{x}^*$. The trajectory is computed in the time interval $[t^* - \tau, t^* + \tau]$.

To perform the numerical integration of Equation (A1) for advecting particles, a fifth-order Runge–Kutta was used. To prevent numerical artifacts getting the velocity values at the untabulated points within the grid cell, we utilized bicubic interpolation in space and third-order Lagrange polynomials in time. This choice of interpolating methods is recommended in [42] to get accurate results at a moderate computational cost.

References

1. Mercado, J.M.; Cortés, D.; García, A.; Ramírez, T. Seasonal and inter-annual changes in the planktonic communities of the northwest Alboran Sea (Mediterranean Sea). *Prog. Oceanogr.* **2007**, *74*, 273–293. [CrossRef]
2. Huertas, I.; Ríos, A.; García-Lafuente, J.; Navarro, G.; Makaoui, A.; Sánchez-Román, A.; Rodriguez-Galvez, S.; Orbi, A.; Ruíz, J.; Pérez, F.F. Atlantic forcing of the Mediterranean oligotrophy. *Glob. Biogeochem. Cycle* **2012**, *26*, GB2022. [CrossRef]
3. Lo Iacono, C.; Gràcia, E.; Ranero, C.; Emelianov, M.; Huvenne, V.; Bartolomé, R.; Booth-Rea, G.; Prades, J. The West Melilla cold water coral mounds, Eastern Alboran Sea: Morphological characterization and environmental context. *Deep Sea Res. Part II Top. Stud. Oceanogr.* **2014**, *99*, 316–326. [CrossRef]
4. REMPEC. *Study of Maritime Traffic Flows in the Mediterranean Sea. Regional Marine Pollution Emergency Response Centre for the Mediterranean Sea (REMPEC)*; Technical Report; Lloyd's Marine Intelligence Unit: Valletta, Malta, 2008.
5. Lacombe, H. Contribution à l'étude du détroit de Gibraltar. I. Etude dynamique. *Cahiers Ocdanographiques* **1961**, *XlII*, 73–107.
6. Stommel, H.; Bryden, H.; Mangelsdorf, P. Does some of the Mediterranean outflow come from great depth? *Pure Appl. Geophys.* **1973**, *105*, 879–889. [CrossRef]
7. Gascard, J.; Richez, C. Water masses and circulation in the western Alborán Sea and in the Straits of Gibraltar. *Prog. Oceanogr.* **1985**, *15*, 157–216. [CrossRef]
8. Sánchez-Garrido, J.; García Lafuente, J.; Álvarez Fanjul, E.; García Sotillo, M.; de los Santos, F. What does cause the collapse of the Western Alboran Gyre? Results of an operational ocean model. *Prog. Oceanogr.* **2013**, *116*, 142–153. [CrossRef]
9. Bryden, H.; Kinder, T. Steady two-layer exchange through the Strait of Gibraltar. *Deep Sea Res. Part A* **1991**, *38*, S445–S463. [CrossRef]
10. Naranjo, C.; Garcia-Lafuente, J.; Sannino, G.; Sanchez-Garrido, J. How much do tides affect the circulation of the Mediterranean Sea? From local processes in the Strait of Gibraltar to basin-scale effects. *Prog. Oceanogr.* **2014**, *127*, 108–116. [CrossRef]
11. García-Lafuente, J.; Naranjo, C.; Sammartino, S.; Sánchez-Garrido, J.; Delgado, J. The Mediterranean Outflow in the Strait of Gibraltar and its connection with upstream conditions in the Alborán Sea. *Ocean Sci. Discuss* **2016**, in review. [CrossRef]
12. Tintoré, J.; Gomis, D.; Alonso, S.; Parrilla, G. Mesoscale Dynamics and Vertical Motion in the Alborán Sea. *J. Phys. Oceanogr.* **1991**, *21*, 811–823. [CrossRef]
13. Viúdez, A.; Haney, R.L.; Vazquez-Cuervo, J. The deflection and division of an oceanic baroclinic jet by a coastal boundary: A case study in the Alboran sea. *J. Phys. Oceanogr.* **1998**, *28*, 289–308. [CrossRef]
14. Renault, L.; Oguz, T.; Pascual, A.; Vizoso, G.; Tintore, J. Surface circulation in the Alborán Sea (western Mediterranean) inferred from remotely sensed data. *J. Geophys. Res. Ocean.* **2012**, *117*, C08009. [CrossRef]
15. Peliz, A.; Boutov, D.; Teles-Machado, A. The Alboran Sea mesoscale in a long term high resolution simulation: Statistical analysis. *Ocean Model.* **2013**, *72*, 32–52. [CrossRef]
16. Sotillo, M.G.; Garcia-Ladona, E.; Orfila, A.; Rodríguez-Rubio, P.; Maraver, J.C.; Conti, D.; Padorno, E.; Jiménez, J.A.; Capó, E.; Pérez, F.; et al. The MEDESS-GIB database: Tracking the Atlantic water inflow. *Earth Syst. Sci. Data* **2016**, *8*, 141–149. [CrossRef]

17. Sotillo, M.; Amo-Baladrón, A.; Padorno, E.; Garcia-Ladona, E.; Orfila, A.; Rodríguez-Rubio, P.; Conti, D.; Madrid, J.J.; de los Santos, F.; Fanjul, E.A. How is the surface Atlantic water inflow through the Gibraltar Strait forecasted? A lagrangian validation of operational oceanographic services in the Alboran Sea and the Western Mediterranean. *Deep Sea Res. Part II Top. Stud. Oceanogr.* **2016**, *133*, 100–117. [CrossRef]

18. Pascual, A.; Faugère, Y.; Larnicol, G.; Le Traon, P. Improved description of the ocean mesoscale variability by combining four satellite altimeters. *Geophys. Res. Lett.* **2006**, *33*, L02611. doi:10.1029/2005GL024633. [CrossRef]

19. Isern-Fontanet, J.; Ballabrera-Poy, J.; Turiel, A.; García-Ladona, E. Remote sensing of ocean surface currents: A review of what is being observed and what is being assimilated. *Nonlinear Process. Geophys.* **2017**, *24*, 613–643. [CrossRef]

20. Held, I.; Pierrehumbert, R.; Garner, S.; Swanson, K. Surface quasi-geostrophic dynamics. *J. Fluid Mech.* **1995**, *282*, 1–20. [CrossRef]

21. Lapeyre, G. Surface Quasi-Geostrophy. *Fluids* **2017**, *2*, 7. [CrossRef]

22. Lapeyre, G.; Klein, P. Dynamics of the Upper Oceanic Layers in Terms of Surface Quasigeostrophy Theory. *J. Phys. Oceanogr.* **2006**, *36*, 165–176. [CrossRef]

23. Isern-Fontanet, J.; Chapron, B.; Klein, P.; Lapeyre, G. Potential use of microwave SST for the estimation of surface ocean currents. *Geophys. Res. Lett.* **2006**, *33*, L24608. [CrossRef]

24. Isern-Fontanet, J.; Shinde, M.; González-Haro, C. On the transfer function between surface fields and the geostrophic stream function in the Mediterranean sea. *J. Phys. Ocean* **2014**, *44*, 1406–1423. [CrossRef]

25. González-Haro, C.; Isern-Fontanet, J. Reconstruction of global surface currents from passive microwave radiometers. *J. Geophys. Res.* **2014**, *119*. [CrossRef]

26. Isern-Fontanet, J.; Olmedo, E.; Turiel, A.; Ballabrera-Poy, J.; García-Ladonaía-Ladona, E. Retrieval of eddy dynamics from SMOS sea surface salinity measurements in the Algerian Basin (Mediterranean Sea). *Geophys. Res. Lett.* **2016**, *43*. [CrossRef]

27. Ponte, A.; Klein, P. Reconstruction of the upper ocean 3D dynamics from high-resolution sea surface height. *Ocean Dyn.* **2013**, *63*, 777–791. [CrossRef]

28. Isern-Fontanet, J.; Lapeyre, G.; Klein, P.; Chapron, B.; Hetcht, M. Three-dimensional reconstruction of oceanic mesoscale currents from surface information. *J. Geophys. Res.* **2008**, C09005. [CrossRef]

29. LaCasce, J.; Mahadevan, A. Estimating subsurface horizontal and vertical velocities from sea surface temperature. *J. Mar. Res.* **2006**, *64*, 695–721. [CrossRef]

30. Davis, R.E. Drifter observations of coastal surface currents during CODE: The statistical and dynamical views. *J. Geophys. Res. Ocean.* **1985**, *90*, 4756–4772. [CrossRef]

31. Arbic, B.; Scott, R.; Chelton, D.; Richman, J.; Shriver, J. Effects of stencil width on surface ocean geostrophic velocity and vorticity estimation from gridded satellite altimeter data. *J. Geophys. Res. Ocean.* **2012**, *117*, C03029. [CrossRef]

32. Olmedo, E.; Taupier-Letage, I.; Turiel, A.; Alvera-Azcárate, A. Improving SMOS Sea Surface Salinity in the Western Mediterranean Sea through Multivariate and Multifractal Analysis. *Remote Sens.* **2018**, *10*. [CrossRef]

33. Isern-Fontanet, J.; Hascoët, E. Diagnosis of high resolution upper ocean dynamics from noisy sea surface temperature. *J. Geopys. Res.* **2014**, *118*, 1–12. [CrossRef]

34. Viúdez, A.; Tintoré, J.; Haney, R.L. Circulation in the Alboran sea as determined by quasi-synoptic hydrographic observations. Part I: Three-dimensional structures of two anticyclonic gyres. *J. Phys. Oceanogr.* **1996**, *26*, 684–705. [CrossRef]

35. Millot, C. Circulation in the Western Mediterranean Sea. *J. Mar. Syst.* **1999**, *20*, 423–442. [CrossRef]

36. Jiménez Madrid, J.A.; Mancho, A.M. Distinguished trajectories in time dependent vector fields. *Chaos: Interdiscip. J. Nonlinear Sci.* **2009**, *19*, 013111. [CrossRef] [PubMed]

37. Mendoza, C.; Mancho, A.M. Review Article: "The Lagrangian description of aperiodic flows: A case study of the Kuroshio Current". *Nonlinear Process. Geophys.* **2012**, *19*, 449–472. [CrossRef]

38. Tintore, J.; Violette, P.E.L.; Blade, I.; Cruzado, A. A Study of an Intense Density Front in the Eastern Alboran Sea: The Almeria–Oran Front. *J. Phys. Oceanogr.* **1988**, *18*, 1384–1397. [CrossRef]

39. Poulain, P.M.; Gerin, R.; Mauri, E.; Pennel, R. Wind Effects on Drogued and Undrogued Drifters in the Eastern Mediterranean. *J. Atmos. Ocean. Technol.* **2009**, *26*, 1144–1156. [CrossRef]

40. Isern-Fontanet, J.; García-Ladona, E.; Font, J. Identification of Marine eddies from Altimetry. *J. Atmos. Ocean. Technol.* **2003**, *20*, 772–778. [CrossRef]

41. Isern-Fontanet, J.; García-Ladona, E.; Font, J. The vortices of the Mediterranean sea: An altimetric perspective. *J. Phys. Oceanogr.* **2006**, *36*, 87–103. [CrossRef]

42. Mancho, A.M.; Small, D.; Wiggins, S. A comparison of methods for interpolating chaotic flows from discrete velocity data. *Comput. Fluids* **2006**, *35*, 416–428. [CrossRef]

 *remote sensing*

Article

A Synergetic Approach for the Space-Based Sea Surface Currents Retrieval in the Mediterranean Sea

Daniele Ciani [1,*], Marie-Hélène Rio [2], Milena Menna [3] and Rosalia Santoleri [1]

[1] Consiglio Nazionale delle Ricerche, Istituto di Scienze Marine (CNR-ISMAR),
    Via del Fosso del Cavaliere 100, 00133 Rome, Italy; rosalia.santoleri@cnr.it
[2] European Space Agency, European Space Research Institute (ESA-ESRIN), Largo Galileo Galilei, 1,
    00044 Frascati, Italy; marie-helene.rio@esa.int
[3] Istituto Nazionale di Oceanografia e Geofisica Sperimentale (OGS), Borgo Grotta Gigante 42/C,
    34010 Sgonico, Trieste, Italy; mmenna@inogs.it
* Correspondence: daniele.ciani@cnr.it

Received: 22 April 2019; Accepted: 24 May 2019; Published: 30 May 2019

Abstract: We present a method for the remote retrieval of the sea surface currents in the Mediterranean Sea. Combining the altimeter-derived currents with sea-surface temperature information, we created daily, gap-free high resolution maps of sea surface currents for the period 2012–2016. The quality of the new multi-sensor currents has been assessed through comparisons to other surface-currents estimates, as the ones obtained from drifting buoys trajectories (at the basin scale), or HF-Radar platforms and ocean numerical model outputs in the Malta–Sicily Channel. The study yielded that our synergetic approach can improve the present-day derivation of the surface currents in the Mediterranean area up to 30% locally, with better performances for the the meridional component of the motion and in the western section of the basin. The proposed reconstruction method also showed satisfying performances in the retrieval of the ageostrophic circulation in the Sicily Channel. In this area, assuming the High Frequency Radar-derived currents as reference, the merged multi-sensor currents exhibited improvements with respect to the altimeter estimates and numerical model outputs, mainly due to their enhanced spatial and temporal resolution.

Keywords: surface currents; mediterranean sea; satellite altimetry; sea surface temperature

1. Introduction

The monitoring of the oceanic surface currents is a major scientific and socio-economic challenge. The ocean currents modulate natural and anthropogenic processes at different space and time scales, from global climate change to local dispersal of tracers and pollutants, with relevant impacts on marine ecosystem services and maritime activities (e.g., optimization of the ship routes, maritime safety, coastal protection). An appropriate monitoring of the oceanic currents necessitates high frequency and high resolution observations of the global ocean, which are achieved using satellite measurements. Nowadays, no satellite mission provides a direct measurement of the sea surface currents and their global-to-regional scale monitoring is mainly provided by satellite altimetry. Being based on the observation of the sea surface height (SSH), satellite altimetry reconstructs the large-scale geostrophic component of the surface motions at an operational level. Indeed, the altimeter-derived currents have spatio-temporal spectral responses from the mesoscale to the basin-scale range [1], corresponding to temporal scales from some weeks to several years. This is not sufficient for many applications, especially in semi-enclosed basins as the Mediterranean Sea, where the most energetic variable signals are found at relatively small scales [2,3]. In fact, the synoptic retrieval of the sea surface dynamics in the mesoscale-to-submesoscale range, i.e., at spatial resolutions of a few kilometres and temporal resolution of a few days, is a crucial topic in physical oceanography. The characterization of the surface processes at these scales has significant impacts on the human activities in the marine context [4,5] and can also provide information on the interior dynamics, e.g., [6]. This can be obtained either by conceiving new satellite sensors and/or by optimizing the present-day space-based observations.

For instance, the incoming NASA/CNES Surface Water and Ocean Topography Satellite (SWOT) mission aims at retrieving the global SSH at 15 km horizontal resolution, providing improvements for the retrieval of the surface geostrophic flow (though the temporal coverage of the global ocean will be around 20 days) [7]. In addition, the potential observations from the sea surface KInematics Multiscale monitoring mission (SKIM, presently in competitive feasibility phase to be the ESA-EE9) aims at providing wide-swath measurements of the sea surface currents at 40 km horizontal resolution, going down to a few kilometers for along-swath data, and errors even lower than 0.1 m·s^{-1} [8,9].

On the other hand, a number of approaches are based on the exploitation of the satellite observations in the radio, thermal or optical band, today available at horizontal resolutions up to 1 km (or higher), e.g., [10,11]. For example, the Surface Quasi Geostrophic theory (SQG) relies on the assumption that the surface motions are mostly driven by the sea surface density gradients. This technique proved successful at reconstructing the surface velocity field from single sea-surface temperature (SST) bi-dimensional images [12] and also includes the possibility of deriving the 3-D structure of the current field from surface observations [6,13]. An SQG based approach was also exploited combining the phase of SST and the amplitude of SSH observations, allowing to implement one of the first applications for the global ocean currents retrieval from the synergy of SSH and SST data [14,15]. One requirement of this approach is the condition that SST and SSH patterns are in phase, which is mostly verified for deep Mixed Layer conditions. Moreover, the SQG is based on the assumption that surface density gradients are mostly driven by SST. In reality, the sea-surface salinity (SSS) can also regulate the surface dynamics by compensating the SST anomalies. In the Mediterranean Sea this was observed in the Algerian basin by Isern-Fontanet et al. 2016 [16] and, at global scale, this condition was reported in 0.12% of the cases based on a three years global scale statistics [14].

The radio, thermal and optical-band satellite observations can be exploited for the analysis of consecutive cloud-free two-dimensional images, e.g., sea surface temperature (SST) and surface Chlorophyll concentration (Chl). For instance, the images can be used to infer the surface currents field via maximum cross correlation techniques or inverting the surface tracer dynamical evolution equation [17–20]. However, these approaches only account for the currents horizontal advection and diffusion, neglecting the effects of the source and sink terms for the tracer evolution. Given a surface tracer distribution, these techniques satisfactorily retrieve the cross-gradient velocity field but the along-gradient component cannot be directly inferred and no information can be retrieved in low tracer gradient conditions. In the past, this limitation was overcome by assuming the existence of stochastic forcings for the statistical derivation of unknown parameters [21]. More recently, Piterbarg 2009 (PIT09 hereinafter) [22] solved this issue via an optimal combination of the tracer information with the one of a background velocity. The PIT09 method, whose dynamical framework considers both the effects of advection and of the source/sink terms on the tracer evolution, was successfully applied to an oil spill monitoring case study [23], in which the background velocities were derived from a numerical simulation. In [23], the authors pointed out the potential of using their method with satellite-derived tracers, as SST or Chl, as the main perspective of their results. A first confirmation of their findings appeared recently. Indeed, Rio et al. 2016 [24] showed the potentiality of combing altimetric and thermal-band satellite observations for improving the oceanic surface currents retrieval (nowadays provided by satellite altimetry) at global scale. This was proved in a model study, via an Observing System Simulation Experiment (OSSE) based on one year of currents data simulated via the MERCATOR operational model [25]. The authors showed that, if the large-scale geostrophic currents are combined with the information coming from the SST, the average global improvements in the currents retrieval can reach 35% locally. The method was further successfully applied to satellite altimetry and SST observations (Rio and Santoleri 2018 (RS18 hereinafter)) [26].

The objective of our study is to further exploit the RS18 method to tackle a more challenging application, i.e., we aim at merging the geostrophic currents (derived from sea surface height, SSH) with the SST observations in the Mediterranean Sea. Indeed, the Mediterranean is characterized by Rossby deformation radii that can go down to 10–20 km, hence, its typical mesoscale features ($\mathcal{O}(10$–100 km)) are only partially captured by classical satellite altimetry [3]. Thus, we attempt to quantify the potential of the RS18 method in unvealing the Mediterranean Sea small scale and ageostrohic circulation from space observations, which is not possible using the altimeter estimates alone. In addition, the interest for improving the surface currents retrieval in the Mediterranean has commercial and environmental implications. In fact, this region represents the main ship route between the Indian Ocean and the main harbours of the European Union and hosts 25% of the world oil trade, with a consequent presence of illegal oil spills of the order of 600 kt per year [27]. This makes high-resolution surface currents a necessary information for monitoring the Mediterranean Sea and improving the knowledge of its surface dynamics.

The aim of this paper is to describe the derivation of the merged SSH/SST currents (Optimal currents hereinafter) and to illustrate their assessment via comparison with several in-situ and model derived surface currents. At the same time, the capability of retrieving the small scale ageostrophic circulation from satellite observations is demonstrated. The paper is structured as follows. In Section 2 we present the datasets involved in our study, i.e., used for the determination and the quality assessment of the Optimal currents. Then, Section 3 details the computation of the optimal currents. In Section 4 we will discuss the Optimal currents performances in two test-cases in the western Mediterranean and the Sicily Channel. We go on assessing the quality of the optimal currents via quantitative comparisons with model as well as in-situ derived surface currents and, in the final section, we will give the main conclusions and perspectives of the study.

2. Data

The following datasets have been used to derive and assess the quality of the optimal currents:

1. The CMEMS (Copernicus Marine Environment Monitoring Service) operational delayed-time sea surface geostrophic velocities for the Mediterranean Sea derived from satellite altimetry [28]. These are gap-free daily data with nominal $1/8°$ horizontal resolution computed by the Sea-Level Thematic Assembly Center (TAC). The 2012–2016 time series was extracted. CMEMS data ID: SST-MED-SST-L4-REP-OBSERVATIONS-010-021;

2. The CMEMS operational reprocessed SST for the Mediterranean Sea [29,30]. These are gap-free daily data with nominal $1/24°$ horizontal resolution computed by the SST-TAC. The 2012–2016 time series was extracted. CMEMS data ID: SEALEVELMED-PHY-L4-REP-OBSERVATIONS-008-051;

3. The CMEMS global operational sea surface currents provided by the MERCATOR model (first output at 0.49 m depth) [25]. These are daily data with nominal $1/12°$ horizontal resolution (data available from January 2016 to present). CMEMS data ID: GLOBAL-ANALYSIS-FORECAST-PHY-001-024;

4. The CMEMS near-surface currents given by the operational forecasting model for the Mediterranean Sea, i.e., the Mediterranean Forecasting System (first output at 1.47 m depth) [31]. These are daily data with nominal $1/24°$ horizontal resolution. The data are available from January 2016 to present. CMEMS data ID: MEDSEA-ANALYSIS-FORECAST-PHY-006-013;

5. The EMODNet (European Marine Observation and Data Network) surface currents in the Malta-Sicily channel given by the operational HF-RADAR observing system (CALYPSO Project, http://oceania.research.um.edu.mt/cms/calypsoweb/). These constitute hourly data with $1/37°$ horizontal resolution. The radar observing system is given by three antennas, two of which are placed on the Malta and Gozo Islands and a third one is located in the southern Sicily coast (Pozzallo) (see also Figure 12 [32]. The data are available from 2016 to present;

6. The OGS (Istituto Nazionale di Oceanografia e di Geofisica Sperimentale) drifting buoys-derived sea surface currents and SST in the Mediterranean Basin. These lagrangian observations have six-hourly temporal resolution and the dataset coverage spans from 1986 to late 2016 (data doi=10.6092/7a8499bc-c5ee-472c-b8b5-03523d1e73e9);

7. The CNR (Consiglio Nazionale delle Ricerche) Sea Forecast operational model for the computation of the three-dimensional surface currents, temperature, salinity, SSH, heat&water fluxes and wind stress in the Central Mediterranean (Tyrrhenian Sea and Sicily Channel). We were kindly provided with the monthly averaged data with nominal 1/48° horizontal resolution by the G3O-Operational Oceanography Group. More details available at http://www.seaforecast.cnr.it/forecast/index.php/en/forecast/sicily-strait-region/);

8. The CMEMS Global Blended Mean Wind Fields. These are gap-free 6-hourly global data with 1/4° horizontal resolution computed at CERSAT/Ifremer. The data are available from January 2015 to present [33]. CMEMS data ID: WIND-GLO-WIND-L4-NRT-OBSERVATIONS-012-004;

9. Two datasets of Sea Surface Salinity (SSS):

- The daily, 1/20° maps of L4 SSS obtained from the multifractal fusion applied to the L3 Soil Moisture and Ocean Salinity (SMOS) observations. The data are distributed by the Barcelona Expert Center (BEC) [34]. The data are available from 2011 to 2016;
- The 4-daily, 1/4° LOCEAN L3 SSS estimates derived from a combination of SMOS ascending and descending orbits measurements [35] and are available from 2010 to 2017.

In the list given above, the datasets labeled with 1 and 2 will be combined according to the method described in PIT09 to obtain the Optimal sea surface currents in the period 2012–2016. The Optimal currents will constitute a series of daily gap-free sea surface currents for the Mediterranean basin mapped onto a regular 1/24° grid (more details will be given in Section 3). On the other hand, the datasets labeled with numbers from 3 to 9 will be used for the quality assessment of the Optimal Currents, as we will show in Section 4.

3. Methods

3.1. Derivation of the Optimal Currents

Following PIT09 and RS18, we derive the surface circulation in the Mediterranean Sea combining the CMEMS geostrophic velocities (GV hereinafter) and the SST fields. The Optimal reconstruction is based on the inversion of the ocean heat conservation equation in the mixed layer. The equation, indicated below, is inverted for inferring information on the sea surface currents:

$$\frac{\partial SST}{\partial t} + u\frac{\partial SST}{\partial x} + v\frac{\partial SST}{\partial y} = F \tag{1}$$

In Equation (1), (u,v) are respectively the zonal and meridional component of the ocean surface flow, (x,y) are the zonal and the meridional directions and F is the forcing term, containing the source and sink terms for the heat conservation equation. The resulting expressions for the optimally reconstructed velocity (u_{opt}, v_{opt}) are the following:

$$u_{opt} = u_{bck} + u_0\sin\phi + v_0\cos\phi = u_{bck} + u_{COR}$$
$$v_{opt} = v_{bck} - u_0\cos\phi + v_0\sin\phi = v_{bck} + v_{COR} \tag{2}$$

The Optimal currents are obtained adding the correction factors u_{COR} and v_{COR} to the background ("bck") geostrophic velocities. In practice, Equation (2) is the formal expression of the synergy between the altimeter-derived SSH and the SST observations for the sea surface currents reconstruction. We provide a brief illustration of the terms appearing in the correction factors of Equation (2), referring the reader to PIT09 for more details. The correction velocities u_0 and v_0 have the following expressions:

$$u_0 = \frac{f(\min(\beta, q)) - f(\max(\alpha, -q))}{g(\min(\beta, q)) - g(\max(\alpha, -q))}$$

$$v_0 = pu_0 \tag{3}$$

In addition, ϕ is a function of the SST spatial derivatives, as expressed by Equation (4):

$$\phi = \text{atan}(A/B) \tag{4}$$

where $A = \partial_x SST$, $B = \partial_y SST$ and (x,y) are the generic zonal and meridional coordinates. Moreover, in Equation (3) the functions f, g (expressed as functions of a generic variable γ) and the coefficients p, q, α and β are given by the set of Equation (5):

$$f(\gamma) = -2(q^2 - \gamma^2)^{3/2}/3;$$
$$g(\gamma) = x(q^2 - \gamma^2) + q^2 \text{asin}(\gamma/q);$$
$$p = \sin\phi\cos\phi(\sigma^2_v - \sigma^2_u)q^{-2};$$
$$q = \sqrt{\sigma^2_u \sin^2\phi + \sigma^2_v \cos^2\phi}; \tag{5}$$
$$\alpha = (Au_{bck} + Bv_{bck} + E - h)/\sqrt{A^2 + B^2};$$
$$\beta = (Au_{bck} + Bv_{bck} + E + h)/\sqrt{A^2 + B^2};$$

where h is the error associated to the forcing term F; σ_u, σ_v are respectively the errors for the zonal and meridional background velocities, and E depends on the temporal derivatives of SST, i.e., $E = \partial_t SST\text{-}F$. The determination of the input parameters included in Equations (1)–(5) is illustrated in Sections 3.2 and 3.3.

3.2. The Forcing Term F and the Computation of the Associated Error h

In RS18, the forcing term *F* was approximated as the satellite-derived, low-pass filtered SST temporal derivatives, only retaining spatial scales larger than 500 km. This choice is due to the following two facts:

- The lack of long and high-resolution time series of observed forcing terms estimates;
- The forcing term *F* of the SST heat conservation equation includes the vertical advection, the diffusion, the atmospheric heat fluxes and the entrainment velocity. At the first order, the main contributors to *F* are the atmospheric heat fluxes. In turn, the fluxes can be estimated as the large scales of the SST temporal derivatives, the smaller scales being related to the currents advection. This is also consistent with the analyses described in Rio et al. 2016 [24], who analyzed one year of MERCATOR model outputs at global scale (see, e.g., Figure 10 of Rio et al. 2016).

For the case of the Mediterranean basin, F was identified as the satellite SST temporal derivatives, $\partial_t SST$, low-pass filtered at 400 km. This was achieved filtering the $\partial_t SST$ at several spatial scales and choosing the one that minimized the in-situ forcing estimates obtained with the OGS drifter-derived observations.

Using $\simeq$ 30 years of drifting buoys-derived SST in the Mediterranean basin (data doi = 10.6092/7a8499bc-c5ee-472c-b8b5-03523d1e73e9), we computed an in-situ derived forcing term (F_{dri}), whose formal expression is given by Equation (6).

$$d_t SST_{ins} = \frac{\partial SST_{ins}}{\partial t} + u\frac{\partial SST_{ins}}{\partial x} + v\frac{\partial SST_{ins}}{\partial y} + w\frac{\partial SST_{ins}}{\partial z} = F_{ins} \tag{6}$$

where the subscripts "ins" and "t" respectively stand for in-situ measured and derivative with respect to time.

For every drogued buoy trajectory, we evaluated the in-situ daily $d_t$sst as

$$d_tSST_{ins} = [SST_{ins}(t + dT/2) - SST_{ins}(t - dT/2)]/dT$$

where dT = 1 day. Moreover, taking advantage of the 1986 to 2017 satellite-derived SST (dataset labeled with "2" in Section 2) we could compute the daily remotely sensed $\partial_t SST_{sat}$, including the filtered maps at scales ranging from 50 to 600 km. This operation enabled to derive the quantity L-$\partial_t SST_{sat}$, with L $\in$ [50 km , 600 km]. Afterwards, each of the filtered L-$\partial_t SST_{sat}$ was spatially and temporally colocated with respect to the drifters estimates, i.e., our benchmark. This allowed to generate two-dimensional maps of Root Mean Square Errors (RMSE) between the satellite and the in-situ-derived SST temporal derivatives. In the end, computing the mean, basin-scale RMSE between each of the L-$\partial_t SST_{sat}$ and the $d_t SST_{dri}$, we found that for L = 400 km such RMSE reached a minimum (Figure 1) letting us identify it as the optimal filtering scale. In addition, we assumed that the RMSE error map between the 400 km-D$_t SST_{sat}$ and the $\partial_t SST_{dri}$ is an estimate of the forcing term error h appearing in Equation (2). A map of h, binned in $1° \times 1°$ boxes, is given in Figure 2.

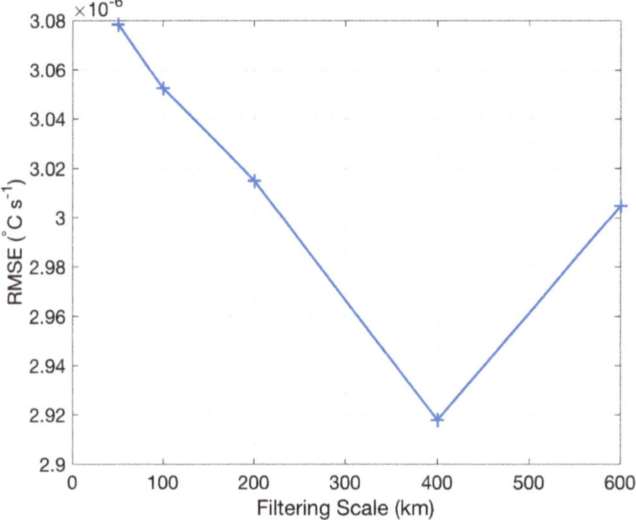

Figure 1. RMSE between the satellite and the in-situ derived forcing term as a function of the $D_t SST_{sat}$ filtering scale.

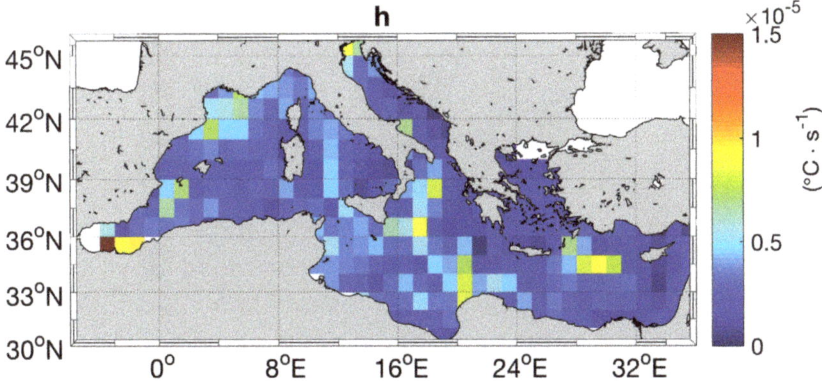

Figure 2. RMSE between F and F_{dri}, binned on $1° \times 1°$ boxes.

The order of magnitude of the RMSE is 10^{-6} °C·s^{-1} over most of the Mediterranean Sea and its spatial variability is mostly correlated with the major dynamical features in the Mediterranean Sea (e.g., the Algerian current, the Liguro-Provenzal, the Atlantic Ionian Stream and Western Adriatic current) where the values can also exceed 1.5 °C·s^{-1}.

3.3. The Uncertainties on the Geostrophic Velocities (σ_u, σ_v)

Another input parameter for the reconstruction method is the error on both the components of the surface geostrophic currents, previously defined as σ_u and σ_v for the zonal and meridional components of the current, respectively. This was estimated in a similar manner as for the forcing term, via comparison with the in-situ derived surface currents in the Mediterranean basin (data doi = 10.6092/7a8499bc-c5ee-472c-b8b5-03523d1e73e9). Using 24 years of altimeter-derived GV (introduced in Section 2) we could colocate the GV along the buoys trajectories and generate two-dimensional maps of the RMSE between the satellite and the in-situ measured currents. This was achieved selecting all the buoys evolving in presence of their drogue and binning the RMSE values in $1° \times 1°$ boxes, as illustrated in Figure 3.

The basin-scale mean error on the geostrophic velocities is of the order of 10 cm·s^{-1} for both the components of the motion and it is slightly larger for the meridional component than for the zonal component of the motion (10.5 cm·s^{-1} versus 9.6 cm·s^{-1}, respectively). The maximum RMSE values can exceed 20 cm·s^{-1} in correspondence of the Alboran gyres for both the surface flow components and their spatial variability is also correlated with the mean positions of the major currents in the basin. Indeed, the largest RMSE values ($\geq$15 cm·s^{-1}) are found in correspondence of the Algerian, Liguro-Provenzal, Libyo-Egyptian and Cilician currents as well as the Atlantic Ionian Stream [3].

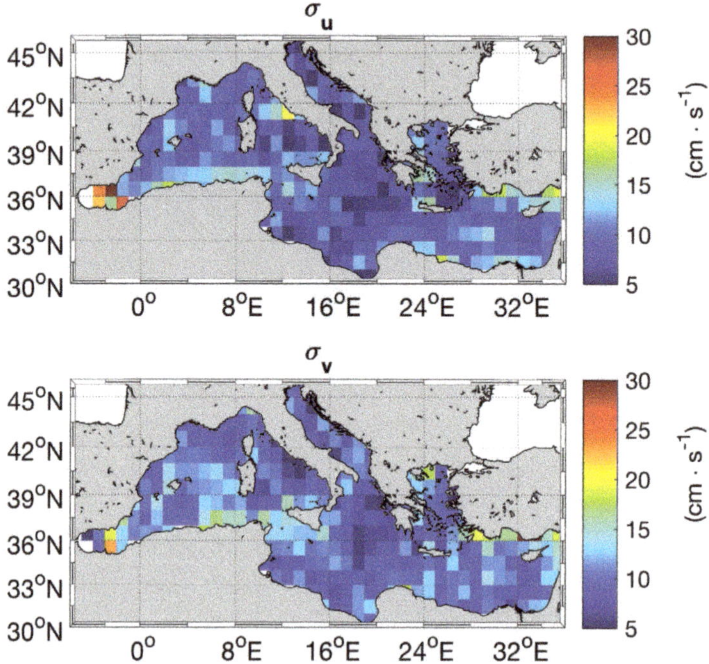

Figure 3. RMSE between the altimeter and the drifter-derived surface currents, binned on $1° \times 1°$ boxes.

4. Results

Relying on the CMEMS altimeter-derived GV, SST and on the error fields described in the previous section, we implemented the RS18 method to generate five years of Optimal currents in the Mediterranean Sea (from 2012 to 2016), a time window compatible with all the datasets used in the study. Such currents have spatial temporal resolution of $1/24°$ and 1 day, respectively. In the following sections we discuss the Optimal currents performances with respect to the geostrophic estimates, mainly via validation against in-situ or model-derived data.

4.1. Case Study 1: Vortex Dynamics

Here we give a first example of the Optimal Currents capabilities in presence of an oceanic vortex (generally referred to as eddy in the oceanographic community). An anticyclonic eddy located halfway between the Balearic Islands and Sardinia is detected on 25 April 2016 and is recognisable by a ring-shaped SST anomaly centered at approximately $39°45'N$ and $5°50'E$ and is associated to a clockwise circulation (Figure 4).

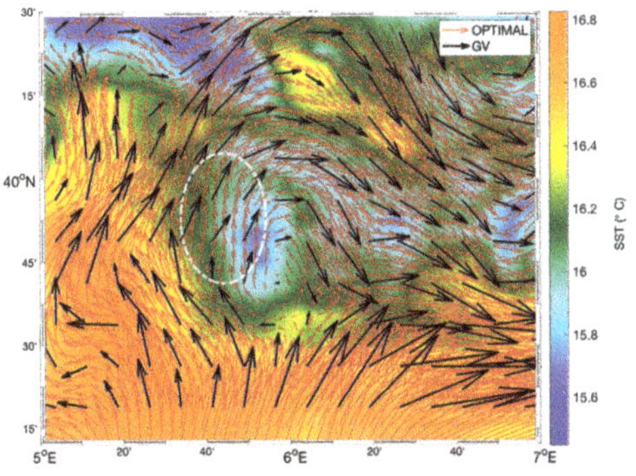

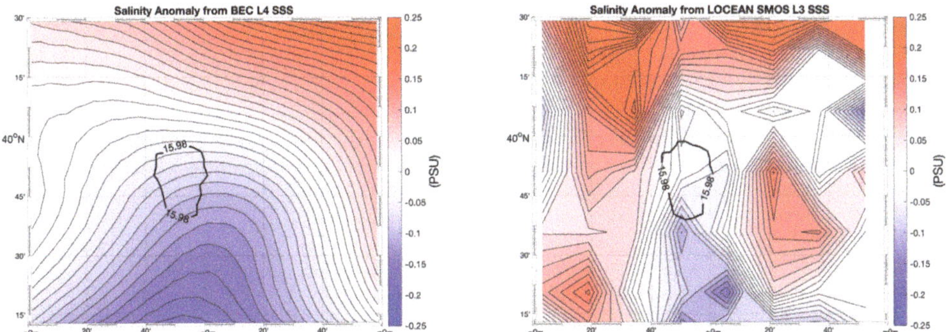

Figure 4. Top: Comparison between the Optimal currents (red arrows) and the GV (black arrows) in presence of anticyclonic circulation in the Western Mediterranean (25th of April 2016). The surface currents fields are superimposed to the SST (in colours, °C); the white ellipse is referenced in Section 4.1. Bottom left: Soil moisture and ocean salinity (SMOS)-derived sea-surface salinity (SSS) salinity anomaly from Barcelona Expert Center (25th of April 2016). The black closed contour is the 15.9 °C level indicating the position of the vortex. Bottom right: Same as bottom left, using the LOCEAN L3 SMOS SSS data, on the 26th of April 2016 .

Cold SST anomalies in presence of surface-intensified anticyclones have already been reported in literature [36,37]. In the Mediterranean sea, they can result from SSS compensation mechanisms, as also described in [16]. Using the Barcelona Expert Center SMOS L4 SSS for the Mediterranean Sea as well as the LOCEAN L3 SSS estimates (the latter being available on 26 April 2016), we found that the eddy lies in a negative salinity anomaly area. This is confirmed by the lower panels of Figure 4. Here, the eddy is highlighted by the 15.9 °C closed contour and the SSS anomaly is evaluated as the SSS in the core of the eddy minus the SSS at the eddy periphery.

Thus, the salinity compensation mechanism is probably the cause of the observed oceanographic structure. However, the detailed investigation of the eddy formation and characteristics is outside the scope of this study. Here, we want to assess the capability of using the SST information to improve the altimeter-derived currents. In the western boundary of the vortex, the GV do not resolve such a dynamical feature, indicating a circulation which crosses the north western section of the eddy with a northeastward flow direction. This effect is a result of the mapping procedure statistics in order to obtain two dimensional maps from along-track altimeter data. On the other hand, if we consider the dynamical evolution of the SST, i.e., we account for the SST spatial-temporal gradients, the Optimal reconstruction method prescribes a correction factor that deviates the flow from the altimeter estimate, resulting in a more accurate description of the eddy circulation. This is particularly evident in the area highlighted by the white ellipse in the left panel of Figure 4.

This case study shows one of the main advantages of RS18, i.e., merging a background currents field with a higher resolution SST map. Other approaches, as SQG, rely on SST images to infer the surface flow. Nevertheless, when salinity anomalies contribute significantly to the sea surface density gradients, such approach may well lead to a current estimate which is in an opposite direction with respect to the actual one, as also found by Isern-Fontanet et al. 2016 [16]. Instead, using the RS18 method this ambiguity is overcome.

4.2. Case Study 2: Coastal Upwelling in the Sicily Channel

The Sicily Channel area, whose approximate coordinates are 11°E to 13°E and 36°N to 38°N constitutes an interesting testbed for the performances of the Optimal Currents. In this region, in proximity of the southern coast of Sicily, coastal upwelling is a very frequent phenomenon and is known to generate cold SST patches associated with an offshore directed current [38,39]. In particular Piccioni et al. 1988 [39], using a statistical approach, found that such events are mostly detectable during summer and that they mostly develop with a delay of three days with respect to the southeasterly wind input (i.e., with a non-zero alongshore component with respect to the southern coast of Sicily).

We show an example of coastal upwelling detected on 23 July 2016. The result of its dynamics is clearly visible from the SST spatial distribution and is consistent with the CERSAT/IFREMER averaged surface winds during the three previous days (20–22 July 2016) (Figure 5).

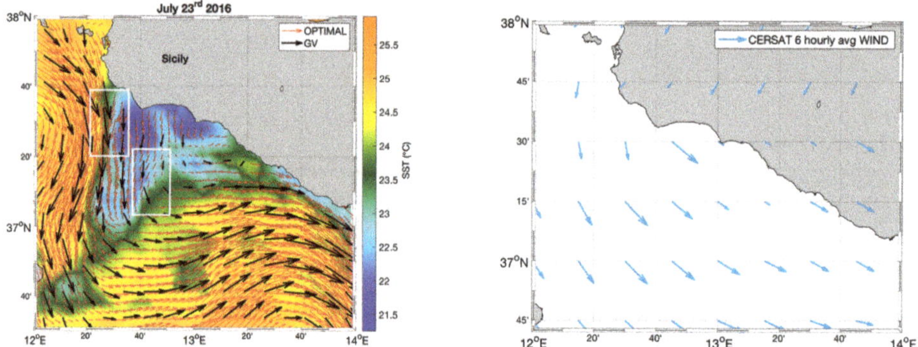

Figure 5. Left panel: Geostrophic Velocities (black arrows, 1/8° horizontal resolution) and Optimal Velocities (red arrows, 1/24° horizontal resolution) over sea-surface temperature (SST) in °C—23 July 2016 (the white boxes are referenced in Section 4.2). Right panel: CERSAT/IFREMER 6 hourly averaged surface winds—20–22 July 2016).

The SST anomaly between the interior of the upwelled waters and the surrounding areas exceeded 4 °C and the shape of the patch (in agreement with the mechanism described in [39]) indicates an offshore current, perpendicular to the southern coast of Sicily (between 12°20′E and 13°E). The offshore circulation was not entirely captured by the GV (Figure 5). Indeed, at the positions 12°40′E–37°30′N and 12°20′E–37°15′N, highlighted by the white boxes, the GV exhibit a cross thermal gradient component which is not consistent with the wind-driven circulation. A cross-thermal gradient geostrophic circulation is also found in correspondence of the lower tip of the cold SST patch. This effect is reduced when the Optimal reconstruction is implemented. The qualitative validation between the GV and the Optimal currents patterns allows to conclude that, enriching the GV with the information contained in the spatial and temporal variability of SST, it is possible to retrieve the ageostrophic circulation (as it is the case for the wind-driven upwelling dynamics) starting from the geostrophic, altimeter-derived currents. In order to further confirm this, we computed the divergence of the surface currents field. The divergence is defined as $\partial_x u + \partial_y v$ (with u and v the generic zonal and meridional components of the motion, respectively). When applied to a geostrophic field, the divergence operator yields zero in the whole surface currents domain, while it can be $\mathcal{O}(10^{-5}\ \mathrm{s}^{-1})$ for a total surface current field, i.e., considering both the geostrophic and the ageostrophic components of the motion [40]. Indeed, choosing the same region of Figure 5, we compared the divergences obtained using the Optimal currents, the model-derived surface currents of the MFS operational model and the MERCATOR global operational model (Figure 6). In addition, a comparative analysis with the monthly outputs of the CNR-Sea Forecast operational model will be given. Such divergences were all computed using a 9-point stencil width technique, as in [41].

The GV, though not exhibiting divergence values strictly equal to zero (mostly due to the numerical scheme of the computation), were mostly $\mathcal{O}(10^{-8}\ \mathrm{s}^{-1})$ everywhere in the analyzed area, (as well as in the rest of the Mediterranean Basin (not shown). Thus, we assumed this order of magnitude to be the lower limit for the surface currents field divergence, that can be referred to as geostrophy. On the other hand, the upper limit for the divergence values is identified with the results given by the MFS operational model. In this case, the divergence could reach the value of $\pm 1 \times 10^{-5}\mathrm{s}^{-1}$ (bottom panel of Figure 6). Quite interestingly, the Optimal currents exhibit divergence values at least two orders of magnitude larger than the GV, up to $\pm 8 \times 10^{-6}\mathrm{s}^{-1}$, hence comparable with the ones of a total surface current. The current field of the MERCATOR operational model also exhibits values $\mathcal{O}(10^{-6}\ \mathrm{s}^{-1})$, though its maxima only reached the 70% of the values observed in both the Optimal and the MFS-model case. Moreover, using the monthly averages of the submesoscale-permitting (resolution $\simeq 2$ km) operational model for the Sicily Channel and Tyrrhenian Sea (dataset labeled with "7" in Section 2), we had further confirmations that the expected value of the surface currents divergence during July 2016 must be $\mathcal{O}(10^{-5}\ \mathrm{s}^{-1})$, as also shown in Figure 7.

In addition, if we compare the spatial distribution of the divergences (DIV hereinafter) of the Optimal, MFS and MERCATOR surface currents (for 23 July 2016), the following similarities can be found:

1. The presence of an overall flow divergence area (DIV > 0) off the western cape of Sicily, around the Egadi Islands, highlighted in light green in Figure 6, with some convergence areas intrusions found in both the MFS and the Optimal Currents estimates;
2. The alternating flow convergence (DIV < 0) and divergence (DIV > 0) patterns inside the cold SST filament extending almost vertically from $37°36'\mathrm{N}$–$12°20'\mathrm{E}$ to $37°00'\mathrm{N}$–$12°20'\mathrm{E}$. Such adjacent patterns are mostly found around $37°06'\mathrm{N}$ and are highlighted in black in Figure 6. They are common to all the current fields, though looking smoother in the MERCATOR estimates.

The most interesting result of this comparison is that the combination of the geostrophic current and the SST data is a promising technique for the retrieval of the total surface current field from an initial geostrophic estimate, i.e., the one provided by satellite altimetry. Indeed, the Optimal current divergence is of the same order of magnitude predicted by high-resolution numerical models and observations, also sharing common features in the spatial patterns given by the MERCATOR and MFS-model estimates (though in the numerical model outputs the structures are slightly smoothed compared to the Optimal case).

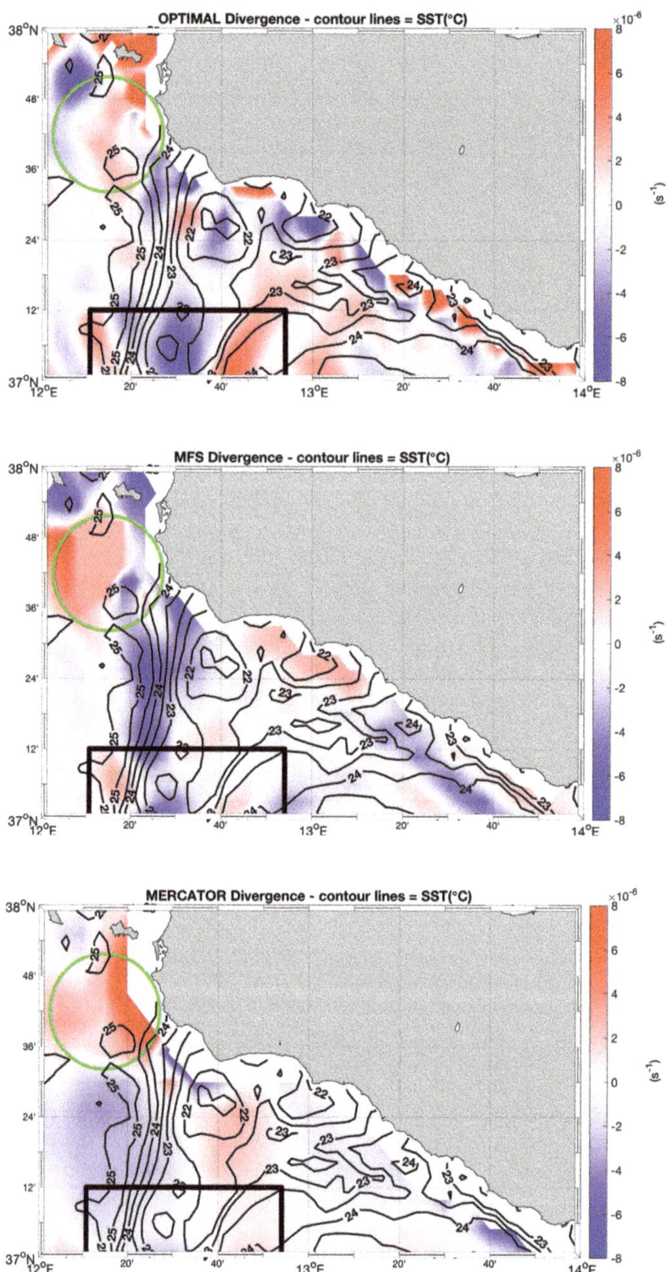

Figure 6. Divergence of the surface currents field in the Sicily Channel (23 July 2016). Top panel: Optimal currents; middle panel: MFS operational model for the Sicily Channel; bottom panel: MERCATOR operational model. Black contours indicate sea surface temperature, in °C. The light green and dark green circles are referenced in Section 4.2.

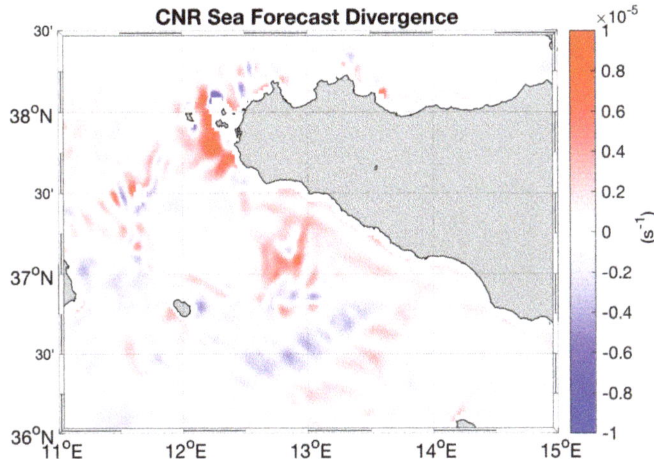

Figure 7. Divergence of the surface currents field in the Sicily Channel (July 2016 monthly average). Prediction of the CNR-Sea Forecast submesoscale-permitting operational model for the Sicily Channel and the Tyrrhenian Sea.

In the following sections we will describe the quantitative validation of the Optimal currents via two approaches: a first one in which the comparison will be performed with respect to the in-situ drifting buoys measurements in the Mediterranean Sea and a second in which the Optimal currents will be compared to the estimate of the High-Frequency Radar (HFR hereinafter) surface currents in the Malta–Sicily channel.

4.3. Validation with the In-Situ Measured Currents

In the period 2012 to 2016, the in-situ sea surface currents have been deduced from the trajectories of 104 15m-drogued drifting buoys at the positions indicated in Figure 8 (data doi = 10.6092/7a8499bc-c5ee-472c-b8b5-03523d1e73e9, [42]).

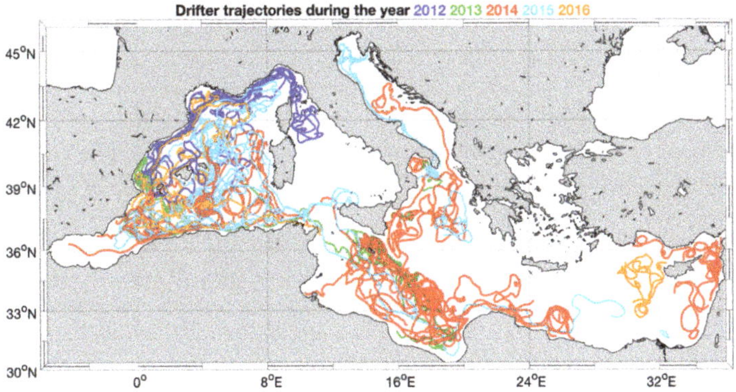

Figure 8. Positions of the sea surface currents measurements given by the surface drifting buoys in the Mediterranean Area during the period 2012–2016 [42].

The buoys have been mostly occupying the Western, the Central Mediterranean Sea, the Adriatic Sea and the area around the Island of Cyprus (eastern Mediterranean), measuring the sea surface currents with 6-hourly temporal resolution. Selecting the totality of the buoys trajectories, we compared the performances of the GV and the Optimal currents against the in-situ-derived measurements. This was achieved creating two match-up databases for the GV and the Optimal currents, which are both gridded fields. The match-up database was generated using cubic interpolation for the spatial colocalization of the gridded currents over the buoys positions while a linear interpolation was used to create the 6-hourly GV and Optimal currents (which are originally daily data) and finally perform the colocalization in time.

Similarly to RS18, the validation is based on the computation of the RMSE, correlation coefficients (CC) and biases between the GV, the Optimal currents and the drifter-derived currents, the latter being our benchmark. The computation of the biases did not evidence significant differences between the GV and the Optimal currents. For both datasets, the basin scale bias of the zonal and meridional component of the motion is around -0.3 cm/s and 0.4 cm/s, respectively. Concerning the RMS, the Optimal currents always exhibited lower values than the ones obtained with the GV, with larger RMSE reduction for the meridional component of the motion. Moreover, we found that the difference between the GV and the Optimal currents RMSE increases with the increasing magnitude of the SST spatial gradients (A,B in Equation (4)), indicating that our reconstruction method gives the best performances in large SST gradient areas. This was found selecting all the GV and Optimal current values laying in areas of progressively increasing $|\nabla SST|$, with $|\nabla SST| = \sqrt{A^2 + B^2}$ comprised between 0.1 and 6×10^{-5} $°C \cdot m^{-1}$ (such interval represents more than 90% of the $|\nabla SST|$ values in the basin). Based on the RMSE computation, we also defined a parameter to estimate the percentage of improvement of our reconstruction method with respect to the GV estimates. The parameter is defined as follows:

$$\text{IMPROVE}_{(U,V)} = 100 \times \left[1 - \left(\frac{\text{RMSE}_{(U,V)}^{\text{Optimal}}}{\text{RMSE}_{(U,V)}^{\text{GV}}} \right)^2 \right] \tag{7}$$

where U,V respectively indicate the zonal and the meridional component of the sea surface currents.

In general, our reconstruction is more satisfying for the meridional component of the motion and, at the basin scale, yields improvements that can exceed 10% (while they are only about 5% for the zonal component of the motion). Such improvements increase almost monotonically with the magnitude of the $|\nabla SST|$ except for the zonal currents, where a linear increasing trend is only observed until 5×10^{-5} $°C \cdot m^{-1}$. In a similar fashion, the Optimal currents show correlations with the in-situ measured currents generally higher than the GV, with a correlation improvement which is larger for increasing $|\nabla SST|$ values. Such results, schematically summarized in Figure 9, indicate that the Optimal currents better represent the sea surface currents variability observed via the in-situ measurements, confirming the potentiality of bringing the spatio-temporal variability of the SST remote observations inside the large-scale geostrophic currents.

During 2012–2016 we also investigated the local variability of the Optimal reconstruction method. This has been done binning the values obtained using Equation (7) in $2° \times 2°$ boxes (Figure 10 upper panels). The available drifter-derived currents allowed to perform the analysis in the sites of validation (SOV) indicated by the black squares in Figure 10 (bottom panel). In general, the best performances are found in the Central and Western Mediterranean, where local improvements can also reach 30%. The results of the validation, both at the basin scale and in terms of local variability, are in agreement with the results found at global scale by RS18. Also, we found that in the Western Mediterranean, the summertime reconstruction can lead to improvements exceeding 40%, which was not found during winter (not shown). Considering our investigation area, the zonal component of the surface currents is improved over 60% of the domain, while this percentage raises up to 70% for the meridional component of the motion. Combining these results, we could also indicate all the SOV where the method can improve both of the surface flow components, schematically given in the bottom panel of Figure 10. As expected, such SOV are mostly located in the Western Mediterranean. Unfortunately, some of the sites of the Optimal currents validation, though never exceeding the 40% of the analyzed domain, indicate that the method can also degrade the quality of the surface currents, when compared to the geostrophic estimates. These sites are mostly located in the eastern section of the Mediterranean basin. Quite interestingly, they are characterized by two properties: either they lay in areas of low $|\nabla\mathrm{SST}|$ ($\simeq 0.1 \times 10^{-5}\ ^\circ\mathrm{C}\cdot\mathrm{m}^{-1}$), which is not the ideal condition for the method applicability, or their statistical properties are computed based on a poor number of observations, affecting their statistical significance. This can be checked comparing the maps of $|\nabla\mathrm{SST}|$ and numbers of available in-situ observations binned in $1^\circ \times 1^\circ$ boxes in Figure 11. In a future study we plan to compute such statistics on a longer time series of optimal currents ($\simeq$20 years) and check the effects on the local variability of the optimal reconstruction method performances.

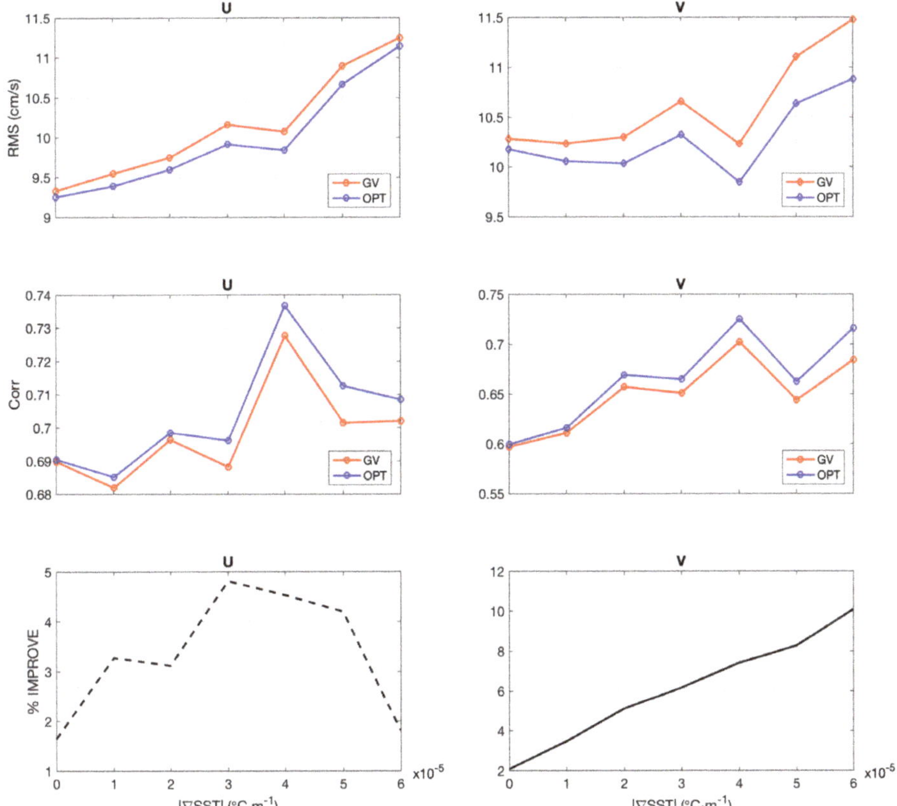

Figure 9. Basin scale improvements of the Optimal reconstruction method expressed via root mean square error (RMSE), Correlation Coefficients (Corr) and percentage of improvement (% IMPROVE) in the period 2012–2016. The Root mean square errors and correlation coefficients are computed for both the Optimal currents (blue lines) and the geostrophic velocities (GV) (red lines) using the drifting buoys-derived surface currents as a benchmark. The results are presented for both the zonal (U) and the meridional component (V) of the surface motion, respectively shown in the left and right column of the figure.

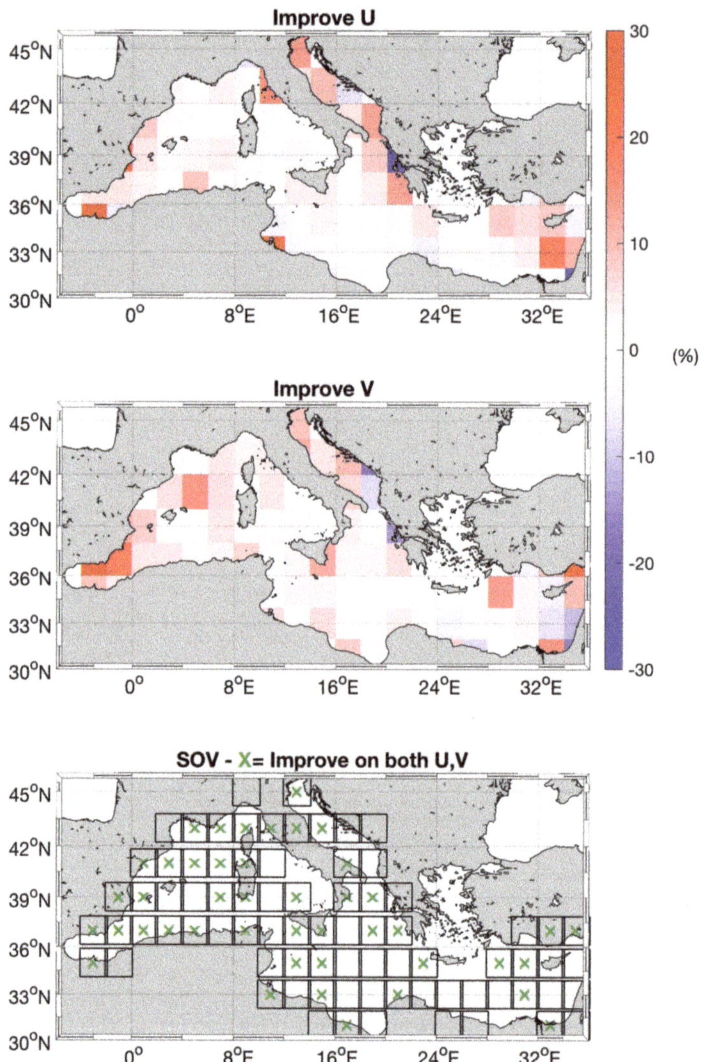

Figure 10. Spatial variability of the percentage of improvement of the Optimal reconstruction method for the zonal (U) and meridional component (V) of the surface motion, respectively shown in the top and middle panel of the figure. The percentage of improvement, binned in $2° \times 2°$ boxes has been computed in the sites of validation (SOV) given by the black squares in the bottom panel. The crosses indicate the SOV where both the zonal and meridional components of the motion are improved.

Nevertheless, it is worth noticing that the validation against the in-situ observations confirmed the validity of the case study analyzed in Section 4.2. Indeed, in the coastal area of the Sicily Channel (close to the southern coast of Sicily), during the period 2012–2016, our surface-currents reconstruction method yielded positive improvements for both the components of the motion. Hence, we wanted to further investigate the quality of the Optimal currents via a comparison with the HFR-derived surface currents in the Malta–Sicily Channel. The results are presented in Section 4.4.

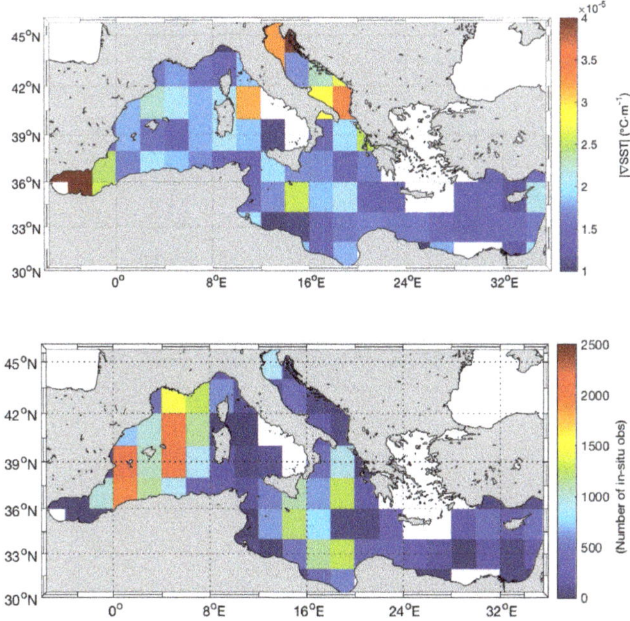

Figure 11. Top panel: $|\nabla\mathrm{SST}|$ binned in $1° \times 1°$ boxes, computed along the positions occupied by the drifting buoys during the period 2012–2016. Bottom panel: Number of in-situ surface-currents observations laying in each of the $1° \times 1°$ boxes.

4.4. Comparative Analysis: Optimal Currents Validation in the Malta–Sicily Channel

In this section we present an additional quality assessment for the Optimal currents via comparison with the daily surface currents derived from the Calypso High Frequency Radar platform in the Malta–Sicily Channel (during the year 2016), our benchmark. The coverage of the Calypso platform, schematically shown in Figure 12, extends from 35.7°N to 36.8°N and from 13.7°E to 15.2°E. As stated in Section 2, such platform provides hourly observations at $1/37°$ horizontal resolution, hence, our validation was realized creating daily maps of HFR currents. For each day, we computed bias and root mean square errors of the differences between the GV, the Optimal, the MERCATOR, the MFS-derived currents and the HFR estimates. Unfortunately, no comparison with the drifting buoys-derived currents was performed as, during the year 2016, the buoys have been circulating in the westernmost and easternmost sections of the Mediterranean basin only (Figure 10).

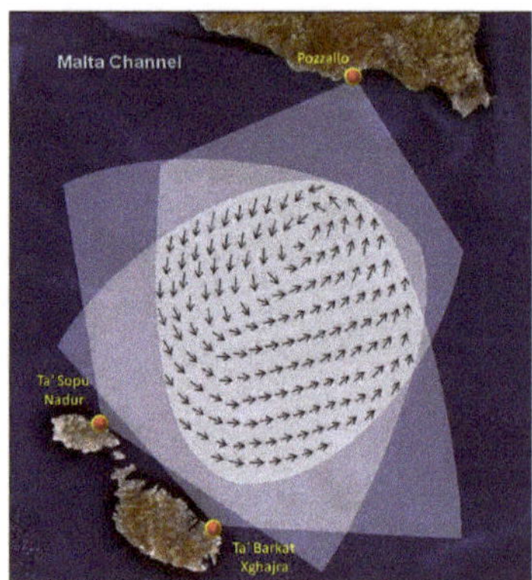

Figure 12. The Calypso High Frequency (HF) Radar platform: Schematics (after the Calypso Project web portal).

Unlike the case of the drifting buoys, the comparison with the High-Frequency Radar (HFR) currents involves two-dimensional gridded data. In order to carry out the comparison, all the analyzed daily currents were remapped over the HFR grid via bi-cubic interpolation. The mean results of the comparative study are summarized in Tables 1 and 2.

Table 1. BIAS and Root Mean Square Error (RMSE) between the Optimal-GV-MERCATOR-MFS and the High-Frequency Radar (HFR) surface currents in the Malta–Sicily Channel. Zonal Component.

ZONAL	RMSE (cm/s)	BIAS (cm/s)
OPTIMAL	9.72	4.04
GV	11.23	5.11
MERCATOR	12.50	4.30
MFS	12.70	2.60

Table 2. BIAS and Root Mean Square Error (RMSE) between the Optimal-GV-MERCATOR-MFS and the HFRsurface currents in the Malta-Sicily Channel. Meridional Component.

MERIDIONAL	RMSE (cm/s)	BIAS (cm/s)
OPTIMAL	9.00	1.40
GV	9.12	1.30
MERCATOR	12.65	3.60
MFS	13.23	3.19

Focusing on both the Optimal and the GV, using the HFR estimates as benchmark, the current retrieval is more satisfying for the meridional component of the motion, exhibiting lower RMSE and BIAS (respectively $\simeq 9.00$ and $\simeq 1$ cm·s^{-1}) compared to the zonal one, where RMSE and BIAS are respectively $\simeq 10.00$ and $\simeq 4$ to 5 cm·s^{-1}. Moreover, if we observe the behaviour of the Optimal currents with respect to the other datasets, we find that the RS18 method brings larger improvements on the retrieval of the zonal component of the motion. Indeed, for such component the Optimal currents have the lowest RMSE of all, 9.72 cm·s^{-1} (the MFS model yielding the larger value, 12.70 cm·s^{-1}) and a lower BIAS (4.04 cm·s^{-1}) than the GV and the MERCATOR estimates. The only exception is given by the BIAS of the MFS model, whose value is $\simeq 3$ cm·s^{-1}.

For the meridional component of the motion, the Optimal currents have RMSE and BIAS in line with the GV and always lower than the numerical model estimates. These results may look surprising. Indeed, at the basin scale the RS18 method brought the largest improvements for the meridional component of the motion. Nevertheless, here we are restricting our analysis to a much smaller coastal area in which our reference, the HFR currents, are known to be more accurate in retrieving the zonal component of the motion [43], probably affecting the statistics on the meridional flow.

Considering these results, we can conclude that the Optimal currents perform globally better than altimeter-derived estimates or model outputs.

5. Discussion and Conclusions

During the last decades, both physical and operational oceanography have been demanding high spatial and temporal resolution observations. Such observations are useful for the validation, optimization and run of the main operational models, especially when data assimilation is required. In the specific case of the sea surface currents retrieval, the high-resolution and repetitive observations are also necessary for several human activities in the marine context (navigation, safety and rescue activities) and for the environmental safeguard (illegal oil spills and pollutants monitoring).

The objective of our study was to improve the spatial and temporal resolution of the remotely sensed, altimeter-derived surface currents in the Mediterranean Sea. Based on the RS18 method, we could successfully introduce the information contained in a high-resolution satellite-derived SST ($\Delta t = 1$ day and $\Delta x = 1/24°$) inside the larger scale altimeter geostrophic currents. The RS18 method has the main advantage to optimize the use of SST to derive information on the surface currents field. This is achieved accounting for the source and sink terms in the SST evolution equation, unlike in maximum cross correlation techniques where only advection and diffusion can be considered. An additional advantage of the RS18 method is the possibility to rely on a background surface currents estimate, i.e., the altimeter measurements. This enables to derive the surface circulation even in areas of low tracer gradients or when the tracer alone is not a direct proxy of the ocean surface dynamics [16]. We applied the RS18 reconstruction method to the CMEMS datasets of geostrophic surface currents and sea surface temperature satellite data for the Mediterranean area. This enabled to estimate an Optimized velocity during five years.

The highlights of our study are listed here:

- During the period 2012–2016, we computed daily gap-free maps of Optimal currents. Using the OGS in-situ measured surface currents as a benchmark, we compared the performances of the Optimal currents and the altimeter-derived estimates. We found that the Optimal reconstruction method yields the largest improvements (locally up to 30%) in the western section of the Mediterranean basin and for the meridional component of the motion. Such improvements generally increase with the increasing SST spatial gradients and proved to be slightly higher during summertime. This is mostly due to the larger spatial SST gradients that can occur during summer, especially in areas of upwelling. This effect is particularly enhanced in the Western Mediterranean and in the Sicily Channel area, where you have very active upwelling regions (not shown);
- The Optimal reconstruction method is able to retrieve both the small scale geostrophic and ageostrophic circulation, obtaining a surface field characterized by a non-zero divergence. This was shown for the Sicily Channel. In this area, during summertime, it is frequent to observe patterns of ageostrophic circulation. Indeed, computing the Optimal currents during a coastal upwelling event (on 23 July 2016), we found that the surface divergence of the Optimal currents is comparable with the predictions of the CMEMS operational models, providing an estimate of the total (i.e., with non-zero divergence) surface-current field. The values of the Optimal currents divergence also proved to be in line with the monthly values predicted by the submesoscale-permitting CNR-Sea Forecast operational model for the Sicily Channel;
- The Optimal reconstruction, unlike the SQG method, allowed to overcome the ambiguity in retrieving the surface circulation when SST is not a direct proxy of the surface circulation. In this study, this was proved for a specific eddy dynamics case study in the Western Mediterranean area.
- In the Malta-Sicily channel, using the HFR surface currents as our benchmark, the Optimal currents improved the mean RMSE and BIAS compared to the numerical model outputs and to

the satellite-derived currents (except for the bias in the retrieval of the zonal flow, where the MFS model exhibited the best performance).

The main results listed here express the capability of the RS18 method to increase the spatio-temporal variability of the altimeter-derived currents even in the Mediterranean Area. Provided the existence of non-zero sea surface temperature spatial gradients, the variability of the high-resolution SST data can be successfully introduced inside the large-scale geostrophic currents estimates. The increase in the spatio-temporal variability explains the satisfying comparison of the Optimal currents with both the in-situ measured currents and the HFR estimates. Moreover, this was confirmed by the spectral analysis shown in Figure 13. In particular, we computed the mean spatial Kinetic Energy (KE) spectrum of the Optimal, MFS and altimeter-derived surface currents (GV) as a function of the spatial wavenumber (here given as the inverse of the wavelenght.) This was done in a land-free and dynamically active area of the Western Mediterranean (37.5°N to 38.0°N and from 0°E to 11°E), delimited by the yellow box in Figure 13. Moreover, we performed this analysis done during summertime (on 23 July 2016) i.e., choosing all the favourable conditions for the Optimal reconstruction method. Looking at the behaviour of the KE power spectral density, we can see that the Optimal currents and GV power spectra are superimposed until a scale of 100 km. Afterwards the GV begin loosing energy and their spectrum falls abruptly. Such a result was expected and is intrinsic with the geostrophic large-scale estimate of the surface currents, whose spectral content is mostly in the mesoscale range. On the other hand, for scales smaller than 100 km, the Optimal Currents spectrum contains more energy than the GV one and maintains the same decreasing trend until a scale of 30 km, confirming that the resolution of the Optimal currents is actually increased with respect to altimetry (notice that in the figure we discarded the areas in which the spectra exhibit a noisy behaviour). Moreover, the Optimal Currents spectrum is in agreement with the predictions of the MFS model, that, based on a fully three-dimensional primitive equations framework, simulates the evolution of the total currents field. Quite interestingly, Figure 13 also confirms that the Optimal and MFS spectra are closer to the predictions of the two dimensional turbulence energy cascade, $K^{-5/3}$ and K^{-3} [44]. Though the K^{-3} slope is not fully recovered at small scales, the Optimal Currents spectral separation from the K^{-3} line is reduced compared to the altimeter estimates.

In the end, we computed the GV and Optimal Currents KE temporal spectra in the Western Mediterranean (from 6°W to 8°E), where the optimal reconstruction exhibited the largest improvements (Figure 14). This was done on a time window of 45 days, which is the maximum temporal correlation scale for building the L4 altimeter-derived currents from the altimetric ground-tracks [28]. In particular, the computation was performed during wintertime (Julian day 1 to 45 of the year 2016), when the basin scale averaged KE was found to be larger. Figure 14 shows the Optimal and GV currents KE temporal spectra given as a function of the temporal wavenumber (here given as the inverse of the wavelength). The Optimal currents spectrum is initially superimposed to the GV one until a timescale of about twenty days. Afterwards, due to a power spectral density drop in the geostrophic velocities, the two spectra separate, indicating that the Optimal currents contain more energy up to scales of one week, when the GV spectrum starts exhibiting a noisy behaviour. This analysis confirms the Optimal Currents enhanced temporal variability.

This work could certainly benefit from the following additional analyses. At first, we plan to extend the Optimal currents validation to the years ranging from 1993 to 2016. This would help increasing the statistical significance of the comparisons with the drifting buoys in all the areas of the basin. In addition, using a $\simeq$20 years long time series, would allow to statistically determine the different contributors to the improved currents retrieval. Indeed, the improvements may come from both a better description of the non geostrophic flow and more accurate retrieval of the eddy dynamics, which were here discussed as single case studies.

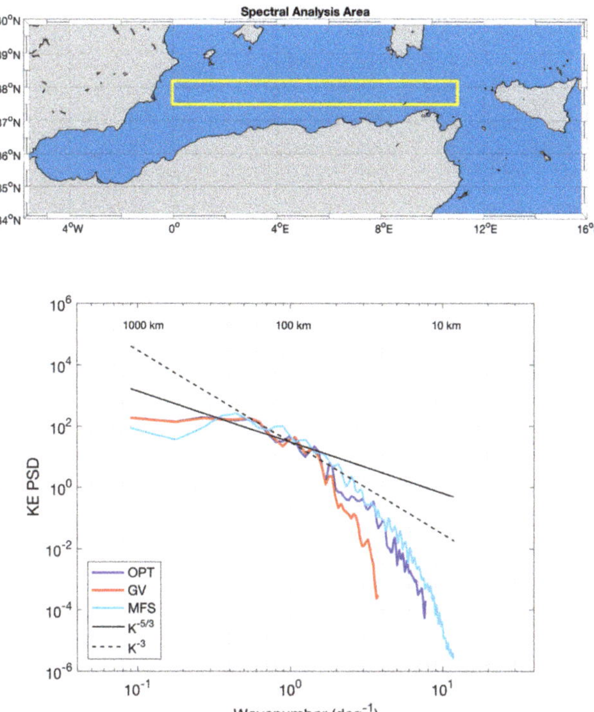

Figure 13. Optimal (blue line), altimeter-derived (red line) and MFS (light blue line) kinetic energy spatial spectra. The mean spectra are computed in the area delimited by the yellow box (top panel). The black continuous and dashed lines represent the surface and internal predictions of the QG turbulence, respectively. The spectra are computed on the date of 23 July 2016 and are represented as a function of the spatial wavenumber.

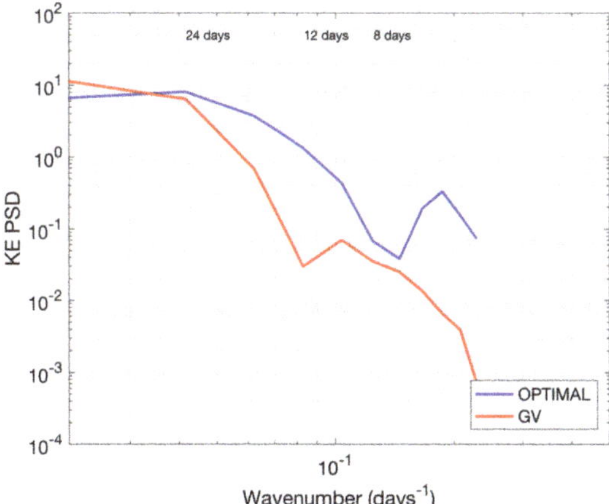

Figure 14. Temporal spectral content of the Optimal currents (blue) and GV (red) kinetic energy (west Mediterranean).

New ways of estimating the forcing term in the SST evolution equation should also be exploited. In the present study, the forcing was approximated by a low pass filtered SST temporal derivative but, in principle, heat fluxes estimates given by model-derived reanalyses, e.g., the ECMWF ERA-5, or estimated using bulk formulae could be tested [45,46]. In addition, the possibility to refine the temporal resolution of the method input parameters (e.g., $\sigma_{u,v}$ and h), presently given by climatological two-dimensional maps, could further improve the reconstruction method capabilities. At present, the definition of $\sigma_{u,v}$ and h is one of the more critical and crucial points of the RS18 method, being strictly dependent on the availability of high quality in-situ measurements of SST and surface currents (as shown in Section 3).

In addition, this study could further benefit from the exploitation of the gap-free SST obtained by geostationary satellites (e.g., METEOSAT [47]). Such an approach could lead to the determination of the sub-daily currents variability, up to the hourly resolution.

In the end, this method could be further applied on other types of satellite derived tracers. For instance, the ocean colour products are available at horizontal resolutions even higher than 1 km, e.g., the observations from the OLCI sensor mounted on board Sentinel-3 or from the MSI sensor mounted on Sentinel-2 and acquiring information in coastal areas. Implementing the method with these observations could allow to further improve the retrieval of the surface and near-surface submesoscale dynamics. Nevertheless, such an approach is not straightforward as it will require to accurately assess the source and sink terms due to biological activity in the tracer conservation equation (Equation (1)). This operation will be crucial for extracting the information related to the surface currents advection.

Author Contributions: Conceptualization: D.C., M.-H.R. and R.S. Formal analysis: D.C. and M.-H.R. Funding acquisition: R.S. Investigation: D.C., M.-H.R. and R.S. Methodology: M.-H.R. and R.S. Supervision: M.-H.R., M.M. and R.S. Validation: D.C., M.-H.R. and M.M. Writing of original draft: D.C. Writing—review and editing: D.C., M.-H.R., M.M. and R.S.

Funding: This research was funded by Progetto Bandiera RITMARE (Ricerca ITaliana per il MARE), the project "Marine Strategy" of the Italian Ministry of Environment, the Consiglio Nazionale delle Ricerche-Collecte Localisation Satellites Service Agreement and the European Copernicus programme.

Acknowledgments: We are grateful for the helpful comments on the manuscript by the three anonymous Reviewers. Moreover, we wish to thank Antonio Olita for providing the CNR-Sea Forecast model outputs and Salvatore Marullo for the fruitful discussions. We also wish to acknowledge Sharon Fan for taking care of the reviewing process.

Conflicts of Interest: The authors declare no conflict of interest. The funders had no role in the design of the study; in the collection, analyses, or interpretation of data; in the writing of the manuscript, or in the decision to publish the results.

References

1. Pujol, M.I.; Dibarboure, G.; Le Traon, P.Y.; Klein, P. Using high-resolution altimetry to observe mesoscale signals. *J. Atmos. Ocean. Technol.* **2012**, *29*, 1409–1416. [CrossRef]

2. Amores, A.; Jordà, G.; Arsouze, T.; Le Sommer, J. Up to What Extent Can We Characterize Ocean Eddies Using Present-Day Gridded Altimetric Products? *J. Geophys. Res. Oceans* **2018**, *123*, 7220–7236. [CrossRef]

3. Malanotte-Rizzoli, P.; The Pan-Med Group. Physical forcing and physical/biochemical variability of the Mediterranean Sea: A review of unresolved issues and directions of future research. In Proceedings of the Report of the Workshop. Variability of the Eastern and Western Mediterranean Circulation and Thermohaline Properties: Similarities and Differences, Rome, Italy, 7–9 November 2011.

4. Falcini, F.; Palatella, L.; Cuttitta, A.; Buongiorno Nardelli, B.; Lacorata, G.; Lanotte, A.S.; Patti, B.; Santoleri, R. The role of hydrodynamic processes on anchovy eggs and larvae distribution in the Sicily Channel (Mediterranean Sea): A case study for the 2004 data set. *PLoS ONE* **2015**, *10*, e0123213. [CrossRef]

5. Corrado, R.; Lacorata, G.; Palatella, L.; Santoleri, R.; Zambianchi, E. General characteristics of relative dispersion in the ocean. *Sci. Rep.* **2017**, *7*, 46291. [CrossRef] [PubMed]

6. Lapeyre, G.; Klein, P. Dynamics of the Upper Oceanic Layers in Terms of Surface Quasigeostrophy Theory. *J. Geophys. Res.* **2006**, *36*, 165–176. [CrossRef]

7. Fernandez, D. *SWOT Project Mission Performance and Error Budget*; NASA/JPL Technical Report; JPL Document D-79084. 2017.

8. Ardhuin, F.; Aksenov, Y.; Benetazzo, A.; Bertino, L.; Brandt, P.; Caubet, E.; Chapron, B.; Collard, F.; Cravatte, S.; Delouis, J.M.; et al. Measuring currents, ice drift, and waves from space: The Sea surface KInematics Multiscale monitoring (SKIM) concept. *Ocean Sci.* **2018**, *14*, 337–354. [CrossRef]

9. Nouguier, F.; Chapron, B.; Collard, F.; Mouche, A.A.; Rascle, N.; Ardhuin, F.; Wu, X. Sea surface kinematics from near-nadir radar measurements. *IEEE Trans. Geosci. Remote Sens.* **2018**, *56*, 6169–6179. [CrossRef]

10. Fu, L.L.; Holt, B. Some examples of detection of oceanic mesoscale eddies by the SEASAT synthetic-aperture radar. *J. Geophys. Res. Oceans* **1983**, *88*, 1844–1852. [CrossRef]

11. Chapron, B.; Collard, F.; Ardhuin, F. Direct measurements of ocean surface velocity from space: Interpretation and validation. *J. Geophys. Res. Oceans* **2005**, *110*. [CrossRef]

12. Isern-Fontanet, J.; Chapron, B.; Lapeyre, G.; Klein, P. Potential use of microwave sea surface temperatures for the estimation of ocean currents. *Geophys. Res. Lett.* **2006**, *33*. [CrossRef]

13. Isern-Fontanet, J.; Lapeyre, G.; Klein, P.; Chapron, B.; Hecht, M. Three-dimensional reconstruction of oceanic mesoscale currents from surface information. *J. Geophys. Res. Oceans* **2008**, *113*, 153–169. [CrossRef]

14. González-Haro, C.; Isern-Fontanet, J. Global ocean current reconstruction from altimetric and microwave SST measurements. *J. Geophys. Res. Oceans* **2014**, *119*, 3378–3391. [CrossRef]

15. Isern-Fontanet, J.; Shinde, M.; González-Haro, C. On the transfer function between surface fields and the geostrophic stream function in the Mediterranean Sea. *J. Phys. Oceanogr.* **2014**, *44*, 1406–1423. [CrossRef]

16. Isern-Fontanet, J.; Olmedo, E.; Turiel, A.; Ballabrera-Poy, J.; García-Ladona, E. Retrieval of eddy dynamics from SMOS sea surface salinity measurements in the Algerian Basin (Mediterranean Sea). *Geophys. Res. Lett.* **2016**, *43*, 6427–6434. [CrossRef]

17. Qazi, W.A.; Emery, W.J.; Fox-Kemper, B. Computing ocean surface currents over the coastal California current system using 30-min-lag sequential SAR images. *IEEE Trans. Geosci. Remote Sens.* **2014**, *52*, 7559–7580. [CrossRef]

18. Bowen, M.M.; Emery, W.J.; Wilkin, J.L.; Tildesley, P.C.; Barton, I.J.; Knewtson, R. Extracting multiyear surface currents from sequential thermal imagery using the maximum cross-correlation technique. *J. Atmos. Ocean. Technol.* **2002**, *19*, 1665–1676. [CrossRef]

19. Warren, M.; Quartly, G.; Shutler, J.; Miller, P.; Yoshikawa, Y. Estimation of ocean surface currents from maximum cross correlation applied to GOCI geostationary satellite remote sensing data over the Tsushima (Korea) Straits. *J. Geophys. Res. Oceans* **2016**, *121*, 6993–7009. [CrossRef]

20. Kelly, K.A. An inverse model for near-surface velocity from infrared images. *J. Phys. Oceanogr.* **1989**, *19*, 1845–1864. [CrossRef]

21. Frankignoul, C.; Reynolds, R.W. Testing a dynamical model for mid-latitude sea surface temperature anomalies. *J. Phys. Oceanogr.* **1983**, *13*, 1131–1145. [CrossRef]

22. Piterbarg, L.I. A simple method for computing velocities from tracer observations and a model output. *Appl. Math. Model.* **2009**, *33*, 3693–3704. [CrossRef]

23. Mercatini, A.; Griffa, A.; Piterbarg, L.; Zambianchi, E.; Magaldi, M.G. Estimating surface velocities from satellite data and numerical models: Implementation and testing of a new simple method. *Ocean Model.* **2010**, *33*, 190–203. [CrossRef]

24. Rio, M.H.; Santoleri, R.; Bourdalle-Badie, R.; Griffa, A.; Piterbarg, L.; Taburet, G. Improving the Altimeter-Derived Surface Currents Using High-Resolution Sea Surface Temperature Data: A Feasability Study Based on Model Outputs. *J. Atmos. Ocean. Technol.* **2016**, *33*, 2769–2784. [CrossRef]

25. Madec, G.; NEMO Team. *NEMO Ocean Engine*; Institut Pierre-Simon Laplace (IPSL): Paris, France, 2015.

26. Rio, M.H.; Santoleri, R. Improved global surface currents from the merging of altimetry and Sea Surface Temperature data. *Remote Sens. Environ.* **2018**, *216*, 770–785. [CrossRef]

27. Pisano, A.; De Dominicis, M.; Biamino, W.; Bignami, F.; Gherardi, S.; Colao, F.; Coppini, G.; Marullo, S.; Sprovieri, M.; Trivero, P.; et al. An oceanographic survey for oil spill monitoring and model forecasting validation using remote sensing and in situ data in the Mediterranean Sea. *Deep Sea Res. Part II Top. Stud. Oceanogr.* **2016**, *133*, 132–145. [CrossRef]

28. Pujol, M.I.; Faugère, Y.; Taburet, G.; Dupuy, S.; Pelloquin, C.; Ablain, M.; Picot, N. DUACS DT2014: The new multi-mission altimeter data set reprocessed over 20 years. *Ocean Sci.* **2016**, *12*, 1067–1090. [CrossRef]

29. Pisano, A.; Nardelli, B.B.; Tronconi, C.; Santoleri, R. The new Mediterranean optimally interpolated pathfinder AVHRR SST Dataset (1982–2012). *Remote Sens. Environ.* **2016**, *176*, 107–116. [CrossRef]

30. Buongiorno Nardelli, B.; Tronconi, C.; Pisano, A.; Santoleri, R. High and Ultra-High resolution processing of satellite Sea Surface Temperature data over Southern European Seas in the framework of MyOcean project. *Remote Sens. Environ.* **2013**, *129*, 1–16. [CrossRef]

31. Clementi, E.; Pistoia, J.; Delrosso, D.; Mattia, G.; Fratianni, C.; Storto, A.; Ciliberti, S.A.; Lemieux-Dudon, B.; Fenu, E.; Simoncelli, S.; et al. A 1/24 degree resolution Mediterranean analysis and forecast modeling system for the Copernicus Marine Environment Monitoring Service. In Proceedings of the Eight EuroGOOS International Conference, Bergen, Norway, 3–5 October 2017.

32. Drago, A.; Ciraolo, G.; Capodici, F.; Cosoli, S.; Gacic, M.; Poulain, P.; Tarasova, R.; Azzopardi, J.; Gauci, A.; Maltese, A.; et al. CALYPSO—An operational network of HF radars for the Malta-Sicily Channel. In *Proceedings of the 7th International Conference on EuroGOOS (Lisbon, Portugal, October 2014)*; Eurogoos Publication: Brussels, Belgium, 2015; pp. 28–30.

33. Bentamy, A.; Grodsky, S.; Carton, J.; Croizé-Fillon, D.; Chapron, B. Matching ASCAT and QuikSCAT winds Abderrahim. *J. Geophys. Res.* **2011**, *117*. [CrossRef]

34. Olmedo, E.; Taupier-Letage, I.; Turiel, A.; Alvera-Azcárate, A. Improving SMOS Sea Surface Salinity in the Western Mediterranean Sea through Multivariate and Multifractal Analysis. *Remote Sens.* **2018**, *10*, 485. [CrossRef]

35. Boutin, J.; Vergely, J.L.; Marchand, S.; D'Amico, F.; Hasson, A.; Kolodziejczyk, N.; Reul, N.; Reverdin, G.; Vialard, J. New SMOS Sea Surface Salinity with reduced systematic errors and improved variability. *Remote Sens. Environ.* **2018**, *214*, 115–134. [CrossRef]

36. Millot, C.; Taupier-Letage, I. Circulation in the Mediterranean sea. In *The Mediterranean Sea*; Springer: Berlin/Heidelberg, Germany, 2005; pp. 29–66.

37. Bashmachnikov, I.; Boutov, D.; Dias, J. Manifestation of two Meddies in altimetry and sea-surface temperature. *Ocean Sci.* **2013**, *9*, 249–259. [CrossRef]

38. Bignami, F.; Böhm, E.; D'Acunzo, E.; D'Archino, R.; Salusti, E. On the dynamics of surface cold filaments in the Mediterranean Sea. *J. Mar. Syst.* **2008**, *74*, 429–442. [CrossRef]

39. Piccioni, A.; Gabriele, M.; Salusti, E.; Zambianchi, E. Wind-induced upwellings off the southern coast of Sicily. *Oceanol. Acta* **1988**, *11*, 309–314.

40. Scherbina, A.Y.; Klymak, J.M.; Molemaker, J.; Novelli, G.; Guigand, C.M.; Haza, A.C.; Haus, B.K.; Ryan, E.H.; Jacobs, G.A.; Huntley, H.S.; et al. Ocean convergence and the dispersion of flotsam. *Proc. Natl. Acad. Sci. USA* **2017**, *115*, 1162–1167.

41. Arbic, B.K.; Scott, R.B.; Chelton, D.B.; Richman, J.G.; Shriver, J.F. Effects of stencil width on surface ocean geostrophic velocity and vorticity estimation from gridded satellite altimeter data. *J. Geophys. Res. Oceans* **2012**, *117*. [CrossRef]

42. Menna, M.; Poulain, P.M.; Bussani, A.; Gerin, R. Detecting the drogue presence of SVP drifters from wind slippage in the Mediterranean Sea. *Measurement* **2018**, *125*, 447–453. [CrossRef]

43. Capodici, F.; Cosoli, S.; Ciraolo, G.; Nasello, C.; Maltese, A.; Poulain, P.M.; Drago, A.; Azzopardi, J.; Gauci, A. Validation of HF radar sea surface currents in the Malta-Sicily Channel. *Remote Sens. Environ.* **2019**, *225*, 65–76. [CrossRef]

44. Vallis, G.K. *Atmospheric and Oceanic Fluid Dynamics*; Cambridge University Press: Cambridge, UK, 2006; p. 745.

45. Kondo, J. Air-sea bulk transfer coefficients in diabatic conditions. *Bound.-Layer Meteorol.* **1975**, *9*, 91–112. [CrossRef]

46. Hersbach, H. The ERA5 Atmospheric Reanalysis. In *AGU Fall Meeting Abstracts*; American Geophysical Union: Washington, DC, USA, 2016.

47. Marullo, S.; Minnett, P.; Santoleri, R.; Tonani, M. The diurnal cycle of sea-surface temperature and estimation of the heat budget of the Mediterranean Sea. *J. Geophys. Res. Oceans* **2016**, *121*, 8351–8367. [CrossRef]

Letter

Synergy between Ocean Variables: Remotely Sensed Surface Temperature and Chlorophyll Concentration Coherence

Marta Umbert [1,2,*], **Sebastien Guimbard** [3], **Joaquim Ballabrera Poy** [1,2] and **Antonio Turiel** [1,2]

[1] Departament of Physical and Technological Oceanography, Institut de Ciències del Mar, CSIC, 08003 Barcelona, Spain; joaquim@icm.csic.es (J.B.P.); turiel@icm.csic.es (A.T.)
[2] Barcelona Expert Center on Remote Sensing, CSIC-UPC, 08003 Barcelona, Spain
[3] Ocean Scope, Plouzané, 29280 Brest, France; sebastien.guimbard@ocean-scope.com
[*] Correspondence: mumbert@icm.csic.es

Received: 27 February 2020; Accepted: 29 March 2020; Published: 3 April 2020

Abstract: The similarity of mesoscale and submesoscale features observed in different ocean scalars indicates that they undergo some common non-linear processes. As a result of quasi-2D turbulence, complicated patterns of filaments, meanders, and eddies are recognized in remote sensing images. A data fusion method used to improve the quality of one ocean variable using another variable as a template is used here as an extrapolation technique to improve the coverage of daily Aqua MODIS Level-3 chlorophyll maps by using MODIS SST maps as a template. The local correspondence of SST and Chl-a multifractal singularities is granted due to the existence of a common cascade process which makes it possible to use SST data to infer Chl-a concentration where data are lacking. The quality of the inference of Level-4 Chl-a maps is assessed by simulating artificial clouds and comparing reconstructed and original data.

Keywords: remote sensing; ocean color; data fusion; data merging; physical oceanography; singularity analysis

---

1. Introduction

Global maps of Chl-*a* concentration at sea surfaces, similar to what happens with many other remote sensing products, suffer from data gaps. This is especially true for high-spatial resolution (few kilometers) daily maps, where orbital voids add to the data loss due to clouds, aerosols or sunglint. For this kind of products, gap-filling and noise-reduction techniques are required to provide uniform products suitable for environmental monitoring or the initialization of ecosystem models. To achieve this goal, data interpolation and noise reduction are commonly performed using univariate methods as, for example, [1]. Multivariate methods have also been applied to generate uniform products that combine information from different sensors; examples of multivariate approaches include optimal interpolation [2], empirical orthogonal functions [3,4], classification methods [5] and more recently using data-diven methods as analog data assimilation [6,7]. The method discussed in this paper is based on the empirical perception that fronts, eddies, and filaments are identifiable in the satellite imagery of different ocean variables (sea level, ocean color, sea surface temperature, sea surface salinity) and, sometimes, even in the raw radiance recorded by the satellites [8,9]. Remote sensing and drifter data have shown that the statistical distribution of surface ocean fields has characteristics of a flow in fully developed quasi-2D turbulence. Owing to the underlying turbulence of the ocean, the application of a technique known as singularity analysis [10] allows to retrieving information about ocean currents from images of any advected scalar.

The analysis of the spatial variability of pigment images found no difference between the wavenumber spectra of SST and ocean color maps, leading to the conclusion that phytoplankton cells in dynamic areas effectively behave as passive scalars [11,12]. In [13] it was shown that sea surface temperature and chlorophyll maps have the same multifractal structure. This fact can be interpreted as a consequence of the turbulent advection at the scales of observation and due to the existence of a common cascade process [14]. The multifractal structure of any scalar can be characterized through singularity analysis [15]. Singularity analysis is any technique capable of calculating a map of the singularity exponents associated with a given scalar. A singularity exponent is a non-dimensional number that characterizes the sharpness, or regularity, of the scalar around a given point [10]. Nieves et al. [13] showed that the singularity exponents extracted from SST, Chl-*a* or brightness temperature maps are very similar, a fact that was interpreted as an effect of the advection term. Notice that the equivalence among singularity exponents does not require advection to be the largest term in the equation of evolution. Advection is a non-linear term in the equation of evolution, and thus it is at the origin of singularities.

That the common singular structure of variables advected by the turbulent ocean has been exploited to reduce the noise in SMOS sea surface salinity maps using data from auxiliary, high-quality SST maps [16,17]. The same algorithm can be used not only to reduce noise but also to extrapolate provided that the template is defined at the missing points. Although the theoretical foundation of the data fusion method is relatively complex, its practical application is rather simple: a non-parametric, weighted local linear regression between Chl-*a* and SST maps are calculated at each point, then the regression parameters obtained at each point are applied to the value of SST at that point to infer a new variable of Chl-*a*.

The goal of this work is to assess the ability of a local linear regression approach to fill remote sensing chlorophyll concentration data gaps. This local relation between SST and Chl-*a* can be built from a single snapshot of both variables, so the implementation of the method is simple, computationally fast, and no training set is required; moreover, the method is well fit to deal with strong, rapid changes in the dynamics of the flow as far as it is always turbulent. Additionally, the analysis of the auxiliary parameters of the fusion algorithm allows characterizing the modulation of the seasonal correlation modulation between Chl-*a* and SST variables.

2. Satellite Data

Two Aqua-MODIS ocean products from the NASA Aqua spacecraft are used in this study: Chl-*a* concentration Level-3 daily product, and standard MODIS Aqua Level 3 SST Thermal IR Daily 4 km Nighttime, both at 4 km × 4 km grid [18]. Our dataset period goes from January to December 2006. The data were downloaded from the Ocean Color web portal, http://oceancolor.gsfc.nasa.gov/. According to many studies, it is assumed that Chl-*a* follows a lognormal distribution [19] and so we work with the logarithm of chlorophyll concentrations for ease of use (theoretically the singularity exponents should not change if a monotonic function is applied to the data [10], but it is numerically more stable to work with a dynamic range comprising fewer orders of magnitude).

3. Method

3.1. Singularity Analysis

Singularity analysis is the key stone of the so-called Microcanonical Multifractal Formalism (MMF) [10]; where the fluid is understood as a hierarchical arrangement of different fractal components, each own with its own fractal dimension and characterized by a particular value of the singularity exponent. Singularity exponents are related to the ocean circulation and thus they are not specific to any particular scalar under study. In fact, it has been verified that with a good approximation singularity lines coincide with the streamlines of the flow [20], confirming that singularity exponents are characterized by the flow and not scalar-specific. The emergence of such a singular structure is the result of the advective forces acting on a quasi-2D turbulence regime, and can be thus observed for scales ranging from kilometers to the planetary scale.

The calculation of singularity exponents in a given noisy, discretized signal requires the use of an appropriate interpolation scheme, the most usual one being wavelet projections. Let s be an arbitrary 2D scalar signal. It will be said that s has a singularity exponent $h(x)$ at the point x if, for any wavelet function Ψ the following relation holds:

$$T_\Psi |\nabla s|(x,r) \equiv \int dx' \, |\nabla s|(x') \frac{1}{r^2} \Psi\left(\frac{x-x'}{r}\right)$$
$$= \alpha(x)\, r^{h(x)} + o\left(r^{h(x)}\right) \tag{1}$$

where r stands for an arbitrary scale parameter (normalized by the integral scale so it is dimensionless and smaller than 1) and $o\left(r^{h(x)}\right)$ is a term that becomes negligible compared to $r^{h(x)}$ when r goes to zero. The amplitude function $\alpha(x)$ does not depend on the particular scale at which the wavelet projection is calculated and has the same units as the gradient of the scalar.

The left hand side, i.e., $T_\Psi |\nabla s|(x,r)$, is called the wavelet projection of the gradient modulus of s over the wavelet Ψ, and represents a local zoom of variable size around the point x. What is important in Equation (1) is the function $h(x)$, which is called the singularity exponent of the function s at the point x. The singularity exponent is, by construction, a dimensionless measure of the regularity or irregularity of the function s around the point x.

3.2. Fusion Method

As discussed in Turiel et al. [10], ocean scalars as chlorophyll and temperature (denoted respectively by c and θ) are multifractal in the microcanonical sense. This implies that a singularity exponent can be assigned at each point and that they are arranged in a particular way, which in turn implies that they are related to a property of the flow (its circulation) rather than to any specificity of the associated scalar.

Assuming that the singularity lines of Chl-*a* and SST coincide as it was shown in Nieves et al. [13], it follows that their gradients must be related by a smooth 2×2 matrix ρ [17] :

$$\nabla c(\vec{x}) = \rho(\vec{x})\, \nabla \theta(\vec{x}), \tag{2}$$

Although a large number of $\rho(\vec{x})$ verifying Equation (2) exist, the indetermination is considerably reduced by imposing that the matrix $\rho(\vec{x})$ must be smooth: a non-smooth $\rho(\vec{x})$ would be a source of singularities, and hence the singularity lines of c and θ would differ. If we further assume that the gradients of c and θ are aligned, then Equation (2) leads to:

$$c(\vec{x}) \approx a(\vec{x})\, \theta(\vec{x}) + b(\vec{x}) \tag{3}$$

with smooth functions a and b (i.e., they have small gradients as compared to those of c and θ).

Let us now suppose that the function c is contaminated by some source of noise n, so in fact our data sets consists of $\theta(\vec{x})$ (assumed noiseless) and $c'(\vec{x}) = c(\vec{x}) + n(\vec{x})$. We can construct a filtered version of c, denoted by c_f, as:

$$c_f(\vec{x}) = \hat{a}(\vec{x})\, \theta(\vec{x}) + \hat{b}(\vec{x}) \tag{4}$$

where $\hat{a}(\vec{x})$ and $\hat{b}(\vec{x})$ are estimates of the actual parameters $a(\vec{x})$ and $b(\vec{x})$ in Equation (3). These estimates are obtained from the values of c' and θ by performing scale-invariant linear regressions weighted around each point. The weight of a specific point x' is defined as $w_x(x') = |x-x'|^{-2}$.

The total weight of a point $\vec{x}$, $N(\vec{x})$, is defined as follows:

$$N(\vec{x}) \equiv \sum_{\vec{x} \neq \vec{x}' \in \text{sea}} \frac{1}{|\vec{x}' - \vec{x}|^2} \tag{5}$$

The weighted regression uses all available data of each field, using the function $N(\vec{x})$ that decay as a power-law with the distance (Equation (5)); thus it is scale-invariant, as it does not introduce any preferred scale [16,17].

With $\hat{a}(\vec{x})$ and $\hat{b}(\vec{x})$, an estimation of the chlorophyll c_f may be obtained applying Equation (4) as soon as the value of θ at that location $\vec{x}$ is available. This algorithm was shown in Umbert et al. [17] to lead to a large increase in the quality of SMOS SSS maps using OSTIA SST maps as a template; besides, the method leads to significant restoration of the structure of singularity exponents.

4. Results

4.1. Scalar Synergy

Figure 1 shows examples of daily maps of MODIS Chl-*a* concentration (top panel) and SST (middle panel) in 1 January 2006. The values of Chl-*a* are derived from measurements in the visible part of the spectrum, which may be affected by artifacts like aerosols, sun glint and high turbidity of the water. Several flags are introduced to characterize Chl-*a* data quality and a result, maps of Chl-*a* usually suffer from a larger incompleteness than those of SST, which are derived from infrared measurements in the 4 µm range. An example of the application of the fusion algorithm in the global ocean for 1 January 2006 is shown in the bottom panel of Figure 1. As stated before, the algorithm allows estimating the local Chl-*a*-SST regression by taking into account all possible couples of data (SST,Chl-*a*) weighted using function $N(\vec{x})$ in Equation (5). Fused Chl-*a* maps integrate the relation between structures present in the SST and Chl-*a* specific structures. In fused daily products we recognize some expected Chl-*a* global patterns, near the ocean surface, where availability of sunlight is not limiting, phytoplankton growth depends on temperature and nutrient levels. High chlorophyll concentrations are found in nutrient-rich, cold polar waters and where ocean currents cause upwelling, which brings nutrient-rich deep-cold water to the surface.

We will focus in the region of the Gulf of California, a narrow sea between mainland Mexico and the Baja California peninsula, where high primary productivity levels are found as a result of an efficient nutrient transport of waters from under a shallow pycnocline into the euphotic zone [21]. Figure 2 shows the input variables of our algorithm (top panel: Chl-*a*, middle panel: SST) and the output fused Chl-*a* (bottom panel). The corresponding singularity exponents are shown in the right column. Rich singularity structures can be recognized in the original chlorophyll concentration maps associated with the fronts mainly caused by the horizontal transport from high primary production areas onto less productive ones. The SST field exhibits also the richness of patterns associated with the circulation in that area. Where both images are not affected by data gaps, the singularity structure (although not the magnitude) is similar between them. Places where the correspondence of both singularity images fails may identify places at which intrinsic dynamics of the variable competes with flow advection.

Once the data fusion is applied, the L4 Chl-*a* is extrapolated to all the pixels where SST was available. The structure of Chl-*a* is well represented (compared to the original image), although a smoothing of the original variable degrades the dynamic range of the variable. The singularity exponents of the fused Chl-*a* data reveal that most of the structures present in the SST map are re-integrated in the chlorophyll map at the same time that Chl-*a* specific magnitude and structures are maintained; for instance, the strong gradient associated with the biological activity present in the Gulf of California, appear delineated in the fused Chl-*a*. This implies that the fusion algorithm can partially integrate SST structures without destroying Chl-*a* ones.

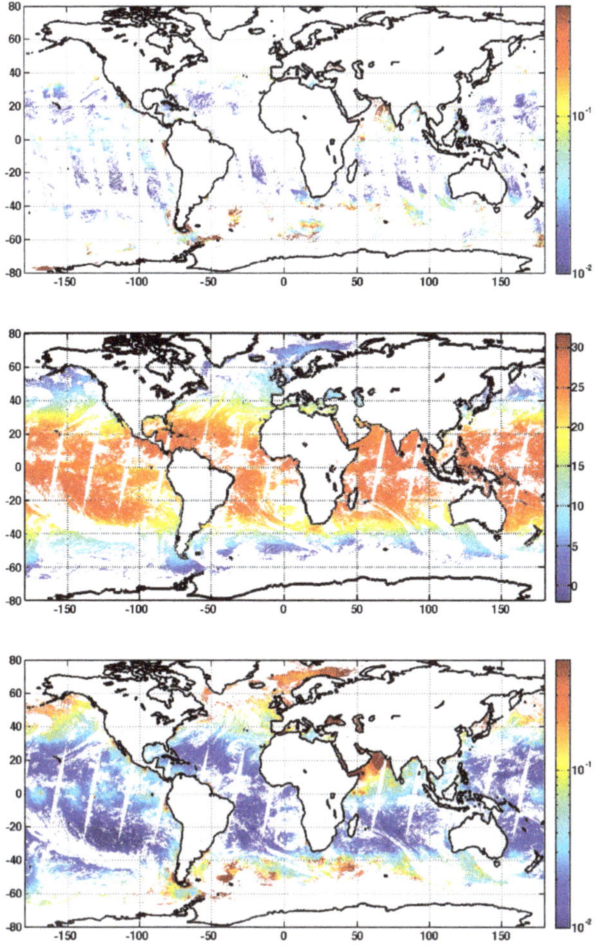

Figure 1. (**top panel**) MODIS Aqua Level 3 Chl-*a* (mg/m^3), (**middle panel**) MODIS Aqua Level 3 SST (°C) and (**bottom panel**) Level 4 Chl-*a* concentration (mg/m^3) extended to the areas where MODIS SST was available, for 1 January 2006.

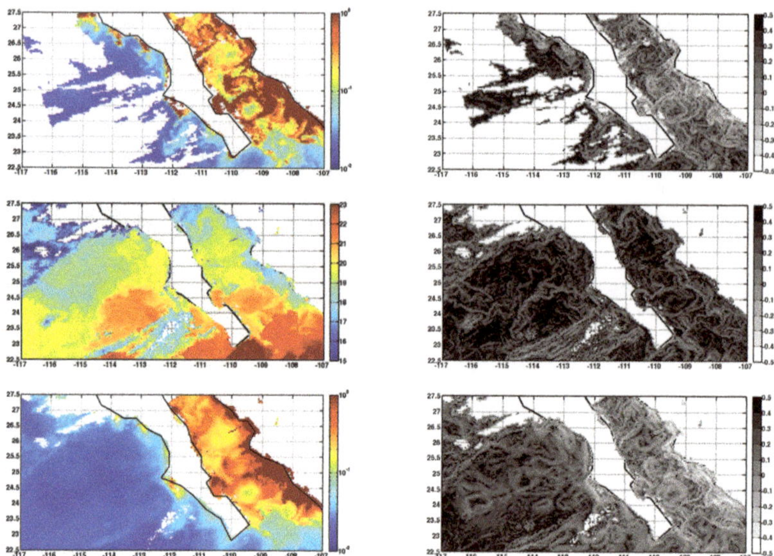

Figure 2. (**top row**) MODIS Chl-*a* (mg/m^3), (**middle row**) MODIS SST (°C) and (**bottom row**) fused Chl-*a* (mg/m^3), for 1 January 2006 (**left column**) and associated singularity exponents (**right column**).

4.2. Interpretation of Auxiliary Parameters

As shown in Equation (4), the functions $\hat{a}(\vec{x})$ and $\hat{b}(\vec{x})$ provide information about the local functional dependence between SST and Chl-*a*. The local slope, $\hat{a}(\vec{x})$, will be negative at those places where SST decreases as Chl-*a* increases in the neighborhood, and the converse. Considering that cold waters tend to have more nutrients than warm waters, phytoplankton is more abundant where surface waters are cold. So, as we move from one given point to another with colder water, Chl-*a* would usually increase and thus the slope $\hat{a}(\vec{x})$ will be negative. However, the relationship changes from point to point and it should be expected to change from one image to the next one. Figure 3 shows the seasonal average of the slope and intercept (considering winter as January-February-March, spring as April-May-June, summer as July-August-Septemebr and fall as October-November-December).

The coherent patterns of the auxiliary parameters of the method delineate areas with different relation between Chl-*a* and SST. In the same spirit, Longhurst [22] introduced the concept of ocean biogeochemical provinces, characterized by their particular physical and biological behavior. Longhurst definition is based on the mixed layer depth lying close to the ocean-atmosphere interface. Specific provinces have common characteristics and can generally be classified as four general biomes: the coastal, polar, westerly and trade winds biomes. A visual comparison between local functions of the fusion method and the Longhurst definition is shown in Figure 3.

As expected, negative slopes are present in the upwelling areas associated with the easternmost currents of the great anticyclonic gyres, corresponding to the Benguela and Canary currents in the Atlantic Ocean and the Peru and California currents in the Pacific Ocean. Notice that this negative slope is present all year long. These oceanic areas, as well as the upwelling zones and continental margins, are rich in Chl-*a* as a result of the proximity to areas where the resurgence of nutrients take place and the local circulation is favorable to nutrient accumulation. A band of cool, chlorophyll-rich water is also apparent all along the equator; the strongest signal at the Atlantic Ocean and the open waters of the Pacific Ocean also leads to negative values of the local intercept $\hat{b}(\vec{x})$.

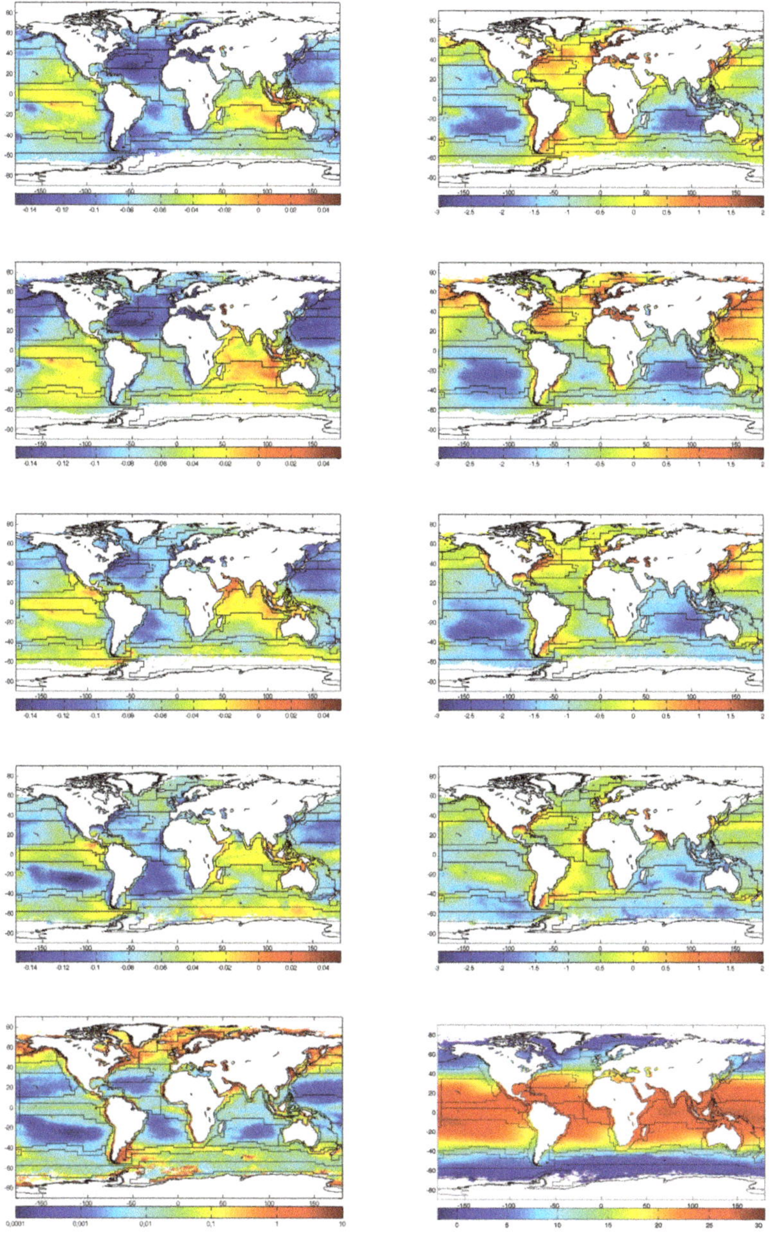

Figure 3. (**left column**) Seasonal mean local slope estimation $\hat{a}(x)$ for spring, summer, fall and winter. (**right column**) Local intercept estimation $\hat{b}(x)$ for spring, summer, fall and winter. These are the mean estimations used in the derivation of the fused maps as the one presented in Figure 1. (**bottom row**) Chl-*a* mean concentration for year 2006 and SST mean for the same year.

Negative values of $\hat{a}(\vec{x})$ are also found in areas where Chl-*a* concentration decreases as SST increases; this situation, which is typically found in the (oligotrophic) subtropical gyres, intensifies in the Atlantic Ocean during the northern hemisphere winter and spring. The Pacific Ocean exhibits an intensified negative pattern in the Northern subtropical gyre during boreal spring and a negative pattern in the southern hemisphere during austral spring. In both cases, such intensification in the oligotrophic subtropical gyres is driven by the seasonal cycle of sea surface temperature.

Subpolar gyres are also characterized by high Chl-*a* concentrations linked to nutrient accumulation during winter, when the mixing layer reaches the deep ocean followed by the stratification of the water column during spring. During the dark winter months, the local slope between SST and Chl-*a* is positive. However, when sunlight returns and nutrients are trapped near the surface during spring and summer, the phytoplankton flourishes in high concentration, seen as negative values of $\hat{a}(\vec{x})$.

The local regression coefficient (not shown) has small values along the Equatorial Pacific, and in the Southern and Indian Ocean indicating that horizontal advection of Chl-*a* cannot be locally explained by SST variability in these regions only. Therefore, either additional variables should be taken into account, or a more sophisticated relation between Chl-*a* and SST should be used. For example, in ocean regions of High Nutrient Low Chlorophyll (HNLC) as the equatorial Pacific and the Southern Ocean, low Chl-*a* concentrations are due to a stoichiometric imbalance of iron which have no link to SST. Another possible cause for the low values of the local regression coefficient in some regions is the lack of enough points to provide a quality reconstruction. For instance, 85% of the data points are missing in the Equatorial Pacific due to the large cloudiness.

4.3. Validation of Reconstruction

To assess the quality of the extrapolation resulting from our data fusion method, we validate it by analyzing 5 different regions (as shown in Figure 4); each region has at least an area of 8×8 degrees (200×200 pixels). For each of these areas a mask representing a cloud structure (of the typical size and shape found in chlorophyll images) is defined (the clouds are generated using a real MODIS SST 9-km daily image of 1 January 2006, and are kept fixed in time). Those masks have a surface of about 30% of the region on which they will be applied. To assess the quality of the fusion algorithm, we proceed in three steps. In the first step, points lying in the masked area of a daily Chl-*a* map are removed (notice that there will be additional missing points in the Chl-*a* map because of the gaps in the original remote sensing product; for instance, in the EP area (Figure 4) the average percentage of missing points is 85%). In the second step, the fusion algorithm is applied to the masked Chl-*a* map using the corresponding daily SST map as a template, extrapolating to the missing values. Finally, the extrapolated values are compared to the available original ones on the masked area for each image during the entire year 2006 (365 daily images). We require a minimum of 5% of original Chl-*a* values existed inside the masked area to perform the cross validation. This strategy allows comparing the original Chl-*a* and the retrieved one at the available masked points, and therefore the extrapolation ability of the fusion method.

The performance of the algorithm is studied in different Chl-*a* regimes: oligotrophic regions (Central Atlantic, CA and Pacific in front of California, CL) and eutrophic regions in higher latitudes (North Atlantic, NA), coastal upwelling areas (Benguela upwelling, BG) and the Equator (Equatorial Pacific, EP). NA region is defined by [19.61°W–9.19°W; 51.65°N–57.91°N] and 22% of the area is masked, EP region is defined by [142.54°W–130.04°W; 5.02°S–0.40°N] (29% of the area is masked), CL region is defined by [125.87°W–117.53°W; 22.48°N–30.82°N] (31% of the area is masked), BG region is defined by [9.56°E–17.90°E; 27.53°S–19.19°S] (35% of the area is masked) and CA is defined by [40.44°W–30.03°W; 22.48°N–30.82°N] with a 34% of the area being masked.

Examples of one daily image validation are shown in Figure 4 for each region. A quantitative measure of the quality of the reconstruction of the chlorophyll is given in terms of four parameters: the root mean square (rms), bias (mean) and standard deviation (std) of the reconstructed error and the correlation coefficient (r) between the masked and reconstructed values. In the case of the central

Atlantic and California regions, the concentration of chlorophyll is smaller, followed by the values found in the North Atlantic, the Equatorial Pacific and the Benguela upwelling where highest values of Chl-*a* concentration are found. Correlations coefficients between L4 and original Chl-*a* decrease for the regions where lower pigment concentrations are found. Better statistics are associated to areas of high primary production.

The validation for the entire year 2006 is summarized in Table 1; the mean seasonal values for each one of the statistical parameters and clouds are included. Our data fusion method succeeds in filling gaps with mean annual correlation coefficients ranging from 0.58 to 0.81 for the studied period and artificial clouds. Oligotrophic regions (North Atlantic and California) have mean regression coefficients which are significantly lower than the mean correlation coefficients values for the equatorial Pacific, the North Atlantic and the Benguela upwelling regions respectively. The smaller performance in oligotrophic regions is probably due to the small horizontal gradients of Chl-*a* there together with a larger signal-to-noise ratio. However, the absolute error of the reconstruction is always moderate. A slightly mean positive bias is systematically found, meaning that the L4 estimates are smaller than the original Chl-*a* values, probably due to the smoothing generated by the weighting function used in the fusion algorithm [16].

The seasonal segregation of validation results highlights a worse performance during the January-February-March period, with mean correlation coefficients ranging from 0.15 to 0.47 (Central Atlantic and Benguela upwelling, respectively). During the rest of seasons, the validation results greatly improve with correlation coefficients ranging from 0.67 to 0.94 in spring (AMJ), from 0.77 to 0.94 in summer (JAS) and from 0.74 to 0.94 in fall season (OND). In the winter season, the strengthening of winds and deepening of the mixed layer in mid and high latitudes create a large supply of nutrients from the deep ocean to the surface, which will be available for phytoplankton to proliferate in the following seasons. In equatorial and coastal upwelling regions (Equatorial Pacific and Benguela, respectively), the upwelling intensity also varies seasonally depending on wind strength and direction and the vertical structure of the water column. Under these circumstances, nutrient availability, vertical mixing and the depth of the mixed layer play crucial roles in the distribution of Chl-*a*, which could not be only explained by horizontal advection and the distribution of surface temperature.

The power density spectra (PDS) is computed as in [23] inside the boxes shown in in Figure 5 (top) for each daily file and then the median for the entire period 2006 is calculated. The spatial spectral analysis of MODIS Chl-*a*, MODIS L3 SST and Level 4 Chl-*a* products under the SPURS and STP regions (red and green boxes in Figure 5 (top)) shows that the actual spatial resolution of MODIS Chl-*a* and MODIS L3 SST is about 10 km and, and 15 km for the L4 Chl-*a* product (Figure 5 (bottom)). The intermittent character of MODIS Chl-*a* and the lower resolution of L4 Chl-*a* explain the vertical shift between their PDS.

Table 1. Mean seasonal results of daily reconstruction validation under artificial clouds for year 2006.

	R				STD				BIAS				RMS			
	JFM	AMJ	JAS	OND	JFM	AMJ	JAS	OND	JFM	AMJ	JAS	OND	JFM	AMJ	JAS	OND
CA	0.15	0.67	0.77	0.74	0.20	0.07	0.08	0.06	0.70	0.04	0.03	0.02	0.73	0.08	0.09	0.06
BG	0.47	0.93	0.91	0.93	0.35	0.14	0.13	0.11	0.27	0.03	0.02	0.02	0.47	0.15	0.13	0.12
CL	0.19	0.89	0.80	0.88	0.22	0.04	0.04	0.04	0.72	0.02	0.03	0.03	0.75	0.05	0.05	0.06
EP	0.25	0.94	0.94	0.94	0.17	0.04	0.03	0.03	0.42	0.01	0.02	0.01	0.46	0.04	0.04	0.03
NA	0.33	0.88	0.91	0.86	0.16	0.09	0.10	0.07	0.47	0.02	0.01	0.01	0.50	0.10	0.10	0.07

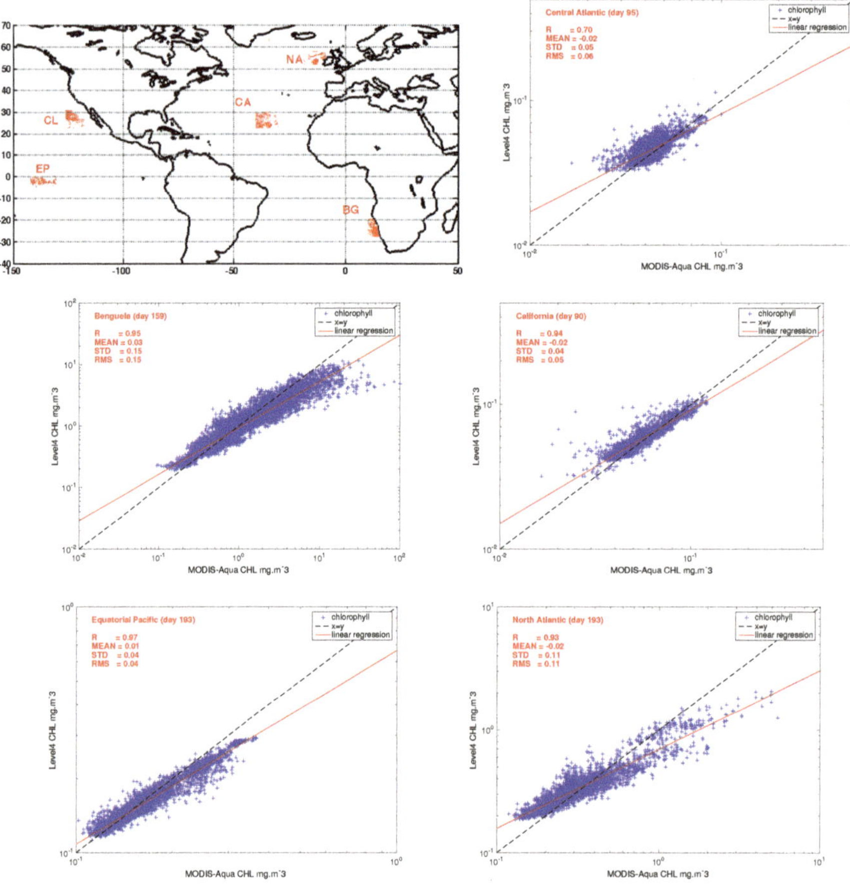

Figure 4. Artificial clouds generated for validation purposes in the Central Atlantic (CA), Benguela upwelling (BG), California Pacific (CL), Equatorial Pacific (EP) and North Atlantic (NA) (**top left panel**). Validation of reconstruction under artificial clouds for MODIS Chl-*a* concentration. For CA region day 95, BG region day 159, CL day 90, EP day 193, and NA day 193 of year 2006.

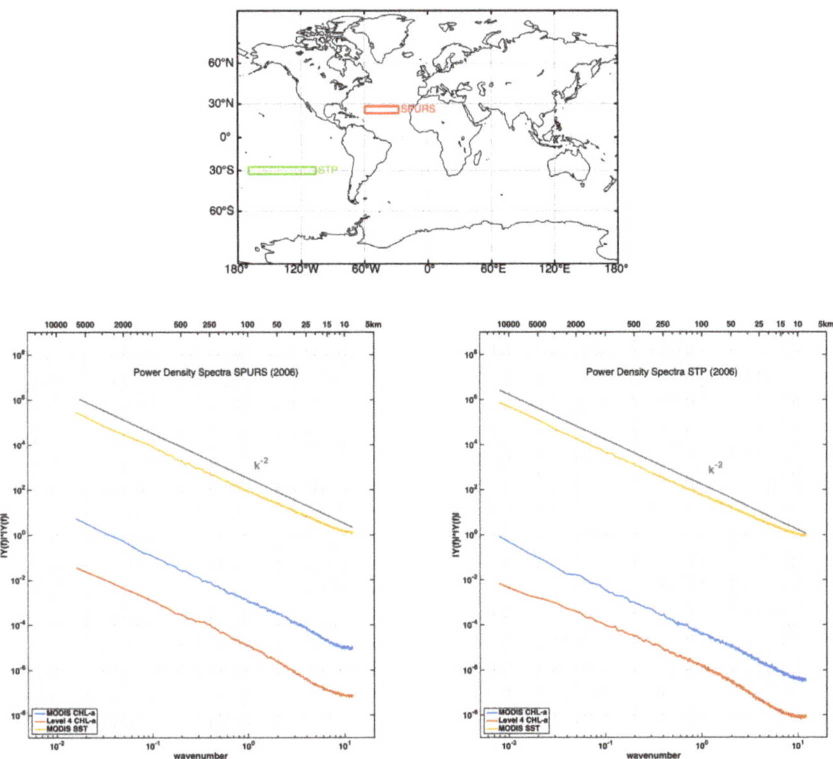

Figure 5. (**top**) Areas for power density spectra (PDS) computation: Red box represent SPURS region (22°–28°N and 28°–60°W) and green box represent STP region (27°–33°S and 106°–170°W). (**bottom**) Median of PDS for the year 2006 of MODIS L3 SST, MODIS Chl-*a*, and Level 4 Chl-*a* for the SPURS region (**left**) and the STP region (**right**).

5. Discussion and Conclusions

Satellite missions that measure chlorophyll concentration offer limited coverage due to orbital gaps, the presence of clouds and aerosols, sun glint, poor retrieval of the geophysical variables, etc. Combining data from several satellites can considerably increase the spatial coverage of daily maps. In this paper, we have used a method to merge the information coming from two different ocean scalars that share common multifractal characteristics. One of the two scalars, denoted as template, is assumed to have higher spatial coverage and signal-to-noise ratios than the other one, denoted as signal. The application presented here uses MODIS SST as template and MODIS Chl-*a* as signal. The template is used to extrapolate the signal by estimating local linear regression coefficients between Chl-*a* and SST, then applying them to the template to restore the signal.

The regression coefficients of the data fusion method exhibit geographical patterns that contain information about the relation between Chl-*a* and SST. The linear approximation, based on the hypothesis that the gradients of the regression coefficients are negligible as compared with the gradients of the variables Chl-*a* and SST, is less accurate on equatorial Pacific, Indian oceans, and during the winter season. In all these cases the present scheme needs to be extended. Besides, auxiliary environmental information, such as surface winds, mixed layer depth or nutrient concentration must be taken into account to fully understand the Chl-*a* behavior.

By exploiting the synergy between Chl-*a* and SST, it has been possible to increase the daily spatial coverage of MODIS Chl-*a*. The extrapolated fields have proved to be consistent with the observed ocean structures. A cross validation of the methodology, which consisted on create artificial gaps, apply the methodology and compare to the original Chl-*a* data, indicates correlation coefficients ranging from 0.67 to 0.94, except in winter season, when the correlation coefficients range from 0.15 to 0.47.

In conclusion, we have demonstrated the efficiency of blending SST and Chl-*a* to improve spatial coverage of chlorophyll products and to study the relation between ocean scalars. The resulting information can help to further characterize the spatio-temporal correlation between SST and Chl-*a*, and the temporal variation and long-term evolution of the biogeochemical provinces.

Author Contributions: Conceptualization, A.T. and M.U.; methodology, A.T. and M.U.; software, M.U., S.G. and J.B.P.; validation, M.U., S.G. and J.B.P.; writing—original draft preparation, M.U.; writing—review and editing, M.U., S.G., J.B.P. and A.T.; visualization, M.U., S.G. and J.B.P.; supervision, A.T. and J.B.P.; funding acquisition, A.T. and M.U. All authors have read and agreed to the published version of the manuscript.

Funding: M. Umbert is funded by the European Union's Horizon 2020 research and innovation programme under the Marie Skłodowska-Curie Individual Fellowship Career Restart Panel (MSCA-IF-EF-CAR Number 840374).

Acknowledgments: The authors acknowledge Ocean Color service website for making publicly available their MODIS-Aqua data at http://oceancolor.gsfc.nasa.gov/. We acknowledge the insightful reviews, comments and suggestions by anonymous reviewers that helped improve the content and readability of the manuscript.

Conflicts of Interest: The authors declare no conflict of interest.

References

1. Smith, T.M.; Reynolds, R.W. Improved extended reconstruction of sst (1854–1997). *J. Clim.* **2004**, *17*, 2466–2477. [CrossRef]

2. Buongiorno, B. A novel approach for the high-resolution interpolation of in situ sea surface salinity. *J. Atmos. Ocean. Technol.* **2012**, *29*, 867–879.

3. Alvera-Azcarate; Barth, A.J.; Weisberg, R. Multivariate reconstruction of missing data in sea surface temperature, chlorophyll and wind satellite fields. *J. Geophys. Res.* **2007**, doi:10.1029/2006JC003660. [CrossRef]

4. Sirjacobs, D.A.; Alvera-Azcarate, A.; Barth, G.; Lacroix, Y.; Park, B.; Nechad, K.; Ruddick, J.-M.B. Cloud filling of ocean colour and sea surface temperature remote sensing products over the southern north sea by the data interpolating empirical orthogonal functions methodology. *J. Sea Res.* **2011**, *65*, 114–130. [CrossRef]

5. Jouini Lévy, M.M.; Crépon, M.; Thiria, S. Reconstruction of satellite chlorophyll images under heavy cloud coverage using a neural classification method. *Remote. Sens. Environ.* **2013**, *131*, 232. [CrossRef]

6. Fablet, R.; Huynh Viet, P.; Lguensat, R.; Horrein, P.H.; Chapron, B. Spatio-Temporal Interpolation of Cloudy SST Fields Using Conditional Analog Data Assimilation. *Remote. Sens.* **2018**, *10*, 2310. [CrossRef]

7. Lguensat, R.; Teo, P.; Ailliot, P.; Pulido, M.; Fablet, R. The analog data assimilation. *Mon. Weather Rev.* **2017**, *145*, 4093–4107. [CrossRef]

8. Crocke, E.R.W.; Matthews, D.; Baldwin, D. Computing ocean surface currents from infrared and ocean color imagery. *IEEE Trans. Geosci. Remote Sens.* **2007**, *45*, 435–447. [CrossRef]

9. Isern-Fontanet; Turiel, J.A.; Garcia-Ladona, E.; Font, J. Microcanonical multifractal formalism: Application to the estimation of ocean surface velocities. *J. Geophys. Res.* **2007**, *112*, C05024.10.1029/2006JC003878. [CrossRef]

10. Turiel, A.H.; Pérez-Vicente, C. Microcanonical multifractal formalism: A geometrical approach to multifractal systems. Part I: Singularity analysis. *J. Phys.* **2008**, *41*, 015501.10.1088/1751-8113/41/1/015501. [CrossRef]

11. Denman, K.L.; Abbott, M.A. Time scales of pattern evolution from cross-spectrum analysis of advanced very high resolution radiometer and coastal zone color scanner imagery. *J. Geophys. Res.* **1994**, *99*, 7433–7442. [CrossRef]

12. Doney, S.C.; Glover, D.M.; McCue, S.J.; Fuentes, M. Mesoscale variability of SeaWiFS satellite Ocean Color: Global Patterns and spatial scales. *J. Geophys. Res.* **2003**, *108*, 6.1–6.15. [CrossRef]

13. Nieves, V.C.; Llebot, A.; Turiel, J.; Solé, E.; García-Ladona, M.; Estrada; Blasco, D. Common turbulent signature in sea surface temperature and chlorophyll maps. *Geophys. Res. Lett.* **2007**, *34*, L23602. [CrossRef]

14. Frisch, U. *Turbulence: The Legacy of A.N. Kolmogorov*; Cambridge University Press: Cambridge, MA, USA, 1995.

15. Turiel, A.; Parga, N. The multi-fractal structure of contrast changes innatural images: From sharp edges to textures. *Neural Comput.* **2000**, *12*, 763–793. [CrossRef] [PubMed]

16. Olmedo, E.J.; Martínez, M.; Umbert, N.; Hoareau, M.; Portabella, J.; Ballabrera-Poy, A. Improving time and space resolution of smos salinity maps using multifractal fusion. *Remote. Sens. Environ.* **2016**, *180*, 246–263. [CrossRef]

17. Umbert, M.; Hoareau, N.; Turiel, A.; Ballabrera-Poy, J. New blending algorithm to synergize ocean variables: The case of SMOS sea surface salinity maps. *Remote. Sens. Environ.* **2014**, *146*, 172–187.10.1016/j.rse.2013.09.018. [CrossRef]

18. Brown; O; Minnett, P.J. Modis Infrared Sea Surface Temperature Algorithm Algorithm Theoretical Basis Document. Version 2.0. Available online: https://modis.gsfc.nasa.gov/data/atbd/atbd_mod25.pdf (accessed on 2 April 2020).

19. Campbell, J.W. The lognormal distribution as a model for bio-optical variability in the sea. *J. Geophys. Res.* **1995**, *100*, 237–254. [CrossRef]

20. Turiel, A.; Nieves, V.; García-Ladona, E.; Rio, M.H.; Larnicol, G. The multifractal structure of satellite temperature images can be used to obtain global maps of ocean currents. *Ocean. Sci.* **2009**, *5*, 447–460. [CrossRef]

21. Valdez-Holguin, J.E.; Lara-Lara, J.R. Primary productivity in the gulf of california effects of el 350 niño 1982-1983 event. *Cienc. Mar.* **1987**, *13*, 34–50. [CrossRef]

22. Longhurst, A.R. *Ecological Geography of the Sea*; Academic Press: Cambridge, MA, USA, 1998.

23. Hoareau, N.; Turiel, A.; Portabella, M.; Ballabrera-Poy, J.; Vogelzang, J. Singularity power spectra: A method to assess geophysical consistency of gridded products—application to sea-surface salinity remote sensing maps. *IEEE Trans. Geosci. Remote. Sens.* **2018**, *56*, 5525–5536. [CrossRef]

 *remote sensing* 

Article

Quantifying Tidal Fluctuations in Remote Sensing Infrared SST Observations

Cristina González-Haro *,†,‡, **Aurélien Ponte** ‡ **and Emmanuelle Autret**

Laboratoire d'Océanographie Physique et Spatiale (LOPS), IUEM, Univ. Brest, CNRS, IRD, Ifremer,
29238 Brest, France; aurelien.ponte@ifremer.fr (A.P.); emmanuelle.autret@ifremer.fr (E.A.)
* Correspondence: cgharo@icm.csic.es; Tel.: +34-932-309-632
† Current address: Physical & Technological Oceanography Department, Barcelona Expert Center Institut de
Ciencies del Mar (ICM), Barcelona, Spain (CSIC).
‡ These authors contributed equally to this work.

Received: 13 September 2019 ; Accepted: 3 October 2019; Published: 4 October 2019

Abstract: The expected amplitude of fixed-point sea surface temperature (SST) fluctuations induced by barotropic and baroclinic tidal flows is estimated from tidal current atlases and SST observations. The fluctuations considered are the result of the advection of pre-existing SST fronts by tidal currents. They are thus confined to front locations and exhibit fine-scale spatial structures. The amplitude of these tidally induced SST fluctuations is proportional to the scalar product of SST frontal gradients and tidal currents. Regional and global estimations of these expected amplitudes are presented. We predict barotropic tidal motions produce SST fluctuations that may reach amplitudes of 0.3 K. Baroclinic (internal) tides produce SST fluctuations that may reach values that are weaker than 0.1 K. The amplitudes and the detectability of tidally induced fluctuations of SST are discussed in the light of expected SST fluctuations due to other geophysical processes and instrumental (pixel) noise. We conclude that actual observations of tidally induced SST fluctuations are a challenge with present-day observing systems.

Keywords: sea surface temperature; satellite observations; tidal currents; internal tides

---

1. Introduction

Internal waves (IW) have recently been shown to significantly contribute to sea level variability at scales smaller than about 100 km [1–3]. The contamination of IW at theses scales limits our ability to infer ocean currents from altimetric data because IW currents are not related to sea level via geostrophy unlike slower/mesoscale balanced structures [4]. Some IW are of tidal origin and stationary with respect to astronomical forcings, such that the signatures of these waves on altimetric data can be predicted harmonically and eventually removed [5]. Unfortunately most IW are either

- from tidal origin but have lost their phase relationship with tidal forcing after interaction with the slower oceanic turbulence (nonstationary internal tides, see Ponte and Klein [6] and Zaron [7]) or
- from non tidal origin (e.g., wind-forced, lee-waves).

In both cases, we have no means today to predict these waves.

A direction of research that has been poorly explored yet deals with the signature of IW on satellite data of non-altimetric nature (e.g., sea surface temperature, color) and potential synergies in order to disentangle IW and balanced signatures in altimetric data. We recently proposed in an idealized context a tentative method in order to carry such synergies [8]. While it is premature to apply this method with real data, it may be time to start collecting remote sensing data that: (1) will allow us to verify critical assumptions for such methods (e.g., smallness of IW signature on SST image) and (2) will ultimately be suitable in order to test such methods. The present work is relevant to the former task. We more precisely focus on the quantification of SST fluctuations induced by stationary baroclinic tides because atlases of their distributions are now available [9]. We will also quantify SST fluctuations induced by barotropic tides because the same methodology may be employed for that purpose.

Sea Surface Temperature (SST) is a challenging parameter to define precisely as the upper ocean (within the first 10 m) has complicated vertical variability that is linked to ocean turbulence and air-sea heat fluxes [10]. Methods for determining SST from satellite remote sensing include thermal infrared (IR) and passive microwave radiometry. Both methods have strengths and weaknesses. Thermal IR SST measurements are derived from radiometric observations at wavelengths of $\sim$3.7 μm and/or near 10 μm. They provide high spatial resolution SST observations ($\sim$1 to 10 km) and good accuracy (0.1–0.8 K) [11,12], however they are affected by cloud coverage and provide observations only for cloud-free pixels. On the contrary, microwave observations (4–10 GHz) provide a better spatial coverage since they are not affected by cloud coverage but their spatial resolution ($\sim$25 to 50 km) is coarser than IR observations ([13] and references therein). In addition, SST measured from space are representative of a depth that is related to the frequency of the satellite instrument. For example, IR instruments measure a depth of about 20 μm, while microwave radiometers measure a depth of a few millimeters [14].

Internal wave modulations of SST from IR aerial observations have been reported at kilometric scales in low-wind conditions [15–19]. Two mechanisms have been considered in order to explain these modulations: fluctuations of the cool-skin temperature (The ocean surface is generally 0.1–0.6 °C cooler than the temperature just below the surface. And this "skin", or ultra-thin region, is less than a 1mm thick. For further details see http://ghrsst-pp.metoffice.com/pages/documents/DocumentFiles/GDS-v1.0-rev1.5.pdf [last access 8 August 2019]) induced by the internal wave straining field and modulations of the upper diurnal surface layer by vertical displacements of the seasonal thermocline. Internal waves imprint their spatial structure on SST in the aforementioned papers. The present work focuses instead on quantifying of Eulerian, i.e., fixed point, modulations of a pre-existing SST distribution induced by tidal currents and the spatial structure of these modulations is thus not expected to reflect the structure of tidal motions.

This study is based on SST satellite observations, tidal current atlases and an atlas of SST gradients which are described in Section 2. The method to quantify internal wave signature on IR SST observations is explained in Section 3 and results are shown in Section 4. Finally, Section 5 discusses to which extent we may capture SST fluctuations of tidal origin in satellite observations and the main limitations for observing these fluctuations (i.e., pixel noise). All the acronyms used in the manuscript are detailed in the Abbreviation table at the end of the manuscript.

2. Data

2.1. Tidal Current Atlases

Barotropic tidal currents are extracted from FES2014 which is the current version of the FES (Finite Element Solution) tidal database [20]. Tidal solutions are obtained from an assimilation of tide gauges and altimetric data and delivered on a 1/16° grid. Tidal currents for the M2 constituents are typically of the order to 2–3 cm/s in the open ocean and exhibit a spatial structure characterized by large spatial scales modulated by topography (Figure 1b).

Baroclinic tidal currents are derived from the High Resolution Empirical Tide (HRET) Models [http://web.cecs.pdx.edu/ zaron/pub/HRET.html]. This database has been created to provide baroclinic tide corrections for sea level measurements collected by the upcoming SWOT mission [21]. It was created from an harmonic analysis of sea level anomalies collected by exact repeat altimetric missions (1992–2015). Maps of tidal sea level are provided on a 0.05-degree near-global grid for the 6 most important constituents (M2, S2, K1, O1, N2, and P1 tides). Tidal currents are derived from sea level maps assuming the following momentum equations [9]:

$$-i\omega u - fv = -g\partial_x\eta - u/\tau, \tag{1}$$

$$-i\omega v + fu = -g\partial_y\eta - v/\tau, \tag{2}$$

where u, v are zonal and meridional currents, η represents sea level, g is gravity, $\tau = 20$ days is damping time scale, and ω is the tidal frequency. Solutions to (1)–(2) are given by:

$$u = \frac{-i\omega_\tau g\partial_x\eta + fg\partial_y\eta}{\omega_\tau^2 - f^2}, \tag{3}$$

$$v = \frac{-fg\partial_x\eta - i\omega_\tau g\partial_y\eta}{\omega_\tau^2 - f^2}, \tag{4}$$

where $\omega_\tau = \omega + i/\tau$. M2 baroclinic currents exhibit a spatial structure characterized by interfering beams originating from well-defined generation spots (e.g., islands archipelagos, sills) and typical amplitudes around 10 cm/s in the open-ocean (Figure 1a)

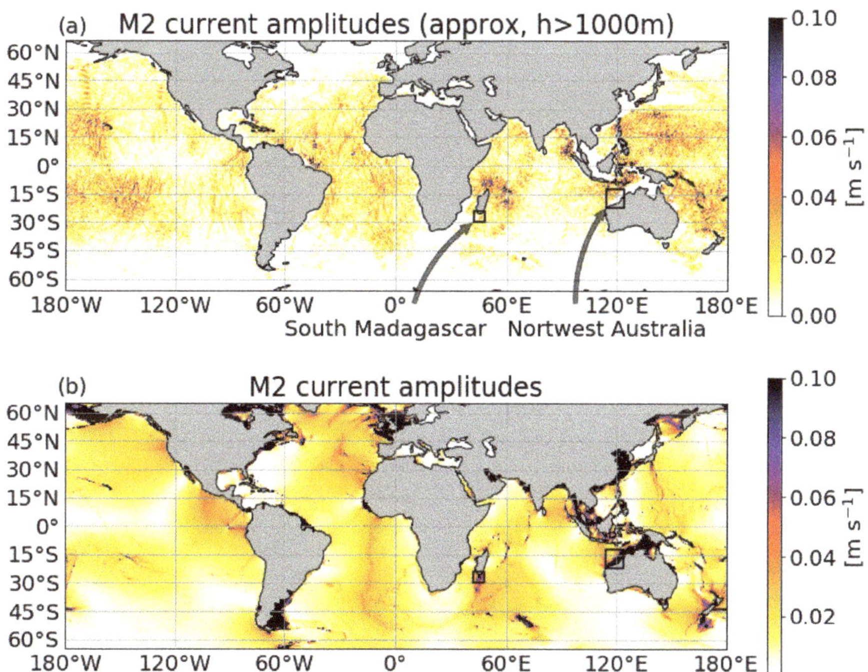

Figure 1. (a) HRET M2 baroclinic current amplitude for regions with bathymetry(h) deeper than 1000 m (b) FES2014 M2 barotropic current amplitude. The colormap is saturated at 10 cm/s on both figures. Two black boxes delimit the regions of the case study presented in Sections 4.1 and 4.2, respectively.

2.2. Atlas of SST Gradients

The climatology of the maximum SST gradient magnitude is courtesy of Peter Cornillon, Graduate School of Oceanography, University of Rhode Island (URI). This climatology is obtained from the entire (1985–1996) Pathfinder 9 km resolution SST dataset and is based on the automated procedure by [22] and [23–25] and developed by [26–28]. Further technical details may be found in this report from the Danish Meteorological Institute (DMI) [29]. The distribution of maximum SST gradient indicates SST spatial gradients exceeding 0.5 °C near western boundary currents, upwelling areas, the antarctic circumpolar current (Figure 2). The coarser resolution of the SST used for this climatology compared to the SST observations described in Section 2.3 may impact the expected amplitudes of tidally induced SST fluctuations using the climatology. The later may be underestimated compare to what would be estimated with higher resolution SST data to an extent that is difficult to anticipate.

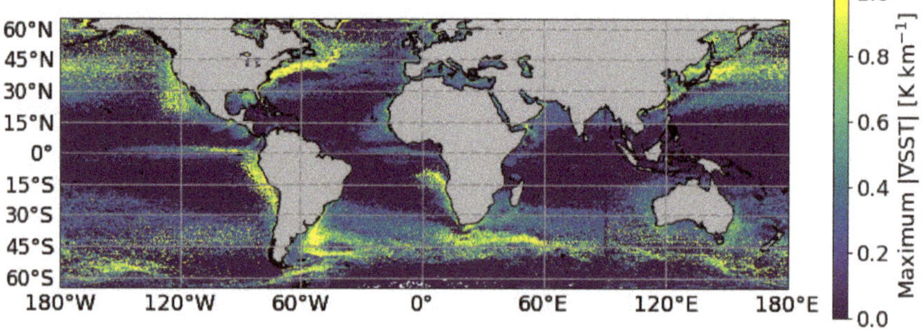

Figure 2. Maximum of the annual climatology of the maximum spatial gradient of SST from University of Rhode Island (URI) Pathfinder 9 km frontal database.

2.3. SST Observations

We use infrared GHRSST SST images captured by several orbiting satellites, particularly, we explored AVHRR-METOP-A (OSISAF) [30], VIIRS and MODIS. In order to find cloud free images, we automatically browsed the entire L2 data catalog for a given region and extracted granules that satisfied a priori selection criteria. The selection procedure consists in dividing the granules in small patches of a fixed size (100×100 pixels) and evaluating the ratio of free cloud pixel for each patch individually. To ensure a good spatial coverage, the granule is selected if cloud-free pixel ratio is equal or higher than 80% in at least 30 patches. Finally, the selected granules are remapped onto a regular grid with a spatial resolution of 0.02°. We have explored the L2 data archive from 1 January 2014 to 31 December 2016 for AVHRR-METOP, VIIRS and MODIS.

3. Method

In order to estimate SST fluctuations associated with tidal currents, we assume that tidal motions follow a dynamics that is linear about other motions and adiabatic. Among both assumptions, that of linearity may be occasionally broken for short solitary-type internal waves of tidal origins but is expected valid for open-ocean low-mode internal tides and barotropic tides. Under the aforementioned assumptions, tidal currents transport SST gradients according to:

$$\partial_t T_w = -u_w \partial_x T_s - v_w \partial_y T_s,$$ (5)

where T_w is the variation of SST due to tidal currents (w stands for wave), u_w and v_w are the zonal and meridional components of tidal currents, respectively, and T_s stands for the SST. Assuming that:

$$T_w = \Re(T_c e^{-i\omega t}), \quad u_w = \Re(u_c e^{-i\omega t}), \quad v_w = \Re(v_c e^{-i\omega t}),$$ (6)

where $T_c = T_r + iT_i$, similarly for u_c and v_c, and ω is the tidal frequency (for M2 constituent $\omega = 1.405 \times 10^{-4}$ rad/s), the amplitude of tidal fluctuations of SST can be estimated as:

$$|T_c| = \left| \frac{u_c \partial_x T_s + v_c \partial_y T_s}{i\omega} \right| \tag{7}$$

The quantification of the SST tidal fluctuations is splitted in three steps:

1. **Computation of the spatial SST gradient**: The Sobel gradient [31,32] of SST is computed from free cloud infrared SST observations. The Sobel operator uses two 3×3 kernels and it emphases pixels that are closer to the center of the mask. (Figures 3b and 4b)
2. **Tidal current amplitudes**: the tidal current amplitudes are either extracted from the data base (barotropic) (Figures 3c and 4c) or derived from the sea level (baroclinic) and then remapped onto the grid of SST observations using bilinear interpolation.
3. **SST tidal fluctuations**: the final step consists in evaluating expression (7) using the SST gradients and the amplitude of tidal currents obtained in the two previous steps. (panels (d) and (f) of Figures 3 and 4)

4. Results

A first attempt at quantifying the expected amplitude of fixed-point SST fluctuations induced by barotropic and baroclinic tidal flows is presented here, by means of two case studies (Sections 4.1 and 4.2) using infrared SST observations, and at a global scale using clymatological analysis of SST gradients (Section 4.3). The two case studies represent a trade-off between clear sky conditions (see Figure A1) and the intensity of tidal currents (Figure 1).

4.1. Northwest Australia

Guided by maps of clear sky probability (Figure A1) and tidal current maps (Figure 1), we focused first on a region located Northwest of Australia, with latitudes between 22°S and 12°S, and longitudes between 114°E and 124°E. We browsed the SST L2 archive (METOP, VIIRS and MODIS) from 2014 to 2016 and selected granules that accomplish at least 30 (100 pixels × 100 pixels) patches with 80% free cloud pixels. Table 1 summarizes the number of available granules for this region. The number of days with more than one L2 granule available is 303 days, which represents a 27% of the analyzed period. The SST corresponding to 8 September 2016 was in particular selected based on this criterium. Gradients of SST were computed and exhibit moderate values around 0.2 °C km^{-1}.

Table 1. Number of L2 granules available from 1 January 2014 to 31 December 2016 in the Northwest Australia case study region.

	Number of Available L2 Granules			
	2014	**2015**	**2016**	**TOTAL**
METOP	181	124	185	490
VIIRS	160	147	120	427
MODIS	59	152	120	331

In this area, semidiurnal internal tide currents (0.08–0.12 m s^{-1}) are typical of what may be found in the open ocean (Figure 1). These currents are masked over the continental shelf because HRET is not reliable there [9]. Barotropic currents are amplified over the shelf with M2 tidal currents up to 0.6–0.8 m s^{-1}.

Figure 3 illustrates the procedure to quantify the signature of M2 SST fluctuations for the Northwest Australia case study region. We first compute the Sobel gradient of the SST field then we retrieve M2 baroclinic and barotropic currents for regions. Finally the amplitude of SST fluctuations associated with baroclinic and barotropic currents are estimated by taking the product of SST gradients and tidal currents (Equation (7)).

Both baroclinic and barotropic tidal SST fluctuations are intensified on fronts and filaments. Barotropic fluctuations are largest on the shelf with values up to 0.3 °C and no significant signal in the open ocean (see Figure 3d). Baroclinic SST fluctuations are one order of magnitude smaller, being 0.03–0.04 °C (see Figure 3f).

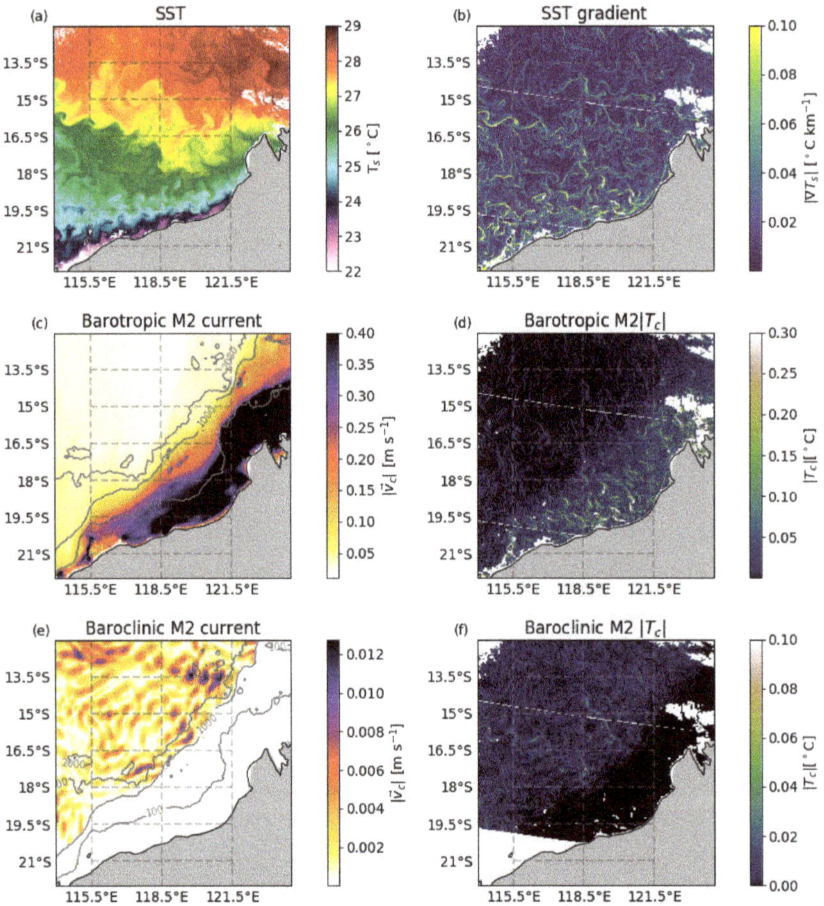

Figure 3. (**a**) SST captured by VIIRS on the 8 September 2016. (**b**) Sobel gradient of SST. Note that the SST field shown here is composed of three L2 granules. The gradient is computed for each granule individually. Thus, the representation of the gradient presents two scan lines with missing data that correspond to the boundaries of the granules. (**c**) **M2 barotropic FES** amplitude current. Gray contour correspond to the bathymetry. (**d**) Estimation of the amplitude of M2 signature on SST. (**e**) **M2 baroclinic HRET** amplitude current for areas with a bathymetry deeper than 1000m. Gray contour lines correspond to the bathymetry. (**f**) Estimation of the amplitude of M2 IW signature on SST.

4.2. South Madagascar

Based again on clear sky maps and tidal currents maps, we selected a second region South of Madagascar (latitude between 30°S and 24°S , longitude between 42°E and 48°E). The number of

L2 granules available between 1 January 2014 and 31 December 2016 are summarized in Table 2. For this region, the number of days with more than one granule is 267 (24 % of the period considered). As shown by the METOP SST observation of 5 June 2014 (Figure 4), this region is characterized by strong SST gradients (0.3-0.4 $^{\circ}$C km^{-1}) see Figure 4b).

Both baroclinic and barotropic tidal currents are weaker in this area with values of the order of 0.07 m s^{-1} and 0.3 m s^{-1}, respectively. Barotropic currents extend to deep areas, which is different than the Northwest Australia case, shown in Figure 3c.

Figure 4 illustrates the procedure to quantify the signature of the semidiurnal M2 baroclinic motion on SST for the South of Madagascar case study region from a SST image captured by METOP on the 5 June 2014. M2 barotropic currents (see Figure 4c) induce SST fluctuations that are up to 0.2 $^{\circ}$C which is less than for the Northwest Australia case. The maximum amplitude of M2 baroclinic tidal SST fluctuations reaches 0.08 $^{\circ}$C for this particular case (Figure 4e), i.e., about twice than for the Northwest Australia region. These two opposite trends emphasizes that SST tidal fluctuations result from the product between SST gradients and tidal currents.

Table 2. Number of L2 granules available from 1 January 2014 to 31 December 2016 in the Madagascar case study region.

	Number of Available L2 Granules			
	2014	**2015**	**2016**	**TOTAL**
METOP	130	147	114	391
VIIRS	70	133	145	348
MODIS	143	111	40	249

4.3. Global Scale

A global perspective on SST tidal fluctuations (Figure 5) is obtained by combining the map of maximum SST gradients (Figure 2) and maps of tidal current amplitudes (Figure 1) as described in Section 3. In this case, since the maximum SST spatial gradient climatology has coarser spatial resolution, it was remapped to onto the grid of the tidal current amplitudes using a bilinear interpolation. Over Northwest Australia and South Madagascar, climatological values of maximum gradients are about 0.3 $^{\circ}$C/km and 0.4 $^{\circ}$C/km respectively, which is between 2 and 6 times more than maximum SST gradients in the two scenes selected (Figures 3 and 4). The same ratio directly reflects into expected tidal SST fluctuation amplitudes.

Baroclinic tidal currents cause largest SST fluctuations around western boundary currents main flow paths, except in the case of the Gulf Stream where the baroclinic tidal current is weak or non-existent (see Figure 1), and nearby upwelling areas. The signal is in general weaker for latitudes within $\pm 15^{\circ}$ equatorial areas. When barotropic currents are large, they induce largest SST signals, i.e., in coastal areas, and with a geographical distributions that thus significantly differs from that associated to baroclinic currents. The amplitude of barotropic tidal SST fluctuations is about three times larger than the one induced by baroclinic currents.

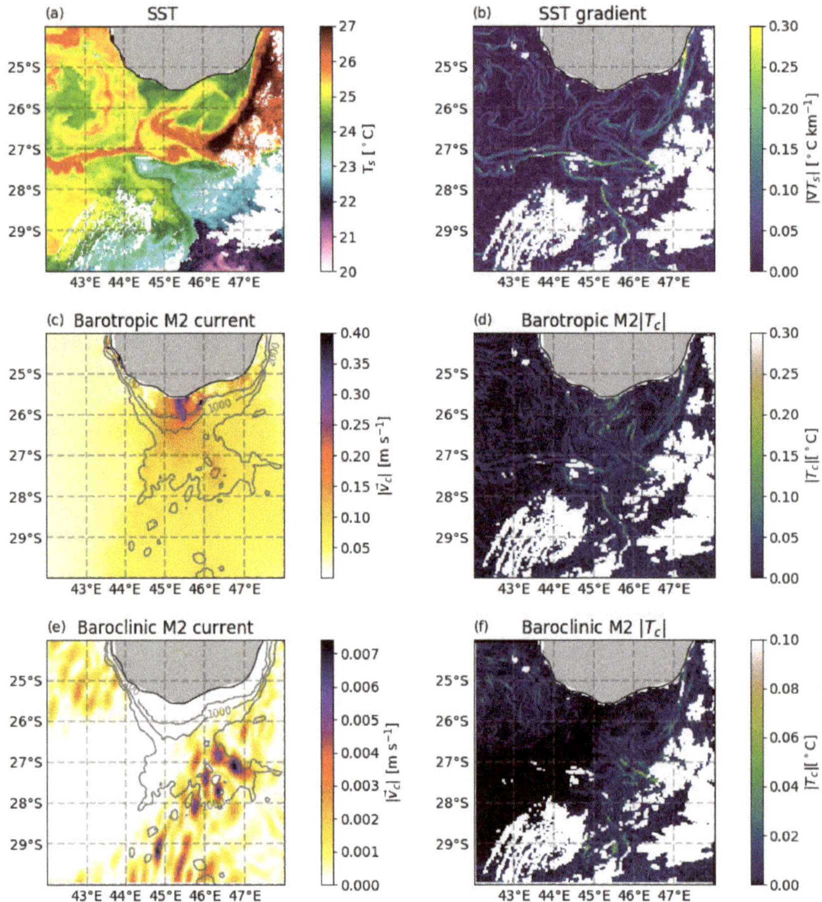

Figure 4. (**a**) SST captured by METOP on the 5 June 2014. (**b**) Sobel gradient of SST. (**c**) **M2 barotropic FES** amplitude current. Gray contour correspond to the bathymetry. (**d**) Estimation of the amplitude of M2 barotropic signature on SST. (**e**) **M2 baroclinic HRET** amplitude current for areas with a bathymetry deeper than 1000m. Gray contour lines correspond to the bathymetry. (**f**) Estimation of the amplitude of M2 IW signature on SST.

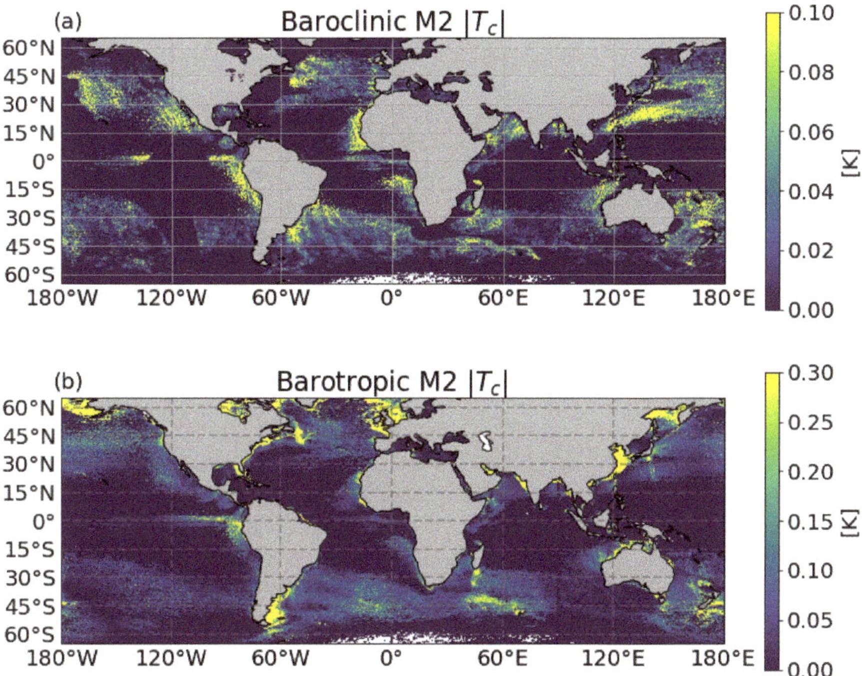

Figure 5. (**a**) Estimation of the amplitude of M2 IW signature on SST. (**b**) Estimation of the amplitude of M2 barotropic signature on SST.

5. Discussion and Conclusions

Baroclinic tidal currents derived from the HRET database correspond to the stationary part of the internal tide signal, i.e., fluctuations that keep a fixed phase relationship with astronomical forcings. These currents represent a fraction of actual baroclinic tidal currents [7]. Therefore, present estimates of baroclinic tidal SST fluctuations may underestimate true fluctuations. This is in particular true over western boundary currents paths (see SST fluctuations minima on Figure 5), where the baroclinic tides are non-stationary to a large extent.

For the sake of brevity, this study has focused on one tidal constituent (M2). Constructive interference with other constituents may therefore modulate present estimates of SST tidal fluctuations. The present analysis may be adapted in order to quantify the amplitude of these modulations. It may also be adapted in order to estimate SST fluctuations by other types of fast oceanic motions such as near-inertial waves [33]. Near-inertial currents are for example at least as energetic as baroclinic tidal ones.

Can we hope to capture SST tidal fluctuations in satellite observations? Consecutive observations of SST may allow the identification of propagating SST fluctuations. This identification may be conditioned by the relative importance of SST tidal fluctuations compared to that induced by other geophysical processes (diurnal cycle, mesoscale, submesoscale motions) and noise. SST tidal fluctuations inherits their spatial structures from frontal features, i.e., of the order of kilometers, which is marginally larger than satellite resolutions. These small scale fluctuations will occur coherently over tidal currents spatial scales (Figure 1) nonetheless, and this may be leveraged in order to identify SST tidal fluctuations. The diurnal cycle of SST is expected to have large spatial scales and thus be clearly distinguishable from SST fluctuations induced by tides [34]. Mesoscale and submesoscale fluctuations have comparable spatial structures on the other hand, but are slower and non-propagating which may be leveraged in order to distinguish their contributions to SST fluctuations. The small scale structure of tidal SST fluctuations indicates that pixel noise is the source of noise that may limit the observation of these fluctuations.

We have estimated the noise present in the granules used, assuming it is Gaussian [35]. Given this assumption, the standard deviation of the noise σ_ϵ can be estimated using the *K-clipping method* [36]. This method exploits the dominance of noise at short wavelengths and can be explained as follows. A first guess of σ_ϵ is obtained as the standard deviation of a high-pass frequency version of the initial field. Then those values with amplitude higher than K times σ_ϵ are rejected, and a new estimation of σ_ϵ is obtained from the standard deviation of the remaining values. This is performed iteratively until σ_ϵ is obtained, then. In practice three iterations ($K = 3$) are enough [36]. The estimated noise standard deviation of the granules considered in the case studies presented in this work is shown in Table 3. It is in general comparable with or smaller than expected SST tidal fluctuations.

Despite the signature of IW on SST exhibit fine-scale spatial structures and thus IR SST observations are more suitable to quantify tidal fluctuation in SST, we explored whether microwave SST observations could be used to extract low-mode stationary internal tides. Figure A2 shows an example of microwave SST captured by AMSR2 for the same case study shown in Section 4.2. Only stronger gradients of SST are captured and its amplitude is halved compared to the IR SST field shown in Figure 4. Similar results have been already reported by [37] in other regions such as the Gulf Stream or California. Despite these weaker values, the larger amount of available microwave data may be beneficial for the extraction of temporally coherent tidal signals and this should be subject for future work.

The following steps are therefore to search for consecutive snapshots of SST in areas of interest. Infrared geostationary satellites may be useful for this purpose (Himawari, SEVIRI). The global maps in the present paper highlights several regions (on top of the two selected here) where tidal variability is important and cloud density is favorable: Patagonian shelf, west Florida shelf, Moroccan shelf, Persian Gulf, East Indian shelf, Strait of Gibraltar.

Table 3. Noise standard deviation of the SST images considered.

Granule ID	Sensor	Region	σ_ϵ (K)
20160908174014	VIIRS	Northwest Australia	0.03
20160908174139	VIIRS	Northwest Australia	0.03
20160908174305	VIIRS	Northwest Australia	1.68
20140605062503	AVHRR	South Madagascar	0.67
20140605184903	AVHRR	South Madagascar	0.06

Author Contributions: The idea of this study was initially designed by Aurélien Ponte. C.G.-H. did most of the work during her PostDoctoral contract at IFREMER. E.A. developed the software to select L2 SST granules from the entired data catalog. All the authors contributed to formal analysis. The paper was written by C.G.-H. and A.P. E.A. contributed to the writing-review and editing of the paper.

Funding: Aurélien Ponte benefited from funding via the ANR project EQUINOx (ANR-17-CE01-0006-01) and CNES TOSCA project "New dynamical tools for submesoscale characterization in SWOT data". Cristina González-Haro benefited from funding by CNES.

Acknowledgments: The authors would like to acknowledge Peter Cornillon (University of Rhode Island) and Stéphane Saux Picart (Météo-France) for providing the atlas of maximum gradient of SST, as well as Ed Zaron for providing the HRET internal tide database. The data from the EUMETSAT Satellite Application Facility on Ocean & Sea Ice used in this study are accessible through the SAF's homepage http://www.osi-saf.org. The MODIS L2P sea surface temperature data are sponsored by NASA. The data from the Naval Oceanographic Office are made available under Multi-sensor Improved Sea Surface Temperature (MISST) project sponsorship by the Office of Naval Research (ONR). The authors want to acknowledge the anonymous reviewers for their valuable and helpful comments.

Conflicts of Interest: The authors declare no conflict of interest. The funders had no role in the design of the study; in the collection, analyses, or interpretation of data; in the writing of the manuscript, or in the decision to publish the results.

Abbreviations

The following abbreviations are used in this manuscript:

ANR	Agence National de la Recherche
AMSR2	Advanced Microwave Scanning Radiometer 2
AVHRR	Advanced Very High Resolution Radiometer
CNES	Centre National d'Études Spatiales
CSIC	Consejo Superior de Investigaciones Científicas
FES	Finite Element Solution
GHRSST	Group for High Resolution Sea Surface Temperature
HRET	High Resolution Empirical Tide
IFREMER	Institut Français pour l'Exploitation de la Mer
IR	Infrared
IW	Internal Wave
MODIS	MODerate resolution Imaging Spectroradiometer
OSISAF	Ocean and Sea Ice Satellite Application Facillity
SEVIRI	Spinning Enhanced Visible and InfraRed Imager
SST	Sea Surface Temperature
SWOT	Surface Water and Ocean Topography
VIIRS	Visible Infrared Imaging Radiometer Suite

Appendix A. Free Cloud Pixel Probability

The availability of IR observations is limited by the cloud coverage. Thus, it may be a key parameter to take into account when selecting the regions to study. Figure A1 is included as supporting information to select studied regions.

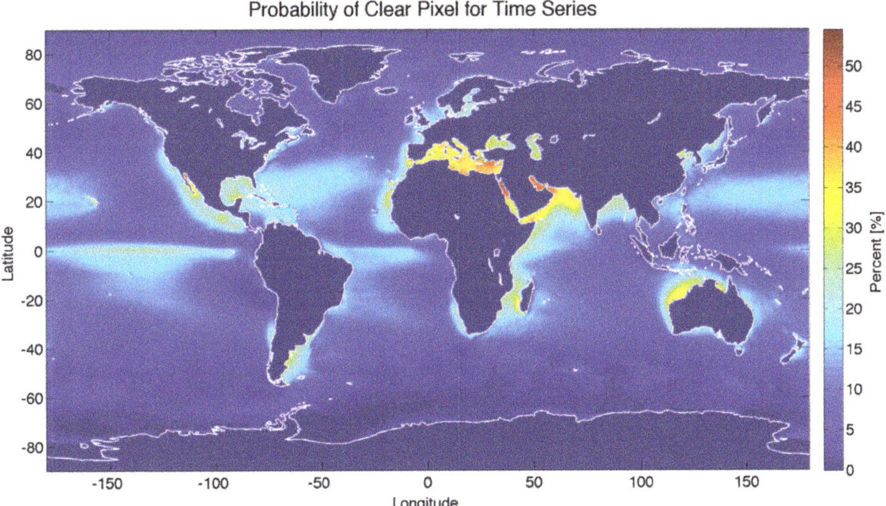

Figure A1. Probability of clear pixel for the time series (1982–2011) considered in the climatology of the maximum gradient of SST from University of Rhode Island (URI) Pathfinder 9 km frontal database (courtesy of Peter Cornillon Graduate School of Oceanography, University of Rhode Island (URI)).

Appendix B. Other SST Observations

Microwave radiometers observations suffer from poor spatial resolution but have great spatial coverage, since they are not affected by the cloud coverage. Their temporal coverage is good for balanced flows but not for IW. We explored to which extent the information about low-modes stationary internal tides (as well as barotropic tides while we're at it) could be extracted from such data. Figure A2 shows an example of a Microwave SST captured by AMSR2 for the same date than the case study in South Madagascar shown in Section 4.2. As it can be observed by comparing Figures 4 and A2, only stronger gradients can be observed in microwave observations, and its amplitude is reduced by half.

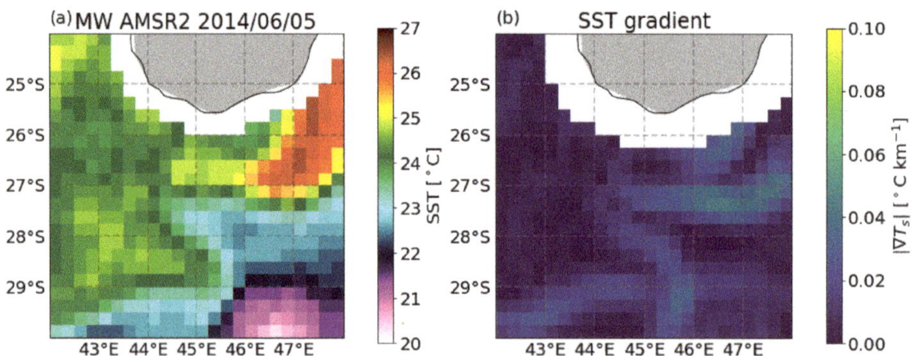

Figure A2. (**a**) SST captured by AMSR2 on the 5 June 2014, same date than SST shown in Figure 4 (**b**) Sobel gradient of SST captured by AMSR2.

References

1. Richman, J.G.; Arbic, B.K.; Shriver, J.F.; Metzger, E.J. Inferring dynamics from the wavenumber spectra of an eddying global ocean model with embedded tides. *J. Geophys. Res.* **2012**, *117*. [CrossRef]
2. Rocha, C.B.; Gille, S.T.; Chereskin, T.K.; Menemenlis, D. Seasonality of submesoscale dynamics in the Kuroshio Extension. *Geophys. Res. Lett.* **2016**, *43*, 11–304. [CrossRef]
3. Savage, A.C.; Arbic, B.K.; Richman, J.G.; Shriver, J.F.; Alford, M.H.; Buijsman, M.C.; Farrar, J.T.; Sharma, H.; Voet, G.; Wallcraft, A.J.; et al. Frequency content of sea surface height variability from internal gravity waves to mesoscale eddies. *J. Geophys. Res. Oceans* **2017**. [CrossRef]
4. Arbic, B.K.; Lyard, F.; Ponte, A.; Ray, R.; Richman, J.G.; Shriver, J.F.; Zaron, E.; Zhao, Z. *Tides and the SWOT mission: Transition from Science Definition Team to Science Team*; Technical Report; CNES: Toulouse, France; NASA: Washington, DC, USA, 2015.
5. Ray, R.D.; Zaron, E.D. M$_2$ Internal Tides and Their Observed Wavenumber Spectra from Satellite Altimetry. *J. Phys. Oceanogr.* **2016**, *46*, 3–22. [CrossRef]
6. Ponte, A.L.; Klein, P. Incoherent signature of internal tides on sea level in idealized numerical simulations. *Geophys. Res. Lett.* **2015**, *42*. [CrossRef]
7. Zaron, E.D. Mapping the Non-Stationary Internal Tide with Satellite Altimetry. *J. Geophys. Res. Oceans* **2016**, *122*, 539–554. [CrossRef]
8. Ponte, A.L.; Klein, P.; Dunphy, M.; Gentil, S.L. Low-mode internal tides and balanced dynamics disentanglement in altimetric observations: Synergy with surface density observations. *J. Geophys. Res. Oceans* **2017**. [CrossRef]
9. Zaron, E.D. Baroclinic Tidal Sea Level from Exact-Repeat Mission Altimetry. *J. Phys. Oceanogr.* **2019**, *49*, 193–210. [CrossRef]
10. Donlon, C.J.; The GHRSST Science Team. *The Recommended GHRSST-PP Data Processing Specification*; Technical Report v1 revision 1.6; GHRSST: Exeter, UK, 2005.

11. Donlon, C.; Rayner, N.; Robinson, I.; Poulter, D.; Casey, K.; Vazquez-Cuervo, J.; Armstrong, E.; Bingham, A.; Arino, O.; Gentemann, C.; et al. The global ocean data assimilation experiment high-resolution sea surface temperature pilot project. *Bull. Am. Meteorol. Soc.* **2007**, *88*, 1197–1213. [CrossRef]

12. Maturi, E.; Harris, A.; Mittaz, J.; Sapper, J.; Wick, G.; Zhu, X.; Dash, P.; Koner, P. A New High-Resolution Sea Surface Temperature Blended Analysis. *Bull. Am. Meteorol. Soc.* **2017**, *98*, 1015–1026. [CrossRef]

13. Hosoda, K. A review of satellite-based microwave observations of sea surface temperatures. *J. Oceanogr.* **2010**, *66*, 439–473. [CrossRef]

14. Donlon, C.; Minnett, P.; Gentemann, C.; Nightingale, T.; Barton, I.; Ward, B.; Murray, M. Toward improved validation of satellite sea surface skin temperature measurements for climate research. *J. Clim.* **2002**, *15*, 353–369. [CrossRef]

15. Walsh, E.J.; Pinkel, R.; Hagan, D.; Weller, R.; Fairall, C.; Rogers, D.; Burns, S.; Baumgartner, M. Coupling of internal waves on the main thermocline to the diurnal surface layer and sea surface temperature during the Tropical Ocean-Global Atmosphere Coupled Ocean-Atmosphere Response Experiment. *J. Geophys. Res. Oceans* **1998**, *103*, 12613–12628. [CrossRef]

16. Marmorino, G.O.; Smith, G.B.; Lindemann, G.J. Infrared imagery of ocean internal waves. *Geophys. Res. Lett.* **2004**, *31*. [CrossRef]

17. Zappa, C.J.; Jessup, A.T. High-Resolution Airborne Infrared Measurements of Ocean Skin Temperature. *IEEE Geosci. Remote Sens. Lett.* **2005**, *2*, 146–150. [CrossRef]

18. Farrar, J.T.; Zappa, C.J.; Weller, R.A.; Jessup, A.T. Sea surface temperature signatures of oceanic internal waves in low winds. *J. Geophys. Res.* **2007**, *112*. [CrossRef]

19. Susanto, R.; Pan, J.; Devlin, A. Tidal Mixing Signatures in the Hong Kong Coastal Waters from Satellite-Derived Sea Surface Temperature. *Remote Sens.* **2019**, *11*, 5. [CrossRef]

20. Carrere, L.; Lyard, F.; Cancet, M.; Guillot, A. FES 2014, a new tidal model on the global ocean with enhanced accuracy in shallow seas and in the Arctic region. In Proceedings of the EGU General Assembly Conference Abstracts, Vienna, Austria, 12–17 April 2015; Volume 17, p. 5481.

21. Fu, L.L.; Alsdorf, D.; Morrow, R.; Rodriguez, E.; Mognard, N. (Eds.) *SWOT: The Surface Water and Ocean Topography Mission*; Jet Propulsion Laboratory: Pasadena, CA, USA, 2012.

22. Cayula, J.F.; Cornillon, P.; Holyer, R.; Peckinpaugh, S. Comparative study of two recent edge-detection algorithms designed to process sea-surface temperature fields. *IEEE Trans. Geosci. Remote Sens.* **1991**, *29*, 175–177. [CrossRef]

23. Cayula, J.F.; Cornillon, P. Edge Detection Algorithm for SST Images. *J. Atmos. Ocean. Technol.* **1992**, *9*, 67–80. [CrossRef]

24. Cayula, J.F.; Cornillon, P. Multi-Image Edge Detection for SST Images. *J. Atmos. Ocean. Technol.* **1995**, *12*, 821. [CrossRef]

25. Cayula, J.F.; Cornillon, P. Cloud detection from a sequence of SST images. *Remote Sens. Environ.* **1996**, *55*, 80–88. [CrossRef]

26. Ullman, D.S.; Cornillon, P.C. Satellite derived sea surface temperature fronts on the continental shelf off the northeast U.S. coast. *J. Geophys. Res. Oceans* **1999**, *104*, 23459–23478. [CrossRef]

27. Ullman, D.S.; Cornillon, P.C. Evaluation of Front Detection Methods for Satellite-Derived SST Data Using In Situ Observations. *J. Atmos. Ocean. Technol.* **2000**, *17*, 1667–1675. [CrossRef]

28. Ullman, D.S.; Cornillon, P.C. Continental shelf surface thermal fronts in winter off the northeast US coast. *Cont. Shelf Res.* **2001**, *21*, 1139–1156. [CrossRef]

29. Andersen, S.; Belkin, I. *Adaptation of Global Frontal Climatologies for Use in the OSISAF Global SST Cloudmasking Scheme*; Technical Report 0614; DMI: Bethesda, MA, USA, 2006.

30. EUMETSAT/OSI-SAF. *Low Earth Orbiter Sea Surface Temperature Product User Manual*, 3.3 ed.; EUMETSAT: Darmstadt, Germany, 2018.

31. Sobel, I.; Feldman, G. *A 3 × 3 Isotropic Gradient Operator for Image Processing*; Stanford Artificial Intelligence Project (SAIL); Stanford University: Stanford, CA, USA, 1968.

32. Danielsson, P.; Seger, O. Generalized and Separable Sobel Operators. In *Machine Vision for Three-Dimensional Scenes*; Academic Press: Cambridge, MA, USA, 1990.

33. Elipot, S.; Lumpkin, R.; Prieto, G. Modification of inertial oscillations by the mesoscale eddy field. *J. Geophys. Res. Oceans* **2010**, *115*. [CrossRef]

34. Kawai, Y.; Wada, A. Diurnal sea surface temperature variation and its impact on the atmosphere and ocean: A review. *J. Oceanogr.* **2007**, *63*, 721–744. [CrossRef]
35. Isern-Fontanet, J.; Hascoët, E. Diagnosis of high-resolution upper ocean dynamics from noisy sea surface temperatures. *J. Geophys. Res. Oceans* **2014**, *119*, 121–132. [CrossRef]
36. Starck, J.L.; Murtagh, F. *Astronomical Image and Data Analysis*; Springer: Berlin/Heidelberg, Germany, 2006.
37. Autret, E. Analyse de champs de température de surface de la mer à partir d'observations satellite multi-sources. Ph.D. Thesis, Université Européenne de Bretagne, Rennes, France, 2014.

Article

Assessment with Controlled In-Situ Data of the Dependence of L-Band Radiometry on Sea-Ice Thickness

Pablo Sánchez-Gámez *, Carolina Gabarro, Antonio Turiel and Marcos Portabella

Institut de Ciències del Mar, Consejo Superior de Investigaciones Científicas (CSIC), Barcelona Expert Center on Remote Sensing (BEC), Passeig Marítim de la Barceloneta 37-49, 08003 Barcelona, Spain; cgabarro@icm.csic.es (C.G.); turiel@icm.csic.es (A.T.); portabella@icm.csic.es (M.P.)
* Correspondence: xauen86@gmail.com

Received: 23 December 2019; Accepted: 13 February 2020; Published: 15 February 2020

Abstract: The European Space Agency (ESA) Soil Moisture and Ocean Salinity (SMOS) and the National Aeronautics and Space Administration (NASA) Soil Moisture Active Passive (SMAP) missions are providing brightness temperature measurements at 1.4 GHz (L-band) for about 10 and 4 years respectively. One of the new areas of geophysical exploitation of L-band radiometry is on thin (i.e., less than 1 m) Sea Ice Thickness (SIT), for which theoretical and empirical retrieval methods have been proposed. However, a comprehensive validation of SIT products has been hindered by the lack of suitable ground truth. The in-situ SIT datasets most commonly used for validation are affected by one important limitation: They are available mainly during late winter and spring months, when sea ice is fully developed and the thickness probability density function is wider than for autumn ice and less representative at the satellite spatial resolution. Using Upward Looking Sonar (ULS) data from the Woods Hole Oceanographic Institution (WHOI), acquired all year round, permits overcoming the mentioned limitation, thus improving the characterization of the L-band brightness temperature response to changes in thin SIT. State-of-the-art satellite SIT products and the Cumulative Freezing Degree Days (CFDD) model are verified against the ULS ground truth. The results show that the L-band SIT can be meaningfully retrieved up to 0.6 m, although the signal starts to saturate at 0.3 m. In contrast, despite the simplicity of the CFDD model, its predicted SIT values correlate very well with the ULS in-situ data during the sea ice growth season. The comparison between the CFDD SIT and the current L-band SIT products shows that both the sea ice concentration and the season are fundamental factors influencing the quality of the thickness retrieval from L-band satellites.

Keywords: L-band radiometry; Soil Moisture and Ocean Salinity (SMOS) mission; Soil Moisture Active Passive (SMAP); sea ice thickness; retrieval model validation; upward looking sonar; Arctic

1. Introduction

The thickness and spatial extent of sea ice are key geophysical parameters, whose retrieval by remote sensors (i.e., L-band passive microwave radiometers) has been carried out a number of times since the late 1970s [1,2]. Thin sea ice is important for climate change due to its dominance over the ocean-atmosphere heat exchange in the Arctic region [3]. The Sea Ice Thickness (SIT) can be indirectly retrieved by measuring the freeboard using laser and radar altimeters [4–7]. The thin SIT can also be estimated using ice surface temperatures from thermal infrared imagery with some limitations [8]. Low frequency (e.g., 0.6 GHz) passive radiometry was first proposed for SIT retrievals by [9], and more recently with L-band passive radiometry [10–13].

SIT retrievals based on freeboard measurements show their merits for developed ice (>1 m) while having limitations for thin ice and reduced spatial and temporal resolutions [6]. The passive microwave radiometric approach at low frequency is better suited for thin ice (<0.5 m) and can produce daily maps of the Arctic regions [11–13]. Nonetheless, passive microwave radiometry observations are limited to the cold season since melting ice produces artefacts in the retrievals [14]. Therefore, these two types of retrieval methodologies are complementary [10].

There are currently two operating L-band passive microwave radiometer satellite missions: the Soil Moisture and Ocean Salinity (SMOS) and the Soil Moisture Active Passive (SMAP) missions. The SMOS satellite was launched in November 2009, carrying a synthetic aperture passive microwave radiometer operating in the L-band at 1.4 GHz (λ = 21 cm) [15]. The SMAP mission was launched in 2015, carrying a radar and an L-band radiometer sharing a 6 m antenna reflector [16]. SMOS provides multi-angular observations, with a spatial resolution of $\sim$30 km $\times$ 30 km at nadir and a swath width of approximately 1200 km [15]. SMAP provides observations at fixed incidence angle of 40°, with a spatial resolution of 36 km $\times$ 47 km and a swath width of 1000 km [16]. These missions were originally designed for inferring soil moisture content and sea surface salinity from the surface emissivity [17–19]. However, [10] described the sensitivity of passive L-band brightness temperature measurements to sea ice properties, mainly to SIT, but also to sea ice temperature (T_{ice}) and salinity (S_{ice}).

There are two types of Passive Microwave Radiometry (PMR) retrieval methodologies for SIT: on the one hand, the methods based on a combination of thermodynamic and radiative transfer models (e.g., University of Hamburg SIT), which account for variations of ice salinity and temperature [12,20]; and on the other hand, empirical algorithms, such as the one developed by the University of Bremen, calibrated using training data from ice growth models [13,21].

The available ground truth limited the assessment and validation of current PMR thickness products [14,22]. Currently used validation datasets are acquired during the late winter and early spring periods [14,23]. Furthermore, they are restricted to specific acquisition dates in different locations, thus limiting the possibility of performing a comprehensive quality assessment of the product. This is due to the very restrictive operational constraints of helicopter-based campaigns. As such, the surveyed ice mostly corresponds to well-developed late-winter (thick) sea ice, not suitable for L-band SIT retrievals. Furthermore, late-winter SIT presents a more disperse distribution as compared with autumn-forming ice. Therefore, at scale of the satellite resolution, the late-winter in-situ SIT measurements are little representative of the satellite-derived SIT scales. The lack of suitable in-situ ground truth data for validation purposes stressed the need for more field campaigns [14]. Authors from [22] and [14] performed comparisons between satellite and airborne Electromagnetic Induction (EM) systems. However, the already mentioned sampling problems (i.e., late-winter acquisitions and lack of a time-continuous record) are still present in all the validation datasets.

The moored Upward Looking Sonar (ULS) from the Beaufort Gyre Exploration Project (BGEP) is considered to be the reference method for SIT estimation [7]. Its year-round sampling of SIT in a fixed location allows for a continuous record of the SIT during the freeze-up period provided that there is enough sea ice drift [24]. The available ULS data, which can be collocated with 7 years of SMOS-coincident measurements, have better temporal coverage than SMOS although with very limited spatial extent (that of the place where the buoy is moored). However, thanks to the drifting sea ice above the mooring, the ULS is actually sampling a wide variety of sea ice conditions, much more representative of sea ice geophysics than the other auxiliary datasets. This enables a spatio-temporal re-sampling of the validation dataset, which makes the ULS data resolution equivalent to that of the satellites. As such, the ULS turns out to be an excellent tool for the assessment of the L-band satellite-derived thin SIT products [12,13].

To the best of our knowledge, this is the first study aiming at comparing ULS with PMR SIT. Due to the limited amount of moorings, the ULS data are insufficient to carry out a thorough analysis of current PMR SIT products. However, available data can be used to verify the agreement between in-situ data (ULS) and modeled data (Cumulative Freezing Degree Days (CFDD)) during the sea ice growth season. Then, the calculated CFDD SIT is used to carrying out a more in-depth analysis of L-band spaceborne SIT products.

This analysis sheds light on the performance and limitations of current PMR ice thickness retrieval algorithms.

The structure of the paper is as follows: in Section 2 the data used in this study are described. Then, in Section 3 the retrieval algorithms for generating SIT from L-band satellite data are presented and discussed. In Section 4 the methodology for the characterization of the different SIT datasets is introduced, while the analysis results are discussed in Section 5. Finally, the concluding remarks can be found in Section 6.

2. Data

2.1. SMOS Data

In this study, the data from the official SMOS Level 1B product version 504 acquired north of $60°$ between 2010 and 2017, are analyzed. The L1B dataset contains the Fourier components of the Brightness Temperatures (TB) at the antenna reference frame. By applying an inverse Fourier transform, we obtain TB snapshots (i.e., an interferometric TB image) [25]. TBs are referenced using an Equal-Area Scalable Earth (EASE) Northern Hemisphere grid of 25 km resolution. The SMOS TB radiometric uncertainty is $\sim$2 K at boresight, although it degrades on the extended alias-free field of view [26].

The TB measurements are corrected for standard contributions such as atmospheric attenuation and geomagnetic and ionospheric rotation [17]. The galactic reflection correction, not significant at high latitudes, was not applied. A 3-σ filtering is applied using the radiometric uncertainty as σ at all the points in the antenna plane. TB measurements are also filtered in regions of the field of view that are known to have low accuracy due to Sun reflections, Sun tails and aliasing effects [27].

The TB measurements from ascending and descending orbits are averaged over periods of 3 days to reduce the noise level. The acquisitions are also averaged by incidence angle bins of $2°$. The SMOS geometry and the presence of interferences can cause missing observations at some incidence angles. A cubic polynomial fit to interpolate TB measurements is used to obtain TB samples over the full range of incidence angles at each grid point [25].

2.2. SMAP Data

The SMAP platform is equipped with an active (Synthetic Aperture Radar or SAR) and a passive (radiometer) microwave system, operating at L-band. The SAR system aimed at improving the quality and resolution of the radiometric signal. However, the radar high-power amplifier had a problem on 7 July 2015 causing a halt in data transmission [28]. Currently, the SMAP products are based on the low-resolution radiometer data [28]. The satellite Equator crossing times are at 6:00 p.m. and 6 a.m., for the ascending and descending nodes respectively, with a revisit of 2–3 days [16]. The global daily SMAP Level 3 V5 gridded TB product on EASE2 grid (36 km resolution) is used in this study. The SMAP *L3_SM_P* product is downloaded from the National Snow and Ice Data Center (NSIDC) [19]. The TBs within the SMAP L3 product are given at the surface level. Therefore, they are corrected, using near surface information, for atmospheric and sky radiation contributions [29].

2.3. Moored Upward Looking Sonar

SIT measurements from in-situ moored ULS are available since the early 1990s [24]. The measuring unit from the BGEP consists of several instruments: an Acoustic Doppler Current Profiler (ADCP), an ice-profiling sonar (IPS) with a pendulum and pressure sensors. The ADCP is an echo sounder that measures motion by using the Doppler shift of a target [30]. The IPS measures the distance to the ice-water interfaces. The sonar is made up of an amount of transducers that shape a fine acoustic beam of 2° at −3 dB [24]. This sharp beam permits sampling a target at 50 m distance with a ∼2 m resolution. The pendulum and pressure sensors allow us to estimate the tilt of the beam and the depth of the moored ULS, respectively. Inferring the depth from the pressure implies the knowledge of the atmospheric pressure which must be subtracted (pressure from a nearby weather station often suffices) [24]. The sampling rate is 0.5 Hz, generating files with more than 40,000 samples per day. The uncertainty of the SIT measurement is of 5–10 cm, and depends on a number of factors [24].

Three in-situ moorings located in the Beaufort Sea (see Figure 1) are used in this study. Data collected between 2010 and 2017 during the freeze-up period of the year; i.e., October–January, are analyzed. The total surveyed track length, assuming an ice drift of ∼8 km/day on average, is ∼20,000 km. This constitutes an unprecedented source of information for validation of current PMR thickness retrievals. Thickness is measured in a Eulerian context (i.e., fixed moorings), thus enabling the analysis of ice thickness development during the freeze-up period. This type of retrievals allow a year-round cost-effective acquisition of ice thickness as opposed to helicopter-based campaigns [14].

2.4. Ancillary Data

Sea Ice Concentration (SIC) maps from the database of the Ocean and Sea Ice Satellite Application Facility (OSI SAF) of the European Organisation for the Exploitation of Meteorological Satellites (EUMETSAT) are used in this study. SIC is estimated from TB observations from the Special Sensor Microwave Imager/Sounder (SSM/I and SSMIS) at 19 and 37 GHz which are corrected for atmospheric effects using the European Centre for Medium-Range Weather Forecasts (ECMWF) model output [31].

Sea Ice drift data from the Polar Pathfinder Sea Ice Motion Vectors [32] available at the NASA NSIDC are also used in this analysis, in particular, Version 4, which includes an improved filtering of the SSMI inputs and updates, i.e., input buoy data and input motion vectors. Ice motion estimates are derived from a number of satellite sensors such as the Advanced Very High Resolution Radiometer (AVHRR), the Advanced Microwave Scanning Radiometer—Earth Observing System (AMSR-E), the Scanning Multi-channel Microwave Radiometer (SMMR), SSMI and SSMI/S sensors, the International Arctic Buoy Programme (IABP) and NCEP/NCAR Reanalysis forecasts [32].

Finally, NCEP/NCAR Reanalysis 1 surface air temperatures [33] at 2.5 degrees resolution in latitude and longitude, with a mean absolute error of ∼0.25 and 1.25 degrees Celsius in summer and winter, respectively [34], are used in this analysis. NCEP data together with the empirical Cumulative Freezing Degree Days (CFDD) model are used to calculate SIT [35,36].

$$SIT[m] = 1.33 * (CFDD[°C])^{0.58} , \tag{1}$$

The following procedure is applied in order to obtain SITs from surface temperatures. Sea ice freezing temperature is subtracted from the available NCEP/NCAR surface air temperatures. We apply the modulus function to the resulting temperatures. The resulting value is added from the beginning of the freezing period (i.e., end of September-beginning of October) to the date under consideration. This cumulative quantity is used as an input in Equation (1) to obtain the desired SITs.

3. L-band Passive Radiometer Ice Thickness

Two algorithms for SIT retrievals have been proposed in the literature: one empirical and another one based on the thermodynamic/radiative-transfer model. The empirical methodology was firstly proposed by [11]. It consists of an exponential function that relates the measured intensity or First Stokes parameter (i.e., the average of the horizontally and vertically polarized TBs) with SIT. The intensity is used because of its better quality and signal preservation (e.g., less affected by Faraday rotation) as compared to the co-polarized TBs.

The L-band SIT empirical retrieval is based on Equation (2):

$$T_{obs} = T_1 - (T_1 - T_0)exp(-\gamma d)$$
$$d \pm \delta d = -\frac{1}{\gamma} ln\left(\frac{T_1 - T_{obs} \pm \delta T_{obs}}{T_1 - T_0}\right),$$

(2)

where d is the inverted ice thickness, γ the attenuation factor, T_1 and T_0 the saturated thick ice and the open water signals respectively, and T_{obs} the observed brightness temperature.

This algorithm has been further developed by [13], including a new fitted function describing the Polarization Difference (PD) dependence on SIT. Also the incidence angle used is different: while [12] uses the incidence angle range of 0–40°, [13] uses the observations retrieved at 40°. Additionally, [13] relies on ice growth models for generating the training dataset. More recently, a method that combines the SMOS and SMAP TBs to produce a state-of-the-art SIT product has been developed [21]. The empirically-based product used in this study is produced and supported by the University of Bremen. They currently distribute a version ingesting the SMOS TBs observations. The retrieval is limited to 50 cm ice thickness [13,21]. Allegedly, the SIT error increases with ice thickness and it could amount up to 50% of the retrieval values [21].

The thermodynamic/radiative-transfer model methodology was first developed by the University of Hamburg [12]. The ice thickness is estimated using an iterative approach due to the dependence of SIT on temperature and ice salinity [12]. The ice thickness estimated by [11], which is calculated with average values of T_{ice} and S_{ice} for the Arctic, is used as a first guess in the initial iteration. The temperature of the ice pack is assumed to be the average of the sea water freezing temperature and the surface air temperature from atmospheric reanalysis data [12]. Sea ice salinity, also required by the radiation model, is estimated with the empirical function defined by [37] using the sea surface salinity based on climatology as input. The thickness retrieval convergence is based on the comparison between the observed intensity at nadir (from 0 to 40° incidence angle) and the modeled intensity from an adaptation of the radiation model by [9]. The obtained thickness field is further refined by applying a lognormal distribution [38] that takes into account the effect of subpixel-scale heterogeneity at SMOS resolution [12]. The limits of this type of retrieval are variable depending on the temperature and salinity of the ice pack. It can reach up to 1.5 m under very special conditions, i.e., low temperatures and low salinities, only attainable at closed seas or water masses with large amounts of fresh water supplies (e.g., Baltic Sea) [12]. The SIT uncertainty increases with thickness and it can be larger than the estimated thickness [12].

The presence of snow over sea ice modulates the L-band signature. The wet snow acts as an absorbent at this frequency, thus suppressing the sea ice thickness sensitivity when wet snow is present. This is the reason why the SMOS SIT maps are not produced from mid April to mid October. On the other hand, the dry snow produces a bias in the TB, especially in the H-polarization TB as already stated in [39].

4. Methodology

The ULS data shows its suitability as SIT reference dataset or ground truth when checking its consistency for the continuous sampling of the ice pack during the freeze-up period. The ground truth data acquisition relies on the drifting nature of the sea ice [24]. The ULS thickness reference is re-sampled into the EASE grid for comparison against L-band radiometer thickness products and CFDD SIT based on NCEP/NCAR surface reanalysis data. To produce a suitable ground truth for comparison against satellite measurements, the ULS data are time averaged over consecutive 24-h spans, i.e., the same time resolution of the daily L-band thickness products.

The total amount of measured samples (which equals the number of measured days during the freeze-up period) was ~2500 which multiplied by the average daily sea ice drift gives a figure of approximately 20,000 km of sampled ice thickness. We imposed a filtering criterion over the time averaged ULS ground truth. The filtering was performed in a two step approach. Firstly, leads were discarded from the statistics whenever their presence in the daily sample was above 5%. Secondly, the daily standard deviation of the ULS dataset with a threshold of 0.2 m was used as a filtering criterion. The latter implied the filtering of approximately $\sim$ 30% of the ULS measured ice thickness. The filtered ULS ice thickness showed its suitability as ground truth given its accuracy, consistency and continuous record of the ice thickness.

To ensure the statistical representativeness of the derived ground truth, we perform a quality control of the ULS data based on three consistency criteria. First, the ice profile sample (dependent on the amount of ice drift) must be long enough; we imposed a minimum of ~100 days of sampling which translated into ~800 km of sampled ice thickness. Second and third, the variability of the SIT profile and SIT field in the perpendicular direction to the profile, respectively, must be low enough; here again, we imposed a threshold of ~0.2 m in the ice thickness standard deviation. A mean ice drift of ~7–8 km day^{-1} for the entire sampling period was estimated on the location of the moorings using both the in-situ ADCP [24] and the NSIDC sea ice velocity record [32]. The variability of the temporally-averaged thickness was calculated with the daily standard deviation of the dataset. Finally, the latter value was used as a proxy for estimating the variability of the ice in the perpendicular direction of the sampled profile.

Figure 1 shows the Buoy C and the CFDD SIT evolution over the freeze-up period in 2015. There is a good match between the ULS thickness data and the CFDD modeled SIT. The temporal evolution of the SIT ground truth shows an increasing standard deviation of the ice thickness daily mean while spring approaches. The thickness values stop increasing at the beginning of spring. As soon as this happens, the ice thickness daily mean begins to change more rapidly. Note also that the SIC increases rapidly at the beginning of the freeze-up period. The former facts indicate that the most suitable time for carrying out a calibration or assessment of L-band ice thickness is precise during the freeze-up period, after the SIC reaches nominal values around 95%, and before the in-situ ice thickness loses its representativeness of satellite resolution scales (i.e., beginning of spring).

We performed the comparison between the daily SIT ground truth and the University of Hamburg (UH) and University of Bremen (UB) SIT products. The UH product provides an uncertainty and a saturation field to inform the users about the quality of the product. As such, the UH SIT data were filtered by using these two parameters following [2]. The filtering was performed using a 1-m uncertainty and a 90% saturation ratio (i.e., the ratio of the estimated thickness to the maximum theoretical retrieval). This filtering procedure reduces the number of ground truth match-ups to approximately one third of its original size, thus triggering the need to extend the validation dataset using model-based SIT data. The comparison between the ground truth and the CFDD SIT (see Equation (1)) shows that the latter is in a first approximation a good estimate of the former as a reference dataset.

The use of CFDD model-based SIT facilitated the extension of the validation. However, this extension was grounded on the homogeneity hypothesis (i.e., the sea ice growth conditions in other regions are considered similar to those found at the buoys' locations). This hypothesis can only be valid during the freeze-up period when the ice growth is fast and the impact of the ice drift is lower on the CFDD modeled ice thickness.

The scatter plot between the ULS-derived ground truth collected over 7 years during the freeze-up periods (from 2010 to 2016) and the co-located CFDD SIT is shown in Figure 2. The scatter indicates that CFDD SIT is a good proxy of the ground truth having only small deviations from the identity line, the regression factor is ~0.99 ± 0.01 at 95% confidence level. Indeed, the CFDD SIT represents quite well ice thickness up to 0.3 m. For larger SIT values, a slight overestimation of the CFDD data is noticeable. We attribute this overestimation to the inability of the model to represent events of strong ice drift that inhibit ice growth for several days. This good correlation between both datasets does not hold for the late winter-early spring months (see Figure 1). Therefore, we only use the CFDD SIT as proxy of the ULS ground truth during the freezing season (i.e., from October to January), and in regions with thin first-year sea ice.

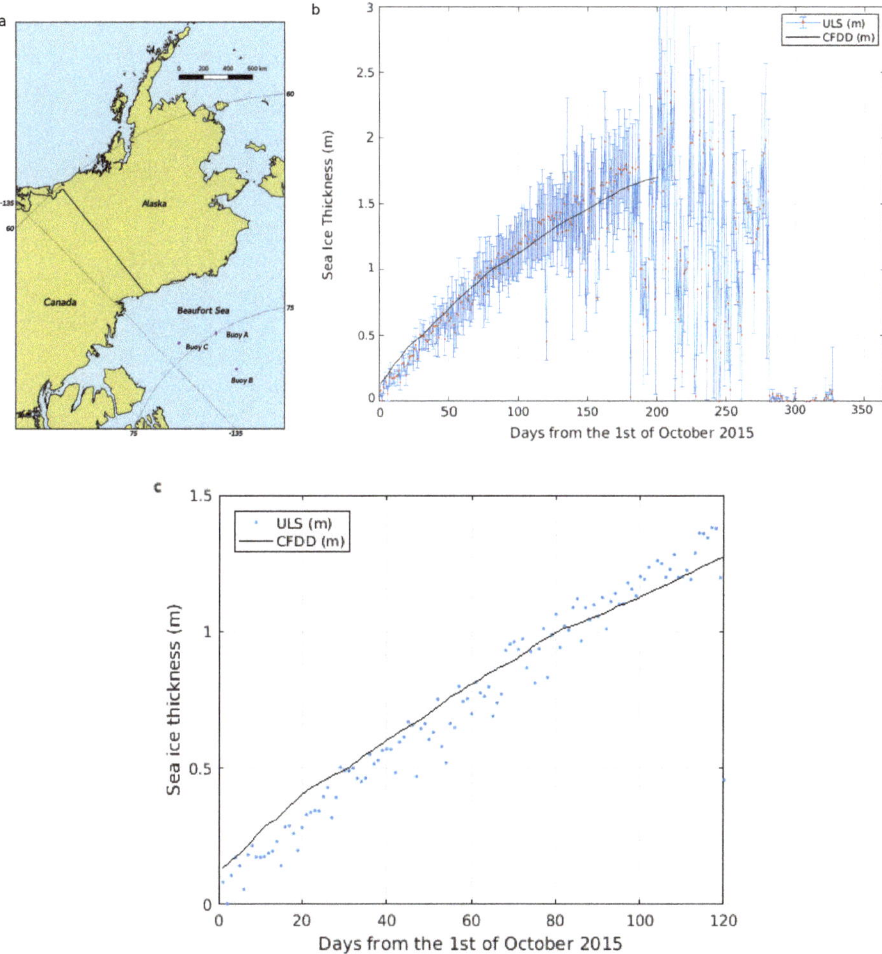

Figure 1. (**a**) Location of Woods Hole Oceanographic Institution (WHOI) buoys in the Beaufort Sea. The map uses the Arctic Polar Stereographic projection (EPSG: 3995) [40]. (**b**) Temporal evolution of the ice thickness measured by Buoy C (red dots) and modeled by Cumulative Freezing Degree Days (CFDD) (black line) over the freeze-up period in 2015. (**c**) Comparison between the modeled CFDD and Upward Looking Sonar (ULS) ground truth gathered during the first three months of the freeze-up period.

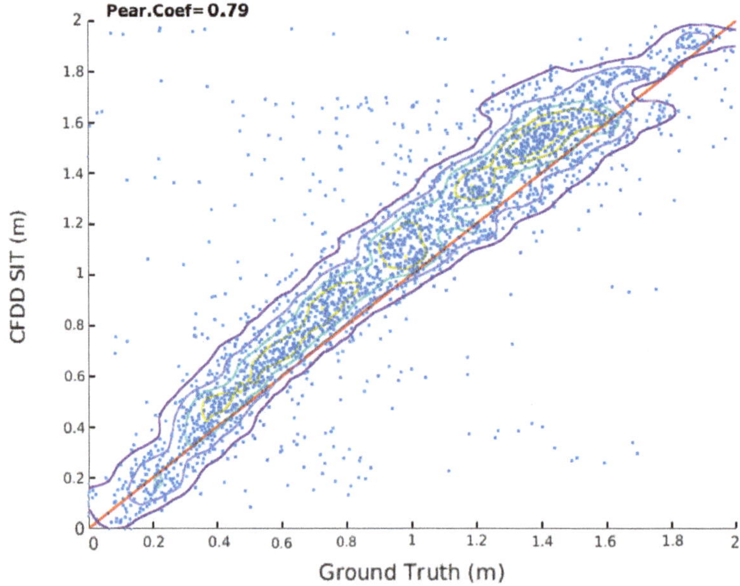

Figure 2. Comparison of the modeled CFDD Sea Ice Thickness (SIT) and the ULS ground truth. Contour lines in scatter plots herein describe the normalized density surface. Contours show a 20% step increase in density in all cases. Pearson's correlation coefficient used herein is defined as the covariance of two variables divided by the product of their standard deviations.

5. Results and Discussion

5.1. L-band Response to SIT

Using the ground truth as described in the previous section allows establishing the relation between SMOS and SMAP L-band First Stokes parameter, the PD and the SIT parameter. Figures 3 and 4 show some L-band parameters as functions of the ground truth, which is filtered with a 95% SIC threshold in all cases. It should be noticed that recent efforts for characterizing the L-band response to ice thickness were hindered by the distribution and characteristics of the ground truth used [2,14,22]. On the one hand, the spatial distribution of the ground truth is limited to those regions sampled by ships. On the other hand, campaigns are normally carried out at the end of winter or in spring, therefore a thickness seasonal bias can be expected [14,22]. The author in [2] uses the Ocean ReAnalysis System 5 (ORAS5) modeled ice thickness for their comparison against SMOS SIT. They notice an overestimation of the ORAS5 with respect to SMOS SIT. This is attributed to the simplified representation of thin ice by the model (see [2] for further details).

Figure 3 displays the relationship between the First Stokes and PD parameters and the daily ULS ground truth. The First Stokes parameter shows a large dispersion for thin ice which decreases with increasing ice thickness. Such a decrease in dispersion reveals a saturation of the response of TB to large SIT values (i.e., above 0.6 m). The signal saturates at a value of ~240 Kelvins. The standard deviation of the TB at large SIT values is in fact an estimation of the radiometric error of the observation system, because at such SIT values, the signal no longer responds to the geophysical parameters and the variability can be mainly attributed to the radiometric noise.

A theoretical relationship between TB and ice thickness based in the incoherent Burke model can be found in [10] and [12]. We note here that larger dispersion is seen for very thin ice while other comparisons between these parameters show that the largest dispersion is found at around 20 cm ice thickness [12,13]. Probably, several physical mechanisms not considered in the theoretical model can be affecting the TB response to thin ice, such as the impact of ice surface temperature and ice surface roughness variations, among others.

The relationship between SMOS PD and the ground truth can be found in Figure 3b. The PD presents a constant but larger standard deviation (with respect to the case of the First Stokes parameter) for ice thinner than 0.6 m, and for thicker ice the PD saturates at a value slightly below ~30 Kelvins. The PD intersects with the ordinate axis at ~70–80 Kelvins. In contrast, [13] indicates an intersection at around 50 Kelvins. We attribute this difference either to a wrongly classified thickness with the CFDD model by [13] or to the inclusion of pixels with different SICs in their analysis.

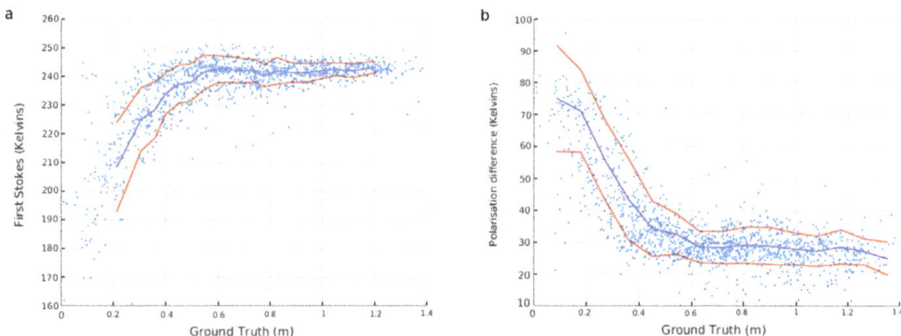

Figure 3. (**a**) Soil Moisture and Ocean Salinity (SMOS) First Stokes parameter calculated for 0–40° incidence angles against ULS derived SIT. (**b**) SMOS Polarization Difference (PD) at 50° incidence angle. The blue line represents the average and the red lines plus and minus one standard deviation.

In order to confirm the L-band response to SIT, as shown by SMOS data (see Figure 3), we perform the same analysis but with SMAP TB. The comparison between SMOS and SMAP TB responses to ice thickness is shown in Figure 4a. The L-band signal response from both sensors presents very similar patterns, including larger dispersion for thin ice and saturation for thick ice. Nonetheless, SMAP TBs are slightly lower for thick ice than those of SMAP, something that can be a consequence of SMOS residual biases [41].

We estimated the standard deviation of the ice thickness retrieval by propagating the TB error to the SIT according to Equation (2). Surprisingly, the standard deviation of ice thickness increases with increasing thickness, as opposed to what happens with the TB. This is due to the formulae used in the thickness retrieval and the larger sensitivity of TB to SIT for thin ice.

The limit of validity of the empirical retrieval is located around the 0.5 m SIT [11,13]. In order to further constrain this limit, we analyzed the factor within the exponential term in Equation (2), γd, which corresponds to a normalized thickness. The scatter plot of Figure 4b evidences the relation between ground truth and the normalized thickness. As expected, we see an almost linear relation for ice thinner than 0.6 m. Besides, the dispersion of the parameter increases with ice thickness and finally, for thicker ice, the sensitivity of TB to SIT is completely lost. The scatter points in that region are distributed around a horizontal line (hence, the attenuation factor γ is meaningless in that region). Obviously, this limitation applies only to the pure empirical retrieval. The thickness retrieval can be extended to thicker ice using a radiative transfer model and assuming a lognormal distribution [12]. However, this approach can be sensitive to the auxiliary fields used, such as SIC, temperature and salinity, whose uncertainties can be difficult to quantify [2].

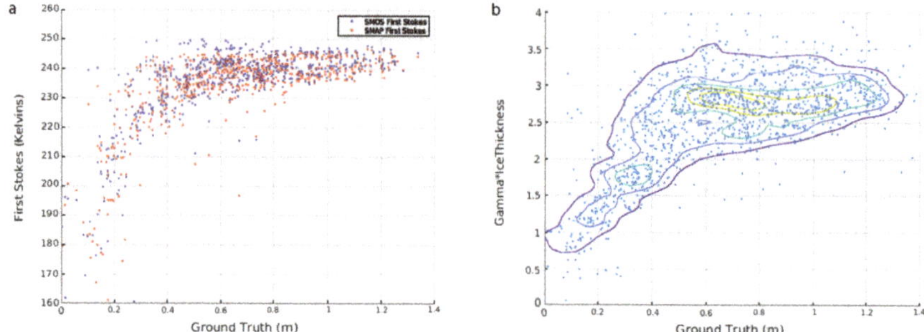

Figure 4. (**a**) SMOS and Soil Moisture Active Passive (SMAP) Brightness Temperatures (TB) First Stokes response to ULS derived ice thickness at 40° incidence angle. (**b**) Normalized thickness from the empirical retrieval against ground truth (see Equation (2)).

5.2. Assessment of UH SIT

The direct comparison between the UH ice thickness and the ground truth reveals that the product underestimates thin sea ice and overestimates thickness larger than 0.3 m (see Figure 5a). This underestimation for thin ice and its relationship with SIC has been widely reported in the literature [2,12]. Considering that CFDD is already overestimating the ULS ground truth for thick ice (see Figure 2), a larger thickness overestimation of the UH product with respect to ULS is expected for thick ice (see Figure 5b and Section 4). The ice thickness standard deviation increases with thickness (see Figure 5b). The observed decrease in the standard deviation for large thickness is due to the decrease of the scatter region analyzed in the calculation.

The CFDD thickness and the ULS scatter plots against the UH SIT reveal with their similarity that the proposed homogeneity hypothesis works fine during the freeze-up periods. This situation provides a solid ground for extending the ULS ground truth with the CFDD ice thickness.

The Probability Density Functions (PDFs) for the freeze-up period for SIC over 90% reveals that the modes values are of similar value for UH and CFDD SITs (see Figure 5). Moreover, the shape of both PDFs are similar. However, UH PDF shows a smoother shape, with gentle slopes and a larger amount of occurrences for thickness above 0.5 m (see Figure 5d). Interestingly, a different PDF is observed for UH SIT during October, at the beginning of the freeze-up period (see Figure 5c). In this case, the modes of both distributions are coincidental as was the case for the whole-winter period. However, the shapes of the distributions are markedly different, even when considering that the CFDD model SIT is around observable values for the SMOS-based L-band SIT retrieval valid range (i.e., from 0.1 to 0.6 m), indicating poor-quality SMOS SIT retrievals, even at these favorable conditions (i.e., for high SIC conditions). A possible explanation for this change on the distribution shape could be the application of the lognormal distribution to the largest thickness values [12]. This change in the shape of the distribution is no longer visible for lower SICs.

The analysis of the distributions with lower SIC values (see Figure 6) indicates that the UH product bias increases with decreasing SIC. Furthermore, the UH SIT distribution is more peaky with decreasing SIC. The increasing negative bias of the product with decreasing SIC has already been reported by [12]. The peaky behavior of the distribution might reveal that the retrieval algorithm is saturated for low SIC values. On the other hand, the thickness distribution for SICs between 70% and 90%, while having a bias, presents a very similar shape to that of the CFDD SIT (see Figure 6c). The latter facilitates a possible readjustment of the product using a simple shift. The distribution of SIT in cases with lower SICs has both a large bias and peaks at lower SIT values (see Figure 6d). In this case, a readjustment would require both a shift and a rescaling provided that the ice thickness signal is still present in the measured TB and is not contaminated by the ocean signal.

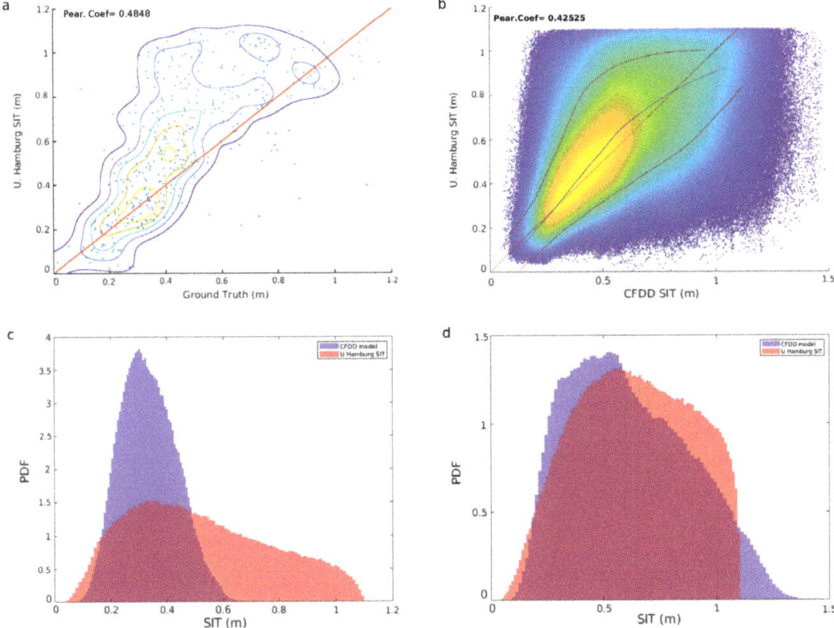

Figure 5. (**a**) Comparison between the ULS ground truth and the University of Hamburg (UH) ice thickness. Soil Moisture and Ocean Salinity (SMOS) thickness uncertainty and saturation ratio are always below 1 m and 90% respectively. The red line represents the identity line. Ground truth was acquired for the freeze-up periods between 2010 and 2017. (**b**) Comparison between the modeled CFDD SIT and UH ice thickness for the freeze-up periods between 2010 and 2017. SMOS thickness uncertainty and saturation ratio are always below 1 m and 90% respectively. The mean and standard deviations are depicted with a blue and red lines respectively. (**c**) Probability Density Functions (PDFs) of CFDD and UH SIT during October between the years 2010 and 2017. (**d**) PDFs of CFDD and UH SIT for case during the freeze-up periods between 2010 and 2017 with SIC over 90% in all cases.

The UH SIT product well reproduces the distribution for the whole freeze-up period for high SIC values (see Figure 5d). However, we recognize the inability of the UH algorithm to represent the real distribution of SITs for high SIC (i.e., above 90%) during specific periods of the year (e.g., October). However, this distribution facilitate that UH product better represents the end of winter fully developed ice for high SICs. During the same period of the year, when lower SICs are considered, UH product departs from CFDD ice thickness. Moreover, we see both the increasing underestimation and kurtosis (peakiness) of the distributions with decreasing SICs (see Figure 6). The satellite sampling which permits a daily coverage for the whole Arctic region [15] contrasts with the limitations of the thickness retrieval for specific periods of the year (see Figure 5c). The inability of the UH product to capture the temporal trend of the ice thickness at high SICs is a significant concern since the satellite temporal resolution is left unexploited (i.e., the retrievals do not properly reflect the real ice thickness distribution for a given time). The latter leaves the door open for future improvements with a temporally enhanced thickness retrieval.

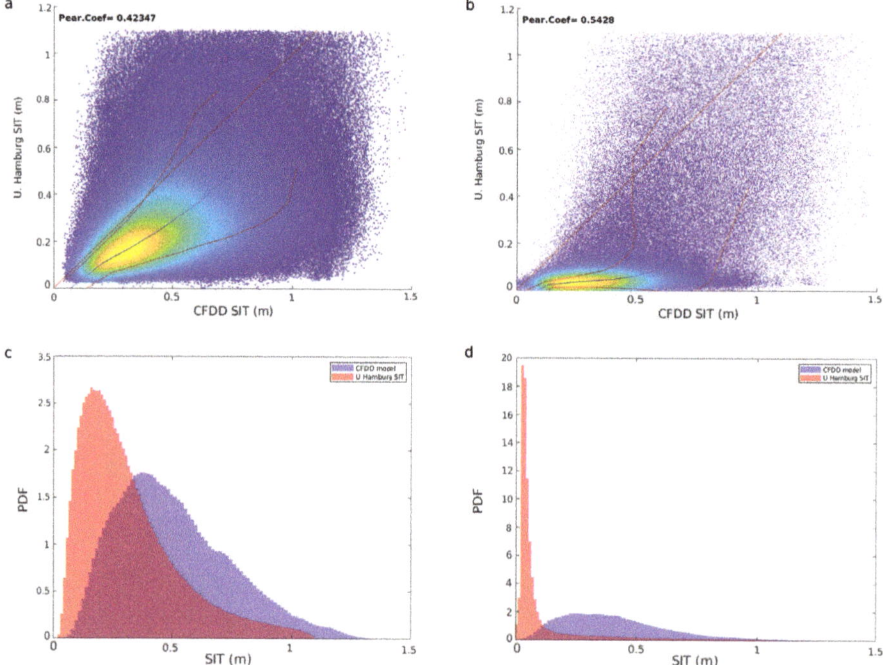

Figure 6. Comparison between the modeled CFDD SIT and the UH ice thickness. (**a**) SIC is between 70% and 90%. (**b**) SIC is between 30% and 50%. SMOS thickness uncertainty and saturation ratio are always below 1 m and 90% respectively. The mean and standard deviations are depicted with a blue and red lines respectively. (**c**) PDFs of CFDD and UH SIT for case a. (**d**) PDFs of CFDD and UH SIT for case b.

5.3. Assessment of UB SIT

The following comparisons were carried out filtering the UB SIT based on the UH restrictions of saturation and uncertainty. The methodology makes the comparison between products and ground truth more consistent and reliable, because we are comparing the same datasets. The analyzed UB SIT cells are sometimes saturated and sometimes fall within the observable thickness range, when saturated they are also filtered (i.e., above 0.5 m). The scatter plot between the buoy ULS thickness and the UB SIT indicates that the UB product underestimates SIT for the whole freeze-up period (see Figure 7b). The UB thickness shows a large amount of occurrences near the saturation limit. We interpret this as an indication that the algorithm is trying to represent in the range of half a meter CFDD thickness values that are clearly over their own retrieval limit (see Figure 7d). We also observe that ice thickness standard deviation increases with thickness (see Figure 7b). In contrast with the UH product, only a slight SIT underestimation is present in the UB product for thickness below 0.4 m, indicating a relatively higher accuracy of the latter within the mentioned thickness range.

Here again, as what happened with the UH SIT, the CFDD thickness and the ULS scatter plots' similarity supports the proposed homogeneity hypothesis. Therefore, we have further support for using the CFDD SIT as an Arctic-wide proxy of SIT ground truth.

Indeed, the UB SIT marginal distribution resembles that of CFDD, for the month of October and SIC above 90% (see Figure 7c). The CFDD thickness values are within the observable L-band thickness range (0–0.5 m), and therefore they are expected to be resolved as they do (as opposed to UH product, see Section 5.2). The good fit comes as no surprise since the UB product is based on the CFDD model and the NCEP/NCAR surface reanalysis auxiliary data [13]. The observed good fit during October is no longer seen for later periods (i.e., from November to January) when UB is not able to resolve CFDD thickness values lying over the retrieval limit (see Figure 7d).

The UB SIT behavior when considering decreasing SIC values is similar to that of the UH product, i.e., an increasing bias and a more peaky distribution on the thinner thickness range (see Figure 8). The distributions reveal that the UB product is wrongly classifying thickness above 0.5 m, by setting them to a fixed value of 0.5 m. The wrongly classified CFDD thickness, calculated after filtering the saturated UB SIT, amount to 50% of the total (this percentage is calculated for the whole freeze-up period, considering values of SIC above 90%). The UB thickness products caps the retrievals at 0.5 m [13] taking into account the limits of the empirical retrieval. By doing so, they are indirectly forcing an artificial saturation visible in all comparisons. On the positive side, we see the related bias of the acquisition for lower SICs, a similar behavior seen for both UB (see Figure 8c,d) and UH (Figure 6a,d) SIT, although the former shows a slightly smaller bias (with respect to ground truth) than the latter.

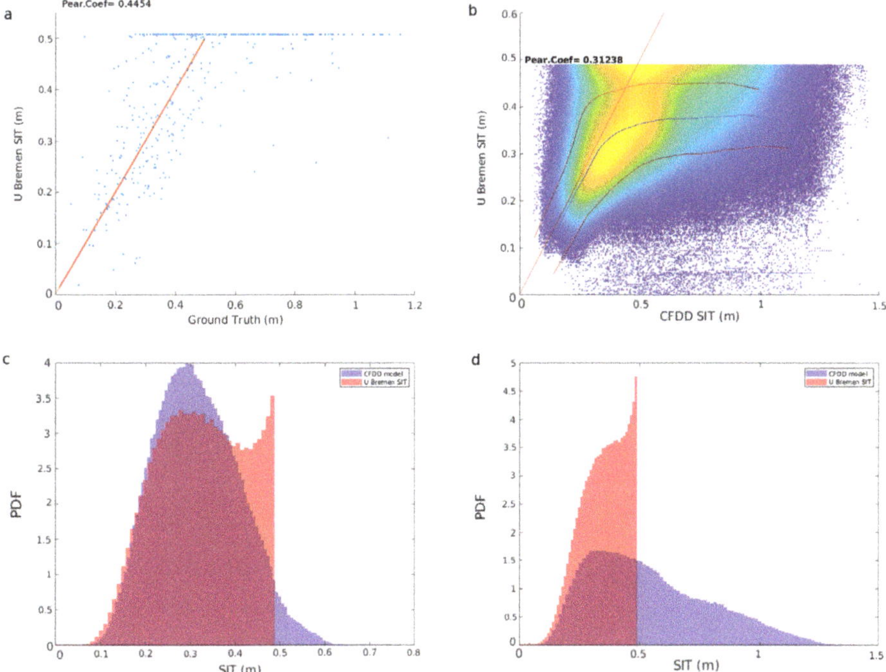

Figure 7. (**a**) Comparison between the ground truth measured by the upward looking sonar and the University of Bremen (UB) ice thickness. SMOS thickness uncertainty and saturation ratio are always below 1 m and 90% respectively. The red line represents the identity line. Ground truth was acquired for the freeze-up periods between 2010 and 2017. (**b**) Comparison between the modeled CFDD SIT and the UB ice thickness for the freeze-up periods between 2010 and 2017. SMOS thickness uncertainty and saturation ratio are always below 1 m and 90% respectively. The mean and standard deviations are depicted with a blue and red lines respectively. (**c**) PDFs of CFDD and UB SIT during October between the years 2010 and 2017. (**d**) PDFs of CFDD and UB SIT for the freeze-up periods between 2010 and 2017 with SIC over 90% in all cases.

5.4. Applicability of the L-band SIT Retrieval Methodology

Assuming that L-band technology is only sensitive to thin ice conditions, the coverage of potentially useful L-band SIT retrievals is analyzed in this section.

Figure 9a shows the sea ice extent for all sea ice cells (i.e., SIC above 15%) and for those cells with specific ice conditions (red curve) and retrievable thin ice (green curve), while Figure 9b shows the latter as percentages with respect to the total sea ice extent. Figure 9b shows that relative percentages of retrievable thickness are highest at the beginning of the freeze-up period (i.e., October–November). Furthermore, when considering the optimal retrievable thickness range (i.e., below 0.5 m), this part of the year has also the largest share of observable ice thickness. Interestingly, as the freezing period progresses, the respective shares of observable thickness drop, but between them, the optimal retrievable thickness presents a further drop below 5% of the total sea ice extent (see Figure 9b).

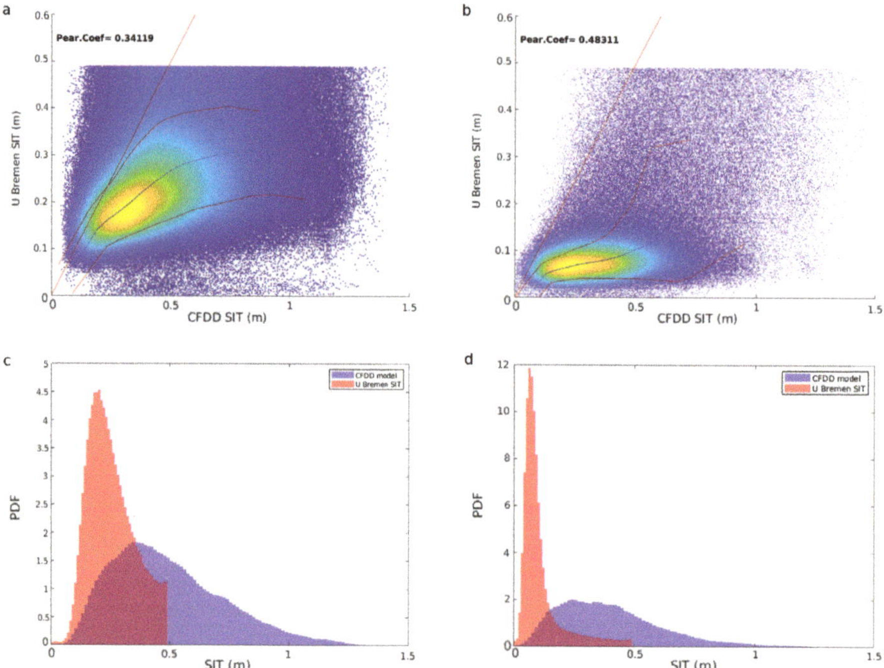

Figure 8. Comparison between the modeled CFDD SIT and the UB ice thickness. (**a**) SIC is between 70% and 90%. (**b**) SIC is between 30% and 50%. SMOS thickness uncertainty and saturation ratio are always below 1 m and 90% respectively. The mean and standard deviations are depicted with a blue and red lines respectively. (**c**) PDFs of CFDD and UB SIT for case a. (**d**) PDFs of CFDD and UB SIT for case b.

It is also relevant to stress the limited temporal coverage of the L-band based SIT products, mainly valid during the freeze-up period and in regions of high SIC values (e.g., during October–November in the Northern Hemisphere). These optimally observable regions are typically located outside the multi-year ice holding half-ringed shape (i.e., a partial ring spanning from the northern coasts of Canada and Russia); eventually, this ring-shaped structure closes when reaching the Bering Strait. These regions shrink in size as the winter progresses (see Figure 9b), thus leading to poor-quality SIT retrievals due to the relatively low spatial resolution of the SMOS satellite. Therefore, optimal retrievable values are expected to be mainly located in these temporally and spatially-varying regions.

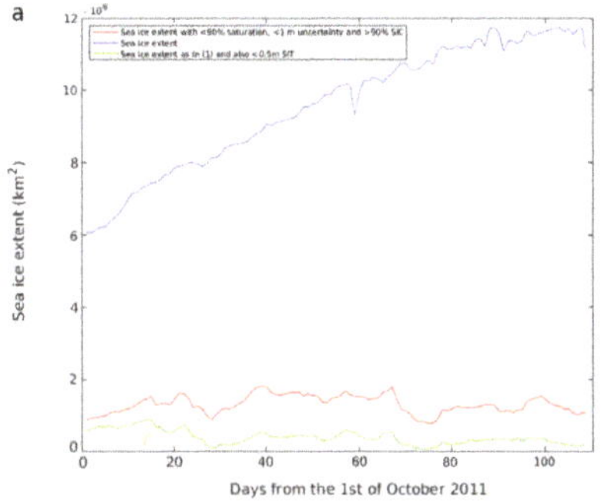 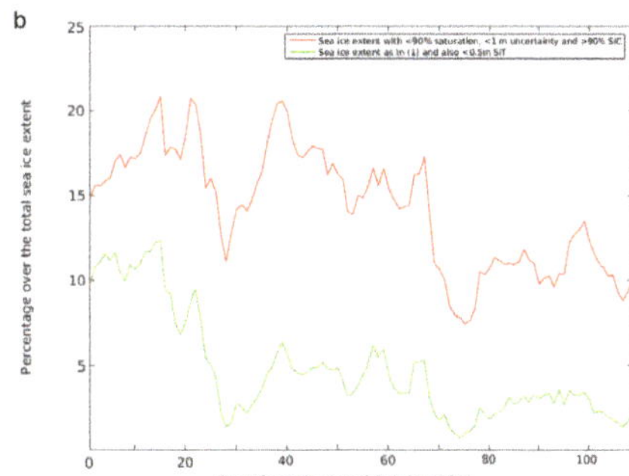

Figure 9. (**a**) Sea ice extent during the freeze-up period of 2011. The blue line depicts the full sea ice extent. The red line for ice that has <90% saturation, <1 m uncertainty and >90% SIC. The green line represents the same conditions as the red line but only for SIT below 0.5 m. (**b**) Same red and green lines as in (**a**) but shown as percentage of the total (blue line).

6. Conclusions

We have characterized the brightness temperature response to SIT using the processed upward looking sonar ground truth during the freeze-up period between October and January from BGEP from WHOI. TB shows a larger dispersion for thin sea ice, increases with SIT and saturates for ice thicker than 0.6 m at approximately 240 Kelvins where the effect of SMOS radiometric uncertainty (radiometric noise) is visible. PD intersects the ordinates axis between 70 and 80 Kelvins and saturates for ice thicker than 0.6 m at approximately 30 Kelvins. The observed dispersion of the TB propagates into the ice thickness estimation with a standard deviation that increases with ice thickness. The analysis of the attenuation parameter of the empirical retrieval reveals the lack of response of the TB to SIT thicker than 0.6 m.

The year-round, cost-effective, accurate record of ULS SIT and the lack of validation data stresses the strong need of deploying and maintaining more measuring units at appropriate locations of the Arctic Ocean. This approach would probably improve the quality and usefulness of validation data with respect to current costly ship-based campaigns performed so far for gathering SIT ground truth with EM [14,22,23], that are also very limited as they can only be done during late winter–early spring months due to the need of helicopter which entails measuring a fully developed ice pack therefore skewing the reference dataset.

The detailed comparison between CFDD-model SIT and ULS ground truth evidenced their strong correlation during the freeze-up period. We have taken advantage of this correlation to use the estimated CFDD SIT as a reference dataset during the freeze-up periods. Furthermore, we saw the similarity of the scatter plots between CFDD and ULS against UH and UB SIT products revealing again the strength of the proposed homogeneity hypothesis. The extension of the ground truth with the model-based thickness allowed characterizing the L-band ice thickness products in different scenarios. The analysis showed that L-band SIT present a clear dependency on SIC while this parameter is not considered in current SIT estimations [12,13].

The comparison between the CFDD SIT and UH SIT product at 90% SIC shows that their modes are coincident during the whole freeze-up period and also partially at other periods. UH SIT slightly underestimates for thicknesses below 0.3 and overestimates for thicker ice. However, the shapes of the PDFs of CFDD SIT and UH SIT, while being similar for the whole winter, becomes notoriously different at specific months. On the other hand, as the freezing period advances UH SIT mode corresponds to the CFDD SIT mode. This change in the shape of the distribution is probably needed to accommodate thickness values outside the L-band observational range and probably relates to the lognormal distribution [12]. As SIC is gradually lowered in the comparison, there is a trend of increasing bias and the UH distribution becomes concentrated at low thicknesses. The shape of both distributions are very similar when the SIC is between 70% and 90%, an encouraging starting point for improving the already visible bias at this concentration range.

The comparison between the CFDD SIT and UB SIT product at 90% SIC indicates that UB SIT fits the ground truth very closely for thickness below 0.4 m. For thicker ice, UB SIT has a tendency to give a saturated response. This saturation stems from the fact that many CFDD thickness values (around 50%) over half a meter are wrongly classified by UB. However, the thickness distribution for October presents a shape very close to that of CFDD distribution. The latter comes as no surprise since [13] product is trained using the CFDD SIT model as a reference. Nonetheless, as the freeze-up period continues, the CFDD SIT distribution displaces to larger thickness outside the empirical retrieval observable range. In these cases UB SIT saturation is more pronounced. As SIC is decreased in the comparison, UB SIT presents an increasing negative bias and its distribution peaks at lower thickness (following exactly the same behavior as the UH product).

The former analysis permits us to understand the current limitations of SMOS-based ice thickness products. These limitations are not only related to the retrieval methodology but also to the geographical extent where L-band based SIT is applicable. There are specific periods of the year when optimal retrieval conditions are only reached for 5% or less of the whole sea ice Arctic coverage. This scarce coverage challenges the consideration of L-band-based SIT as an Arctic wide product. Regarding the methodological limitations, we indicate a minimum of $\sim$90% SIC for a successful thickness retrieval during the freeze-up period when the ice pack thickness remains below half a meter.

The observed limitations of the retrieval methodologies strongly suggest future lines of improvement. Considering the daily temporal resolution of the satellites, a temporally enhanced SIT retrieval would be a step ahead. This temporal adjustment would allow to present a temporally consistent SIT product with no variations in its behavior during the winter. The second line of improvement stems from the need of considering SIC as a required variable for SIT retrieval. This would permit adjusting the thickness inversion depending on the considered SIC. If the latter is not possible, indicating at least the ice concentration for the specific retrieved values as a quality flag is an absolute must. This would give the user fundamental information about the quality of the estimated ice thickness.

Author Contributions: P.S.-G. processed the data, wrote the manuscript and designed the research. C.G., M.P. and A.T. design the research and provide feedback and advice on the paper. All authors have read and agreed to the published version of the manuscript.

Funding: L-BAND project, funded by Spanish R+D Plan (ESP2017-89463-C3-1-R).

Acknowledgments: The data of SIT from ULS were collected and made available by the Beaufort Gyre Exploration Program based at the Woods Hole Oceanographic Institution (https://www.whoi.edu/beaufortgyre) in collaboration with researchers from Fisheries and Oceans Canada at the Institute of Ocean Sciences. Sea ice concentration product of the EUMETSAT Ocean and Sea Ice Satellite Application Facility (OSI SAF, www.osi-saf.org).

References

1. Ivanova, N.; Pedersen, L.T.; Tonboe, R.T.; Kern, S.; Heygster, G.; Lavergne, T.; Sørensen, A.; Saldo, R.; Dybkjær, G.; Brucker, L.; et al. Inter-comparison and evaluation of sea ice algorithms: Towards further identification of challenges and optimal approach using passive microwave observations. *Cryosphere* **2015**, *9*, 1797–1817. [CrossRef]

2. Tietsche, S.; Alonso-Balmaseda, M.; Rosnay, P.; Zuo, H.; Tian-Kunze, X.; Kaleschke, L. Thin Arctic sea ice in L-band observations and an ocean reanalysis. *Cryosphere* **2018**, *12*, 2051–2072. [CrossRef]

3. Maykut, G.A. Energy exchange over young sea ice in the central Arctic. *J. Geophys. Res.* **1978**, *83*, 3646. [CrossRef]

4. Kwok, R.; Cunningham, G.F.; Zwally, H.J.; Yi, D. Ice, Cloud, and land Elevation Satellite (ICESat) over Arctic sea ice: Retrieval of freeboard. *J. Geophys. Res.* **2007**, *112*. [CrossRef]

5. Kwok, R.; Cunningham, G.F. ICESat over Arctic sea ice: Estimation of snow depth and ice thickness. *J. Geophys. Res.* **2008**, *113*. [CrossRef]
6. Zwally, H.J.; Yi, D.; Kwok, R.; Zhao, Y. ICESat measurements of sea ice freeboard and estimates of sea ice thickness in the Weddell Sea. *J. Geophys. Res.* **2008**, *113*. [CrossRef]
7. Laxon, S.W.; Giles, K.A.; Ridout, A.L.; Wingham, D.J.; Willatt, R.; Cullen, R.; Kwok, R.; Schweiger, A.; Zhang, J.; Haas, C.; et al. CryoSat-2 estimates of Arctic sea ice thickness and volume. *Geophys. Res. Lett.* **2013**, *40*, 732–737. [CrossRef]
8. Mäkynen, M.; Cheng, B.; Similä, M. On the accuracy of thin-ice thickness retrieval using MODIS thermal imagery over Arctic first-year ice. *Ann. Glaciol.* **2013**, *54*, 87–96. [CrossRef]
9. Menashi, J.D.; Germain, K.M.S.; Swift, C.T.; Comiso, J.C.; Lohanick, A.W. Low-frequency passive-microwave observations of sea ice in the Weddell Sea. *J. Geophys. Res.* **1993**, *98*, 22569. [CrossRef]
10. Kaleschke, L.; Maaß, N.; Haas, C.; Hendricks, S.; Heygster, G.; Tonboe, R.T. A sea-ice thickness retrieval model for 1.4 GHz radiometry and application to airborne measurements over low salinity sea-ice. *Cryosphere* **2010**, *4*, 583–592. [CrossRef]
11. Kaleschke, L.; Tian-Kunze, X.; Maaß, N.; Mäkynen, M.; Drusch, M. Sea ice thickness retrieval from SMOS brightness temperatures during the Arctic freeze-up period. *Geophys. Res. Lett.* **2012**, *39*, L05501. [CrossRef]
12. Tian-Kunze, X.; Kaleschke, L.; Maaß, N.; Mäkynen, M.; Serra, N.; Drusch, M.; Krumpen, T. SMOS-derived thin sea ice thickness: Algorithm baseline, product specifications and initial verification. *Cryosphere* **2014**, *8*, 997–1018. [CrossRef]
13. Huntemann, M.; Heygster, G.; Kaleschke, L.; Krumpen, T.; Mäkynen, M.; Drusch, M. Empirical sea ice thickness retrieval during the freeze-up period from SMOS high incident angle observations. *Cryosphere* **2014**, *8*, 439–451. [CrossRef]
14. Kaleschke, L.; Tian-Kunze, X.; Maaß, N.; Beitsch, A.; Wernecke, A.; Miernecki, M.; Müller, G.; Fock, B.H.; Gierisch, A.M.; Schlünzen, K.H.; et al. SMOS sea ice product: Operational application and validation in the Barents Sea marginal ice zone. *Remote Sens. Environ.* **2016**, *180*, 264–273. [CrossRef]
15. Kerr, Y.; Waldteufel, P.; Wigneron, J.P.; Martinuzzi, J.; Font, J.; Berger, M. Soil moisture retrieval from space: The Soil Moisture and Ocean Salinity (SMOS) mission. *IEEE Trans. Geosci. Remote Sens.* **2001**, *39*, 1729–1735. [CrossRef]
16. Entekhabi, D.; Njoku, E.G.; O'Neill, P.E.; Kellogg, K.H.; Crow, W.T.; Edelstein, W.N.; Entin, J.K.; Goodman, S.D.; Jackson, T.J.; Johnson, J.; et al. The Soil Moisture Active Passive (SMAP) Mission. *Proc. IEEE* **2010**, *98*, 704–716. [CrossRef]
17. Zine, S.; Boutin, J.; Font, J.; Reul, N.; Waldteufel, P.; Gabarro, C.; Tenerelli, J.; Petitcolin, F.; Vergely, J.L.; Talone, M.; et al. Overview of the SMOS Sea Surface Salinity Prototype Processor. *IEEE Trans. Geosci. Remote Sens.* **2008**, *46*, 621–645. [CrossRef]
18. Font, J.; Camps, A.; Borges, A.; Martín-Neira, M.; Boutin, J.; Reul, N.; Kerr, Y.H.; Hahne, A.; Mecklenburg, S. SMOS: The Challenging Sea Surface Salinity Measurement From Space. *Proc. IEEE* **2010**, *98*, 649–665. [CrossRef]
19. O'Neill, P.; Chan, S.; Njoku, E.; Jackson, T.; Bindlish, R. SMAP L3 Radiometer Global Daily 36 km EASE-Grid Soil Moisture. *Version* **2016**, *4*, R14010. [CrossRef]
20. Mecklenburg, S.; Drusch, M.; Kaleschke, L.; Rodriguez-Fernandez, N.; Reul, N.; Kerr, Y.; Font, J.; Martin-Neira, M.; Oliva, R.; Daganzo-Eusebio, E.; et al. ESA's Soil Moisture and Ocean Salinity mission: From science to operational applications. *Remote Sens. Environ.* **2016**, *180*, 3–18. [CrossRef]
21. Paţilea, C.; Heygster, G.; Huntemann, M.; Spreen, G. Combined SMAP–SMOS thin sea ice thickness retrieval. *Cryosphere* **2019**, *13*, 675–691. [CrossRef]
22. Maaß, N.; Kaleschke, L.; Tian–kunze, X.; Mäkynen, M.; Drusch, M.; Krumpen, T.; Hendricks, S.; Lensu, M.; Haapala, J.; Haas, C. Validation of SMOS sea ice thickness retrieval in the northern Baltic Sea. *Tellus A Dyn. Meteorol. Oceanogr.* **2015**, *67*, 24617. [CrossRef]
23. Haas, C.; Lobach, J.; Hendricks, S.; Rabenstein, L.; Pfaffling, A. Helicopter-borne measurements of sea ice thickness, using a small and lightweight, digital EM system. *J. Appl. Geophys.* **2009**, *67*, 234–241. [CrossRef]
24. Melling, H.; Johnston, P.H.; Riedel, D.A. Measurements of the Underside Topography of Sea Ice by Moored Subsea Sonar. *J. Atmos. Ocean. Technol.* **1995**, *12*, 589–602. [CrossRef]

25. Gabarro, C.; Turiel, A.; Elosegui, P.; Pla-Resina, J.A.; Portabella, M. New methodology to estimate Arctic sea ice concentration from SMOS combining brightness temperature differences in a maximum-likelihood estimator. *Cryosphere* **2017**, *11*, 1987–2002. [CrossRef]
26. Corbella, I.; Torres, F.; Duffo, N.; González-Gambau, V.; Pablos, M.; Duran, I.; Martín-Neira, M. MIRAS Calibration and Performance: Results From the SMOS In-Orbit Commissioning Phase. *IEEE Trans. Geosci. Remote Sens.* **2011**, *49*, 3147–3155. [CrossRef]
27. Camps, A.; Vall-llossera, M.; Duffo, N.; Torres, F.; Corbella, I. Performance of sea surface salinity and soil moisture retrieval algorithms with different auxiliary datasets in 2-D L-band aperture synthesis interferometric radiometers. *IEEE Trans. Geosci. Remote Sens.* **2005**, *43*, 1189–1200. [CrossRef]
28. Chan, S.K.; Bindlish, R.; O'Neill, P.E.; Njoku, E.; Jackson, T.; Colliander, A.; Chen, F.; Burgin, M.; Dunbar, S.; Piepmeier, J.; et al. Assessment of the SMAP Passive Soil Moisture Product. *IEEE Trans. Geosci. Remote Sens.* **2016**, *54*, 4994–5007. [CrossRef]
29. Lannoy, G.J.M.D.; Reichle, R.H.; Peng, J.; Kerr, Y.; Castro, R.; Kim, E.J.; Liu, Q. Converting Between SMOS and SMAP Level-1 Brightness Temperature Observations Over Nonfrozen Land. *IEEE Geosci. Remote Sens. Lett.* **2015**, *12*, 1908–1912. [CrossRef]
30. Brumley, B.; Cabrera, R.; Deines, K.; Terray, E. Performance of a broad-band acoustic Doppler current profiler. *IEEE J. Ocean. Eng.* **1991**, *16*, 402–407. [CrossRef]
31. Tonboe, R.; Lavelle, J.; Pfeiffer, R.H.; Howe, E. *Product User Manual for OSI SAF Global Sea Ice Concentration*; OSI-401-b & EUMETSAT; Danish Meteorological Institute: Copenhagen, Denmark, 2017.
32. Tschudi, M. *Polar Pathfinder Daily 25 km EASE-Grid Sea Ice Motion Vectors*; NASA National Snow and Ice Data Center Distributed Active Archive Center: Boulder, CO, USA, 2019. [CrossRef]
33. Kalnay, E.; Kanamitsu, M.; Kistler, R.; Collins, W.; Deaven, D.; Gandin, L.; Iredell, M.; Saha, S.; White, G.; Woollen, J.; et al. The NCEP/NCAR 40-Year Reanalysis Project. *Bull. Am. Meteorol. Soc.* **1996**, *77*, 437–472. [CrossRef]
34. Mooney, P.A.; Mulligan, F.J.; Fealy, R. Comparison of ERA-40, ERA-Interim and NCEP/NCAR reanalysis data with observed surface air temperatures over Ireland. *Int. J. Climatol.* **2011**, *31*, 545–557. [CrossRef]
35. Bilello, M.A. Formation, Growth, and Decay of Sea-Ice in the Canadian Arctic Archipelago. *ARCTIC* **1961**, *14*. [CrossRef]
36. Weeks, W.F. *On Sea Ice*; University of Alaska Press: College, AK, USA, 2010.
37. Ryvlin, A.I. Method of forecasting flexural strength of an ice cover. *Probl. Arct. Antarct.* **1974**, *45*, 79–86.
38. Key, J.; McLaren, A.S. Fractal nature of the sea ice draft profile. *Geophys. Res. Lett.* **1991**, *18*, 1437–1440. [CrossRef]
39. Maaß, N.; Kaleschke, L.; Tian-Kunze, X.; Drusch, M. Snow thickness retrieval over thick Arctic sea ice using SMOS satellite data. *Cryosphere* **2013**, *7*, 1971–1989. [CrossRef]
40. Wessel, P.; Smith, W.H.F. A global, self-consistent, hierarchical, high-resolution shoreline database. *J. Geophys. Res. Solid Earth* **1996**, *101*, 8741–8743. [CrossRef]
41. Schmitt, A.; Kaleschke, L. A Consistent Combination of Brightness Temperatures from SMOS and SMAP over Polar Oceans for Sea Ice Applications. *Remote Sens.* **2018**, *10*, 553. [CrossRef]

remote sensing

Article

Assessment of Multi-Scale SMOS and SMAP Soil Moisture Products across the Iberian Peninsula

Gerard Portal [1,2,*], Thomas Jagdhuber [3], Mercè Vall-llossera [1,2], Adriano Camps [1,2], Miriam Pablos [2,4], Dara Entekhabi [5] and Maria Piles [6]

[1] CommSensLab–UPC Unidad de Excelencia María de Maeztu, Department of Signal Theory and Communications, Universitat Politècnica de Catalunya (UPC) and IEEC-CTE/UPC, Jordi Girona 1-3, 08034 Barcelona, Spain; merce@tsc.upc.edu (M.V.); camps@tsc.upc.edu (A.C.)

[2] Barcelona Expert Center (BEC), Passeig Marítim de la Barceloneta 37-47, 08003 Barcelona, Spain; mpablos@icm.csic.es

[3] Microwave and Radar Institute, German Aerospace Center (DLR), Münchener Strasse 20, 82234 Weßling, Germany; thomas.jagdhuber@dlr.de

[4] Institute of Marine Sciences, Spanish National Research Council (ICM-CSIC), Passeig Marítim de la Barceloneta 37-49, 08003 Barcelona, Spain

[5] Department of Civil and Environmental Engineering, Massachusetts Institute of Technology (MIT), 15 Vassar Street, Cambridge, MA 02139, USA; darae@mit.edu

[6] Image Processing Laboratory, Universitat de València (UV), Catedrático José Beltrán 2, 46010 València, Spain; maria.piles@uv.es

* Correspondence: gerard.portal@tsc.upc.edu

Received: 20 December 2019; Accepted: 6 February 2020; Published: 8 February 2020

Abstract: In the last decade, technological advances led to the launch of two satellite missions dedicated to measure the Earth's surface soil moisture (SSM): the ESA's Soil Moisture and Ocean Salinity (SMOS) launched in 2009, and the NASA's Soil Moisture Active Passive (SMAP) launched in 2015. The two satellites have an L-band microwave radiometer on-board to measure the Earth's surface emission. These measurements (brightness temperatures T_B) are then used to generate global maps of SSM every three days with a spatial resolution of about 30–40 km and a target accuracy of $0.04~\mathrm{m^3/m^3}$. To meet local applications needs, different approaches have been proposed to spatially disaggregate SMOS and SMAP T_B or their SSM products. They rely on synergies between multi-sensor observations and are built upon different physical assumptions. In this study, temporal and spatial characteristics of six operational SSM products derived from SMOS and SMAP are assessed in order to diagnose their distinct features, and the rationale behind them. The study is focused on the Iberian Peninsula and covers the period from April 2015 to December 2017. A temporal inter-comparison analysis is carried out using in situ SSM data from the Soil Moisture Measurements Station Network of the University of Salamanca (REMEDHUS) to evaluate the impact of the spatial scale of the different products (1, 3, 9, 25, and 36 km), and their correspondence in terms of temporal dynamics. A spatial analysis is conducted for the whole Iberian Peninsula with emphasis on the added-value that the enhanced resolution products provide based on the microwave-optical (SMOS/ERA5/MODIS) or the active–passive microwave (SMAP/Sentinel-1) sensor fusion. Our results show overall agreement among time series of the products regardless their spatial scale when compared to in situ measurements. Still, higher spatial resolutions would be needed to capture local features such as small irrigated areas that are not dominant at the 1-km pixel scale. The degree to which spatial features are resolved by the enhanced resolution products depend on the multi-sensor synergies employed (at T_B or soil moisture level), and on the nature of the fine-scale information used. The largest disparities between these products occur in forested areas, which may be related to the reduced sensitivity of high-resolution active microwave and optical data to soil properties under dense vegetation.

Keywords: soil moisture; moisture variability; temporal dynamics; moisture patterns; spatial disaggregation; Soil Moisture Active Passive (SMAP); Soil Moisture and Ocean Salinity (SMOS); REMEDHUS

---

1. Introduction

Soil moisture (SM) is an essential climate variable (ECV) which plays a crucial role in the interplay between the Earth's land and atmospheric processes [1]. It is involved in the energy flux partition into latent and sensible heat from the land to the atmosphere. SM is closely linked to the soil evaporation, plant transpiration, and the allocation of precipitation into runoff, subsurface flow, and infiltration. Advancing our physical understanding of these land-atmosphere processes and interactions [2] is key for several climate and hydrological applications, such as drought and flood prediction, and weather and climate forecasting. Passive and active microwave sensors (radiometers and radars, respectively) are sensitive to the soil dielectric constant and allow estimation of surface soil moisture (SSM) [3]. Among microwave frequencies, measurements at L-band (1–2 GHz) have a higher soil penetration depth and are less affected by soil roughness, vegetation, and atmospheric effects than at higher frequencies (e.g., C- or X-bands) [4,5].

Currently, there are two L-band missions in orbit which were specifically devoted to measure SSM: (i) SMOS (Soil Moisture and Ocean Salinity) launched by the ESA (European Space Agency) in November 2009, and (ii) SMAP (Soil Moisture Active and Passive) launched by the NASA (National Aeronautics and Space Administration) in January 2015. Both systems have antennas with about a 6-meter aperture. The resulting brightness temperature measurements have about 40 km resolution using the half-power or −3 dB definition.

The spatial resolution of SMOS and SMAP brightness temperatures (T_B) and derived SSM maps are in the order of tens of kilometers. However, to fulfill the needs of a growing number of applications, such as monitoring the evolution of insect pests [6], the prevention of wild fires [7,8], and the early detection of forest decline [9], among others, a higher spatial detail (<1 km) is required. To bridge this gap and improve the spatial resolution of the SSM maps, a variety of spatial enhancement or spatial (sub-pixel) disaggregation approaches have been proposed [10]. They generally differ in the ancillary information they use and the physical assumptions they rely on [11]. Consequently, the performance of these disaggregation algorithms depends mainly on the multi-sensor synergies employed and on the nature of the fine-scale information used which, in turn, may also depend on the season, climate, and land cover. This makes a direct comparison very challenging, since their performance is intrinsically linked to the method and rationale, and can also be time and region dependent.

This paper focuses on the in-depth analysis of SMAP and SMOS radiometer-only based products (SMAP at 9 and 36 km, SMOS at 25 km) and on their enhanced products which are now operational. They are based on two well-known satellite-based downscaling techniques: the active/passive microwave data fusion (SMAP/Sentinel-1 at 1 km and 3 km) [12], and the optical/thermal and microwave data fusion (SMOS/ERA5/MODIS at 1 km) [13,14].

The active/passive microwave data combination aims at obtaining an optimal blend of the high accuracy of passive sensors and the high spatial resolution of active sensors. Microwave radiometers have a high radiometric sensitivity (leading to soil moisture accuracies on the order of 0.04 m^3/m^3) and a high revisit time (three days), but coarse spatial resolution, typically 30–40 km. Therefore, microwave radars, especially Synthetic Aperture Radars (SARs) step in, as their spatial resolution is significantly higher, in the range of some meters. However, the backscatter commonly has a low temporal resolution (around one week) and may be significantly affected by soil roughness and the soil-covering vegetation canopy, which complicates the active-only soil moisture retrieval.

High-resolution maps can be obtained by combining information from the active and passive sensors. For this reason, some studies carried out before and after the SMAP launch, analyzed the covariation between passive and active microwave observations. This covariation is mostly driven by soil moisture dynamics, but also depends on changes on vegetation cover and soil roughness conditions [15–17], as occurs with the backscatter. When the SMAP radar failed, about 4-months after its launch, a method to disaggregate the L-band radiometer T_B using the C-band Sentinel-1 radar backscatter was developed [12,18]. This approach, based on the active/passive covariation, is now the baseline to provide high-resolution SMAP SSM maps at 1 and 3 km [18]. However, the Sentinel-1 measurements are at C-band which have reduced sensitivity for moderate to dense vegetation coverage (up to ~3 kg/m^2).

The optical/thermal and microwave fusion technique takes advantage of the high spatial resolution of optical and thermal remote sensing and on the inverse relationship between the land surface temperature (LST), and the vegetation status, which can be related to the soil moisture content [19]. Note that optical and thermal electromagnetic waves have the drawback of being masked by clouds, whereas microwaves can provide continuous monitoring regardless of atmospheric and illumination conditions. Here we use the latest version [13] of the optical/thermal and microwave algorithm firstly developed by Piles et al. [20,21]. It is an integrative model that holds at the coarse and fine spatial scales. Information of a vegetation index (Normalized Difference Vegetation Index, NDVI) from the optical and LST from the thermal bands of MODIS (moderate resolution imaging spectroradiometer instrument, MODIS) instrument, together with SMOS data, are used to obtain the model coefficients at low resolution. These coefficients are then applied to obtain the SSM fields at high resolution. Since the presence of clouds masking the MODIS LST information resulted in a loss of spatial coverage, a cloud free version of the algorithm [14] was developed in which MODIS LST was replaced with modelled ERA5 climate reanalysis skin temperature from the European Centre for Medium-Range Weather Forecast (ECMWF). Although the spatial resolution of the ERA5 LST is degraded with respect to MODIS LST (33 km vs. 1 km, respectively), the coverage increases dramatically; a comparison study carried out over Australia and Spain showed that the results were consistent for both versions of the algorithm [14]. This cloud-free version of the algorithm is now in operations at the Barcelona Expert Center (BEC) [22].

The aim of this paper is to analyze the temporal and the spatial characteristics of low-resolution (native) and high-resolution (disaggregated) SSM products provided by the SMAP and SMOS missions, with special emphasis on the most recently developed high-resolution ones. The temporal analysis has been carried out in the central part of the Duero basin, Spain, where the dynamics of SMAP and SMOS products at different spatial scales are compared against the data provided by the REMEDHUS in situ network, and their spatial representativeness as well as their correspondence is assessed. A comparison of spatial patterns has been conducted for the whole Iberian Peninsula, with focus on the analysis of their differences and distinct features, as well as on understanding the possible impact of the physical assumptions and multi-sensor synergies in the fine-scale estimates.

The SMAP and SMOS-derived SSM data products as well as the hydrological and climatic variables used in this study are presented in Section 2. Section 3 explains briefly the methodology followed to conduct the temporal and spatial analyses on the different products. The results of these comparisons are shown in Section 4. Section 5 discusses the possible reasons for the mismatch found among the different SSM products. Finally, Section 6 provides main conclusions and perspectives from this study.

2. Data Description

This section introduces four SSM products derived from SMAP, two SSM products derived from SMOS, the in situ SSM measured by REMEDHUS network, and other ancillary information that has been used in this work. The data products used are summarized in Table 1 and described in the following subsections.

Table 1. Summary of the data products used in this study.

Data	Acronym	Grid	Availability
BEC			
SMOS L3	SMOSL3	25 km	3-day
SMOS/ERA5	SMOSL4	1 km	3-day
NASA			
SMAP L2 Radiometer	SMAPL2	36 km	3-day
SMAP Enhanced L2 Radiometer	SMAPL2_E	9 km	3-day
SMAP/Sentinel-1 L2 Radiometer/Radar	SMAP_AP3	3 km	12-day
SMAP/Sentinel-1 L2 Radiometer/Radar	SMAP_AP1	1 km	12-day
REMEDHUS			
In situ SSM		Point	Hourly
Ancillary Data			
Land Cover	LC	300 m	1-year

2.1. Soil Moisture Data

2.1.1. NASA SMAP Products

SMAP is a NASA mission within the Earth System Science Pathfinder (ESSP) program. The mission was launched in January 2015 with the main goal of measuring the SSM and the freeze/thaw state of the soil with high spatio-temporal resolution and global coverage [2]. The data products of this mission serve applications in many disciplines, including hydrology, weather and climate, meteorology, environmental sciences, agriculture, human health, and security [2,23]. Its scientific requirements are to provide estimates of soil moisture of the soil top 5 cm with a target accuracy of 0.04 m^3/m^3 and a spatial resolution of 10 km every 3 days over continental land, excluding areas with standing water, high vegetation content (>5 kg/m^2) or frozen ground as well as urban or mountainous areas.

Three SMAP SSM products were investigated in this study: the SMAP L2 Radiometer (SMAPL2) with a spatial resolution of 36 km [24], the SMAP Enhanced L2 Radiometer (SMAPL2_E) with a gridding of 9 km [25] but still at the radiometer resolution (~40 km) and the SMAP/Sentinel-1 L2 Radiometer/Radar (SMAP_AP) with a spatial resolution of 3 km (SMAP_AP3) and also at 1 km (SMAP_AP1) [26].

The SMAPL2 is a radiometer-only based SSM product derived directly from the SMAP Level-1C T_B (L1CTB) product in a 36 km Equal-Area Scalable Earth Grid 2.0 (EASEv2) grid. To obtain the SMAPL2 from the L1CTB, the Single Channel Algorithm at vertical polarization (SCA-V) is used [27]. In addition to SSM and T_B observations, the ancillary data required to apply the retrieval algorithm is included in the product, namely surface temperature, vegetation opacity, vegetation single scattering albedo, surface roughness, land cover information, soil texture, together with data flags for identification of land, water, precipitation, radio frequency interference, urban areas, mountainous terrain, permanent ice, snow, and dense vegetation [27–29].

The SMAPL2_E is derived from the SMAP Level-1C T_B Enhanced (L1CTB_E) product and contains SSM and T_B data, which are previously interpolated using Backus-Gilbert at T_B level. This optimal interpolation technique takes advantage of the SMAP radiometer oversampling to generate an enhanced version of the T_B that is posted on a 9 km grid. The SCA-V is applied to these T_B data to obtain the SSM retrievals [30].

The SMAP_AP is generated by merging the SMAP radiometer with Sentinel 1A/1B data through a recently developed active/passive downscaling Algorithm [12] (1). It allows to disaggregate the SMAP T_B from a resolution of 36 km to 3 km or 1 km (depending on filtering speckle noise) [31].

$$T_{B_p}(M_j) = \left[\frac{T_{B_p}(C)}{T_S} + \beta'(C)\cdot\left\{ \left[\sigma_{pp}(M_j) - \sigma_{pp}(C) \right] + \Gamma\cdot\left[\sigma_{pq}(C) - \sigma_{pq}(M_j) \right] \right\} \right]\cdot T_s \tag{1}$$

where M (medium) and C (coarse) are the different spatial resolutions at which the variables are used, T_s is the land surface temperature, β' is the active-passive microwave covariation parameter [12], σ_{pp} and σ_{pq} are the radar backscatter with co-pol and cross-pol, respectively, and Γ represents the vegetation heterogeneity within a pixel with C resolution. The SSM at 3 km (or 1 km) is retrieved after applying the SCA-V to the disaggregated T_B.

Descending orbits (06:00 am) of all the SMAP products were used in this study, since they have the same local time of ascending orbits of the SMOS products.

2.1.2. BEC SMOS Products

The SMOS satellite was launched in November 2009, and it is the second Earth observation mission of ESA's Living Planet program [32,33]. After 10 years in orbit, many studies have contributed to understand and improve the quality of SMOS soil moisture products. This mission was designed to observe both soil moisture and ocean salinity, as required by climatological, meteorological, hydrological, and oceanographic applications. The SMOS instrument, the Microwave Imaging Radiometer with Aperture Synthesis (MIRAS), is the first L-band (1.4 GHz) interferometric radiometer on space. It provides global views of the Earth at multiple incidence angles (from 0° to 65°) with a spatial resolution of 35–40 km and a temporal resolution of 3 days [34].

The SMOS Level 3 (L3) and 4 (L4) SSM products used in this study are provided by the BEC [35], an ESA Expert Support Laboratory (ESL) of SMOS L1 and L2 ocean salinity. The BEC SMOS L3 SSM product (SMOSL3) is generated directly from the L2 SSM after discarding invalid retrievals by means of applying quality filters to each grid point. Later, a weighted average based on a data quality index is used to bin the data from the Icosahedral Snyder Equal Area (ISEA) to the 25 km EASEv2 grid [22].

The BEC SMOS L4 SSM (SMOSL4) product is derived from the SMOSL3 using a semi-empirical downscaling algorithm (2) which links the SSM with the T_B, a vegetation index, and the LST [20,21]

$$SSM = b_0 + b_1 \cdot LST + b_2 \cdot NDVI + \frac{b_3}{3} \cdot \sum_{i=1}^{3} T_{BH\theta_i} + \frac{b_4}{3} \cdot \sum_{i=1}^{3} T_{BV\theta_i} \tag{2}$$

where T_{BH} and T_{BV} are the T_B at horizontal and vertical polarizations, respectively, at three different incidence angles (32.5°, 42.5°, and 52.5°). The b parameters represent the downscaling factors associated to each variable. The downscaling is applied daily and the resulting L4 SSM maps are posted on the MODIS 1 km grid.

Ascending orbits (06:00 am) were selected for all the SMOS products used in this study.

2.1.3. REMEDHUS Network

The Soil Moisture Measurements Station Network of the University of Salamanca (REMEDHUS) is an in situ network located in the central part of the Duero basin (41.1° to 41.5°N; 5.1° to 5.7°W). It contains 20 soil moisture monitoring stations that provide information at different depths (here we are using exclusively the topsoil data at 5 cm depth), and four automatic weather stations that measure precipitation, air temperature, relative humidity, wind speed, and solar radiation [36]. These stations are located within a nearly flat area of 1300 km^2 in a semi-arid Continental-Mediterranean agricultural region. This area receives an average annual precipitation of 385 mm, and it has a mean temperature of 12 °C [37]. Most of the region is dedicated to grow rainfed cereals, as shown in Figure 1. Other land uses within this area: irrigated crops, fallow, vineyards, or forest-pasture. The stations record the SSM data every hour, aggregated to a daily average [38] for this study.

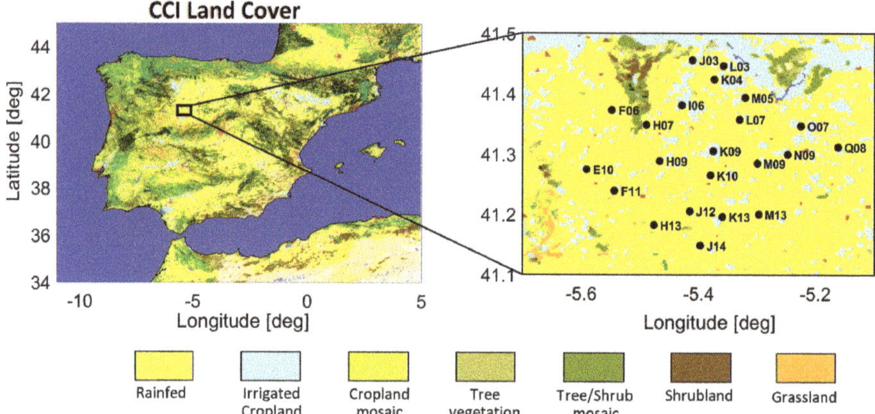

Figure 1. CCI land cover map (at 300 m) over the Iberian Peninsula (left) and a close-up of the REMEDHUS area (right). Black dots depict the 20 in situ SSM stations of the REMEDHUS network available for the study period (from April 2015 to December 2017). The distribution of the land cover within the REMEDHUS area is: agriculture, 95.45% (cropland, 75.44%; irrigated, 16.11%; other, 3.90%); forest, 2.70%; grassland, 0.63%; wetland, 0%; settlement, 0.26%; and other, 0.95%.

2.2. Ancillary Data

Climate Change Initiative: Land Cover

The ESA Climate Change Initiative (CCI) program includes a variety of biological, physical, and chemical variables known as ECV. Here the CCI land cover (LC) is used, which provides information of the geographical distribution of global land cover at a resolution of 300 m [39,40]. The CCI LC from year 2015 will be used in this study to characterize the dominant land cover within each SMOS/SMAP pixel. Minimal differences were observed on the CCI LC over the study region during the period 2013–2017.

3. Methodology

3.1. Statistical Analysis of SSM Time Series at the Network Scale

Ground-based SSM measurements from REMEDHUS have been selected as a benchmark for a cross-validation of the multi-scale remotely sensed SSM products. REMEDHUS stations were placed by the Water Resources Research group of the University of Salamanca (responsible for the maintenance of the network) in areas in which the land use, during the years from 2015 to 2017, were the following: fallow, rainfed, forest-pasture, vineyard and irrigated. A thorough analysis of the 20 operational in situ stations available during the study period and their comparison to satellite data was performed. For the sake of clarity and simplicity, in this work we will focus on 11 of them (see Table 2). They cover the five land uses -and therefore allow studying the impact of land use on the downscaling products- and also provide a good spatial representation when averaged at the network scale.

In this first step of the analysis, we used the data provided by six stations (H13, H9, J3, K13, N9, and O7) representative of the five different land uses over the REMEDHUS network. The SMAP and the SMOS time series of the pixels overlapping these stations have been statistically evaluated with the in situ SSM at two spatial levels, at low resolution (from 9 km up to ~40 km) and at high resolution (3 km and 1 km). Performance metrics, such as the Pearson's correlation (R), the root mean square error (RMSE), the unbiased root mean square error (uRMSE) and the bias, together with the number of available samples (N), have been computed for each station-pixel pair. These performance metrics have been calculated exactly as described in [41].

Table 2. Land use of the region where 11 in situ stations, of the REMEDHUS network, were located, Figure 2015. 2016, and 2017 (provided by the Water Resources Research group of the University of Salamanca). The land uses are: fallow (F), rainfed (R), forest-pasture (FP), vineyard (V), irrigated (I).

	H13	H9	J3	K13	N9	O7	F11	J12	J14	K10	M9
2015	F	FP	V	I	R	R	R	R	R	R	R
2016	F	FP	V	I	R	F	F	F	F	F	F
2017	F	FP	V	I	R	R	R	R	R	R	R

Since rainfed is the most common land cover type within the REMEDHUS area (see Figure 1), the second step of the analysis consisted in reproducing the same statistical evaluation, but using only the dataset of the stations located over rainfed/fallow land uses (F11, H13, J12, J14, K10, M9, and O7). The average value of all these rainfed/fallow stations were compared to the average of the respective SMAP and SMOS pixels covering these stations.

Additionally, statistical scores have been obtained for all seasons (DJF: December, January, February; MAM: March, April, May; JJA: June, July, August; SON: September, October, November). This analysis is needed to evaluate whether the precision (R), accuracy (bias) and quadratic errors (RMSE/uRMSE) of the studied products/methodologies have any seasonal dependence.

3.2. Analysis of the SSM Spatial Patterns

To consistently analyze the spatial features of the SMAP and SMOS SSM maps at 1 km, their maps of daily differences were computed (SMAP_AP1 minus SMOSL4) along the entire study period and the histogram of these daily SSM difference maps has been obtained, together with its mean and standard deviation (std). In addition, daily SSM difference maps have been temporally averaged and compared to the spatial distribution of the most common land cover types over the Iberian Peninsula.

Besides, taking into account different ancillary data (e.g., soil roughness, vegetation indices, skin temperature, or albedo) [24], high-resolution SMAP and SMOS SSM maps are derived from their respective T_B, as described in Sections 2.1.1 and 2.1.2. However, there are noteworthy differences related to T_B polarizations and incidence angles. While the SMAP disaggregation methodology in (1) uses one specific polarization (vertical) with a single incidence angle (40°), the SMOS downscaling algorithm in (2) employs two polarizations (horizontal and vertical), and the average of three incidence angles (32.5° ± 5°, 42.5° ± 5°, and 52.5° ± 5°) over the same target. To analyze the influence of T_B data on the high-resolution SSM maps, the vertical SMAP L1C T_B has been compared to the vertical SMOS L1C T_B at the Earth's surface, using exclusively the central SMOS angle (42.5°). To do so, the SMOS T_B has been corrected by the geometry of the antenna, the ionospheric and atmospheric effects, linearly interpolated to the angles range 42.5 ± 5° and binned to a 25 km EASEv2 grid. The SMAP T_B has been interpolated from the initial 36 km EASEv2 to the same grid of SMOS, using the nearest neighbor. Then, daily differences (SMAP minus SMOS T_B) have been computed from April 2015 to December 2017. The coastal areas of the Iberian Peninsula were discarded to screen out the effect of sea-land contamination.

A low- vs. high-resolution study has also been performed to assess the variations, in volumetric units, between the original and the downscaled SSM maps of the same sensor. In this way, we assessed the impact of the different downscaling methods on spatial soil moisture patterns. To do this, coarse-resolution SMAP and SMOS maps were firstly interpolated to a 1 km grid using the nearest neighbor. The comparison was done by separately calculating the daily differences between the SMAP_AP1 and the SMAPL2 maps, as well as the daily differences between SMOSL4 and SMOSL3 maps along the entire study period.

4. Results

4.1. Statistical Analysis of SSM Time Series at the Network Scale

Interestingly, SMAP and SMOS satellite products agree reasonably well among them, capturing the marked wet up and dry down variations along time. Nevertheless, a strong dependence of results from comparison to in situ on land use is found. Both, SMAP and SMOS products are overestimating the in situ measurements in vineyards (Figure 2a). Instead, satellite data underestimate the in situ SSM for irrigated crops (Figure 2b), while they almost match up with in situ observations for fallow/rainfed crops (Figure 2c), which are the most common land uses in the REMEDHUS area.

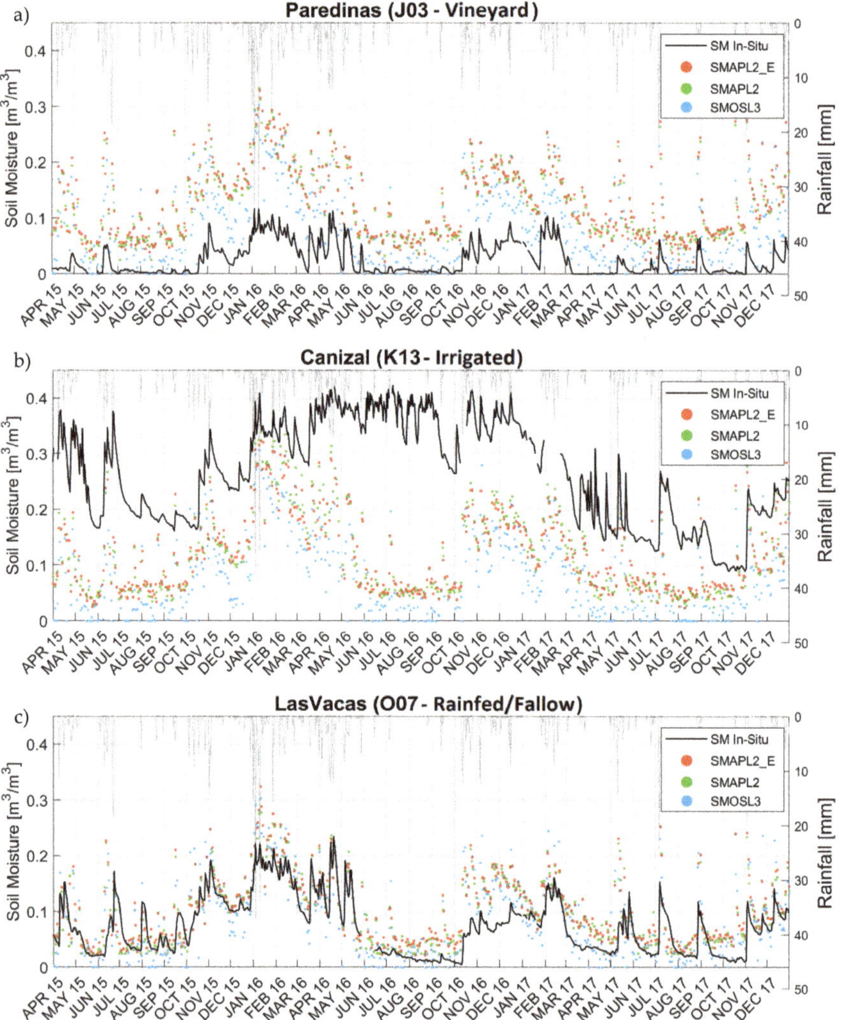

Figure 2. Daily evolution of the in situ SSM (black) and the three low-resolution (radiometer-only) SSM (SMAPL2_E, red; SMAPL2, green; and SMOSL3, blue) at three REMEDHUS stations with different land use: (**a**) J3 (vineyard), (**b**) K13 (irrigated), and (**c**) O7 (rainfed/fallow).

The statistics derived from the temporal inter-comparison of low-resolution SSM products with in situ data using all the concurrent samples available for each dataset are summarized in Table 3. Comparing the different instruments, results show that the two SMAP products have the same or a slightly higher correlation ($\Delta R \leq 0.12$) and similar unbiased errors ($\Delta uRMSE \leq 0.01$ m^3/m^3) than the SMOS product in all the study cases. There are no significant differences between the metrics obtained for the SMAP SSM (9 km vs. 36 km). Regarding the different land uses, the worst results were obtained for K13, an irrigated station, with a R (and a bias) of 0.46 (and -0.142 m^3/m^3) for SMAPL2_E, 0.48 (-0.143 m^3/m^3) for SMAPL2, and 0.46 (-0.183 m^3/m^3) for SMOSL3. Besides, the uRMSE of K13 is twice as high the objective accuracy of both space missions (SMOS and SMAP). This underperformance probably comes from the fact that irrigated land is not the most representative land use within the low-resolution SMAP/SMOS pixels, which is mostly covered by rainfed crops (see Table 4). On the contrary, the best results are obtained for the stations located over rainfed/fallow land cover (H13 and O07), with a R between 0.79 and 0.83 for both SMAP products (SMAPL2_E and SMAPL2), and between 0.70 and 0.80 for SMOS. Their bias is low, between 0.027 and 0.035 m^3/m^3 for SMAP, and between 0.004 and 0.068 m^3/m^3 for SMOS, in absolute values. The uRMSE of these two stations are around 0.04–0.05 m^3/m^3, meeting or almost meeting required accuracy of both missions. In the case of vineyards (J03), intermediate results are obtained. The highest R is obtained for two SMAP products (0.85), but a high R is also obtained for SMOSL3 (0.73). The uRMSE is very similar (from 0.045 to 0.048 m^3/m^3). However, the bias of J03 is similar for SMOSL3 (0.057 m^3/m^3), but the SMAP ones are up to two or three times higher (0.106 and 0.103 m^3/m^3) than the aforementioned ones for rainfed/fallow. At these spatial scales, the number of available samples of SMAP and SMOS with in situ samples are of the same order (around 500 days), ensuring a robust statistical analysis.

Table 3. Statistics obtained from the comparison of *in situ* SSM against the concurrent low-resolution (radiometer-only) pixels SSM time series: of SMAPL2_E (left), SMAPL2 (center) and SMOSL3 (right), from April 2015 to December 2017. The *in situ* stations used (and their respective land use) are: H13 (fallow), H9 (forest-pasture), J3 (vineyard), K13 (irrigated), N9 (rainfed) and O7 (rainfed/fallow).

	In situ vs. SMAPL2_E					*In situ* vs. SMAPL2					*In situ* vs. SMOSL3				
	N	R	RMSE	uRMSE	Bias	N	R	RMSE	uRMSE	Bias	N	R	RMSE	uRMSE	Bias
	[-]	[-]	[m^{3}m^{-3}]	[m^{3}m^{-3}]	[m^{3}m^{-3}]	[-]	[-]	[m^{3}m^{-3}]	[m^{3}m^{-3}]	[m^{3}m^{-3}]	[-]	[-]	[m^{3}m^{-3}]	[m^{3}m^{-3}]	[m^{3}m^{-3}]
H13	540	0.83	0.052	0.044	−0.028	492	0.83	0.056	0.044	−0.035	497	0.80	0.086	0.052	−0.068
H09	524	0.64	0.136	0.075	−0.114	506	0.62	0.137	0.077	−0.113	504	0.58	0.167	0.081	−0.146
J03	550	0.85	0.115	0.046	0.106	537	0.85	0.112	0.045	0.103	516	0.73	0.075	0.048	0.057
K13	502	0.46	0.166	0.086	−0.142	483	0.48	0.167	0.086	−0.143	510	0.46	0.201	0.083	−0.183
N09	502	0.67	0.087	0.052	−0.069	536	0.65	0.076	0.055	−0.052	512	0.60	0.117	0.057	−0.102
O07	490	0.79	0.048	0.038	0.030	486	0.79	0.047	0.038	0.027	510	0.70	0.048	0.048	0.004

Table 4. Percentage of rainfed and irrigated croplands (the two most common land covers over the REMEDHUS network) within the SMOS and SMAP pixels (36 km, 25 km, 9 km, 3 km and 1 km) enclosing the *in situ* stations J3 (vineyard), K13 (irrigated) and O7 (rainfed/fallow).

	J3 (Vineyard)		K13 (Irrigated)		O7 (Rainfed/Fallow)	
	Rainfed (%)	Irrigated (%)	Rainfed (%)	Irrigated (%)	Rainfed (%)	Irrigated (%)
SMAPL2 (36 km)	67.81	20.83	80.27	17.26	67.97	24.68
SMOSL3 (25 km)	61.06	30.51	92.54	6.47	61.06	30.51
SMAPL2_E (9 km)	39.71	52.16	93.08	6.66	68.69	23.27
SMAP_AP3 (3 km)	43.80	42.98	79.55	20.45	66.94	33.06
SMOSL4 (1 km)	56.25	43.75	68.75	31.25	75.00	25.00

When analyzing the metrics derived from the validation of SMAP and SMOS at high resolution (see Table 5), the irrigated station K13 keeps showing the worst results as in the low-resolution case: a R (and a bias) of 0.45 (-0.142 m^3/m^3) for SMAP_AP1, 0.51 (-0.129 m^3/m^3) for SMAP_AP3, and 0.42 (-0.186 m^3/m^3) for SMOSL4. This indicates that irrigated areas are not even spatially representative at the scales of 3 km to 1 km, which denotes the small extent of these areas within the satellite footprint (see Table 4).

Table 5. Statistics obtained from the comparison of *in situ* SSM against the concurrent high-resolution pixels SSM time series: of SMAP_AP1 at 1 km (left), SMAP_AP3 at 3 km (center) and SMOSL4 at 1 km (right), from April 2015 to December 2017. The *in situ* stations used (and their respective land use) are: H13 (fallow), H9 (forest-pasture), J3 (vineyard), K13 (irrigated), N9 (rainfed) and O7 (rainfed/fallow).

	In Situ vs. SMAP_AP1					*In Situ* vs. SMAP_AP3					*In Situ* vs. SMOSL4				
	N	R	RMSE	uRMSE	Bias	N	R	RMSE	uRMSE	Bias	N	R	RMSE	uRMSE	Bias
	[-]	[-]	$[m^3m^{-3}]$	$[m^3m^{-3}]$	$[m^3m^{-3}]$	[-]	[-]	$[m^3m^{-3}]$	$[m^3m^{-3}]$	$[m^3m^{-3}]$	[-]	[-]	$[m^3m^{-3}]$	$[m^3m^{-3}]$	$[m^3m^{-3}]$
H13	100	0.81	0.062	0.040	−0.048	100	0.86	0.046	0.038	−0.025	489	0.80	0.089	0.045	−0.076
H09	96	0.56	0.164	0.086	−0.139	96	0.60	0.155	0.084	−0.131	443	0.59	0.175	0.079	−0.156
J03	98	0.70	0.093	0.046	0.081	98	0.83	0.121	0.043	0.114	513	0.72	0.085	0.054	0.066
K13	97	0.45	0.172	0.097	−0.142	97	0.51	0.156	0.088	−0.129	493	0.42	0.205	0.085	−0.186
N09	101	0.45	0.120	0.071	−0.097	101	0.57	0.101	0.058	−0.082	503	0.63	0.119	0.056	−0.105
O07	98	0.66	0.076	0.063	0.042	99	0.78	0.076	0.050	0.056	501	0.71	0.047	0.047	−0.001

Similarly, the best results are obtained for the stations H13 and O07, with R between 0.66 and 0.86 for SMAP_AP1 and SMAP_AP3, and between 0.71 and 0.80 for SMOSL4. The lowest bias is precisely observed in H13 and O07, ranging between 0.025 and 0.056 m^3/m^3 for SMAP, and from 0.001 to 0.076 m^3/m^3 for SMOS, in absolute values. Again, the reason for that is the predominance of rainfed crops and fallow regions over REMEDHUS (see Table 4). Therefore, both satellites mostly see the land cover types leading to a cover-characteristic signal at low- as well as at high-resolution. As previously observed in Table 3, the metrics for vineyard are in a well acceptable range. On the one hand, taking into account both SMAP and SMOS, R varies between 0.70 and 0.83, and the uRMSE is always around 0.04–0.05 m^3/m^3. On the other hand, the bias of J03 is doubled or even tripled (0.081 and 0.114 m^3/m^3) with respect to the stations H13 and O07. All the SMAP and SMOS products are overestimating the in situ measurements of J03. One reason could be that grapevines are settled on very fine sand, which causes the water not to be retained and it quickly percolates into deeper layers. Additionally, the vineyard areas of REMEDHUS are not spatially representative at scales of 1 km and beyond.

Due to the missing synchronization of SMAP and Sentinel-1 acquisition orbits, the number of samples is much lower in the SMAP_AP1 and SMAP_AP3 (96 to 101 days) than in the SMOSL4 time series (443 to 513).

Similar statistical scores are obtained for the SMOS products when the same number of samples is used at high and low resolution, in line with the results obtained in a previous study [13]. When the same analysis is conducted for the SMAP products, only slightly worst performances are obtained for the SMAP_AP1 product.

Table 6 shows the statistics obtained between the average SSM values of the stations located over a rainfed/fallow land use (F11, H13, J12, J14, K10, M9, and O7) and the average of the concurrent SMAP and SMOS products at high-resolution (see Figure 3). Lower correlations are obtained during summer season (0.62, 0.64 and 0.65, for SMAP_AP1, SMAP_AP3 and SMOSL4, respectively). This is consistent with the results of previous studies [13]. Slightly better results are obtained for SMAP in terms of R (and bias) 0.88 (0.014 m^3/m^3), against SMOS, 0.79 (0.04 m^3/m^3) calculated as an average of DJF, MAM, and SON.

Table 6. Statistics obtained from the comparison of *in situ* SSM against the high-resolution pixel SSM of SMAP_AP1 at 1 km (left), SMAP_AP3 at 3 km (center) and SMOSL4 at 1 km (right) from April 2015 to December 2017 for the different seasons of the year and also for the entire study period (ESP). Statistics are obtained after averaging all-time series of rainfed/fallow stations (F11, H13, J12, J14, K10, M9 and O7) and the pixels that contain these stations.

	In situ vs. SMAP_AP1					*In situ* vs. SMAP_AP3					*In situ* vs. SMOSL4				
	N	R	RMSE	uRMSE	Bias	N	R	RMSE	uRMSE	Bias	N	R	RMSE	uRMSE	Bias
	[-]	[-]	$[m^3m^{-3}]$	$[m^3m^{-3}]$	$[m^3m^{-3}]$	[-]	[-]	$[m^3m^{-3}]$	$[m^3m^{-3}]$	$[m^3m^{-3}]$	[-]	[-]	$[m^3m^{-3}]$	$[m^3m^{-3}]$	$[m^3m^{-3}]$
DJF	17	0.87	0.056	0.053	0.018	17	0.92	0.060	0.048	0.035	88	0.87	0.056	0.047	−0.031
MAM	22	0.91	0.037	0.033	−0.017	22	0.90	0.027	0.026	−0.008	119	0.72	0.071	0.041	−0.058
JJA	26	0.62	0.037	0.035	−0.012	27	0.64	0.035	0.035	−0.006	128	0.65	0.073	0.030	−0.067
SON	33	0.85	0.034	0.033	0.008	33	0.86	0.034	0.032	0.011	125	0.78	0.052	0.041	−0.032
ESP	98	0.86	0.040	0.040	−0.002	99	0.87	0.039	0.038	0.006	460	0.82	0.064	0.043	−0.048

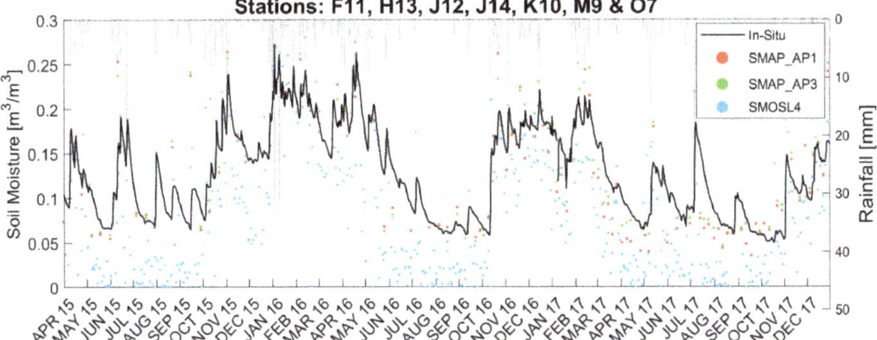

Figure 3. Daily evolution of in situ SSM (black) and the three high-resolution SSM products (SMAP_AP1 at 1 km, red; SMAP_AP3 at 3 km, green; and SMOSL4 at 1 km, blue) after averaging time series of rainfed/fallow stations (F11, H13, J12, J14, K10, M09, and O07) and the pixel time series that contain these stations.

4.2. Analysis of the SSM Spatial Patterns

The study carried out in the previous section shows a general agreement between the temporal dynamics of all the considered SSM products regardless of their spatial resolution (low resolution: ~40 km, 9 km vs. high resolution: 3 km, 1 km). However, as it can be seen from the maps shown in Figure 4, there are clearly visible differences in the spatial patterns attained by the downscaled SMAP and SMOS high-resolution products. In this section, we examine these differences and conduct specific analyses to test two hypothesis: (i) that they are due to differences in the multi-sensor synergies they are built upon (optical-microwave or active-passive) and (ii) that they are due to the rationale of the approach (e.g., whether the downscaling is conducted in brightness temperature- or in the soil moisture-space).

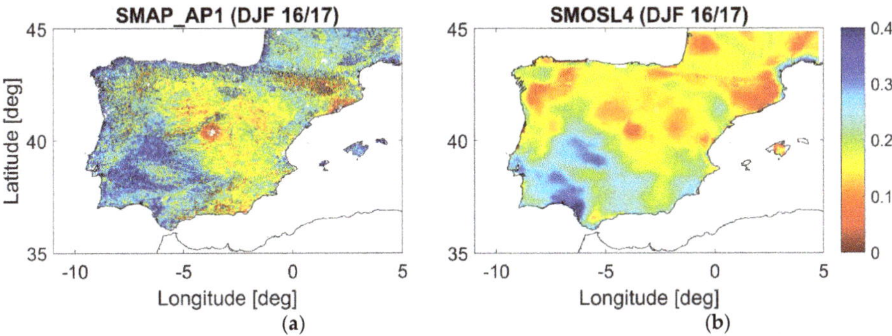

Figure 4. Temporally-averaged map of daily SMAP (**a**) and SMOS (**b**) products at 1 km over the Iberian Peninsula for the period December 2016 to February 2017.

4.2.1. Comparison of SSM Enhanced Resolution Products

Figure 5 shows the daily differences (map and histogram) between the SMAP and SMOS products at 1 km (SMAP_AP1 minus SMOSL4) for the whole study period. The mean of these differences is minimal (of 0.03 m^3/m^3), less than their target accuracy, and their std is also low (of 0.09 m^3/m^3). The same behavior is observed when this study is performed on a year-to-year basis (not shown).

In addition, we conducted the same analysis per season, and we obtained that daily differences ranged between 0.03 and 0.04 m^3/m^3 in mean and from 0.05 to 0.07 m^3/m^3 in std (not shown). These results affirm that the differences cannot be explained by seasonal or yearly differences (e.g., dry or wet year). Yet the temporally-averaged map of daily SSM differences (Figure 5a) reveals that there is a geographic spatial pattern that persists over time when comparing the two high-resolution products, with higher differences located in the north, northwest and west of the Iberian Peninsula, in close correspondence to forested areas (see land cover maps on Figure 6).

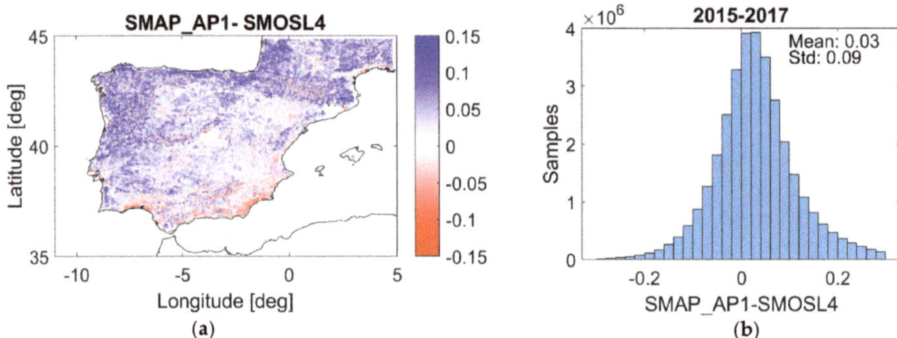

Figure 5. (**a**) Temporally-averaged map of daily SSM differences between SMAP and SMOS at 1 km (SMAP_AP1 minus SMOSL4) and (**b**) histogram of daily SSM differences maps, for the period April 2015 to December 2017.

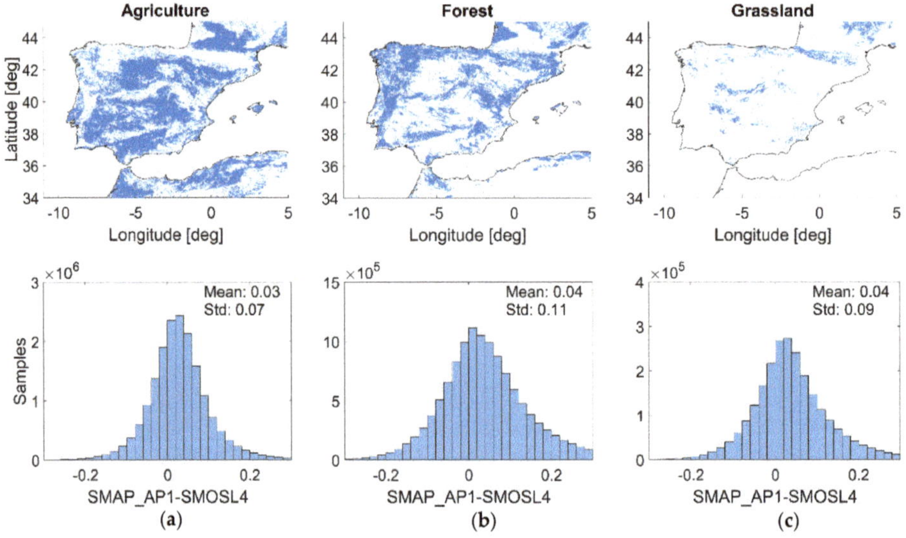

Figure 6. (First row) The three most common land covers types over the Iberian Peninsula (**a**), agriculture; (**b**) forest; and (**c**), grassland) according to the CCI LC map. (Second row) Histograms of the daily SSM differences (SMAP_AP1 minus SMOSL4) for the respective land covers.

The possible dependence of the differences between the two downscaled products on the land cover was further examined. Pixels from the temporally-average map of SSM differences (Figure 5a) were grouped for the most common land cover classes (agriculture, forest and grassland) and their histograms were analyzed (see Figure 6). In general, pixels with SSM differences (SMAP minus SMOS) equal or above 0.1 m^3/m^3 are located within forests (66.71% of the pixels) or agriculture (22.47% of the pixels). The largest SSM differences are observed for forest land cover, with a mean of 0.04 m^3/m^3 and a std of 0.11 m^3/m^3.

Land cover only partially explains the SSM differences between the SMAP and SMOS products at 1 km. Results show that soil moisture values provided by SMAP over the Iberian Peninsula are systematically higher than the ones provided by SMOS (see Figure 5), although the absolute difference is minimal (mean difference of 0.03 m^3/m^3). The SSM values of SMOS exceed those of SMAP less often and with lower intensity, but this effect is mostly occurring in coastal areas. The same SSM difference pattern is found in the temporally-averaged map of daily T_B differences (SMAP minus SMOS T_B) shown on Figure 7. Since the spatial pattern is already present at T_B level, we can conclude it was not introduced by neither SMAP nor SMOS downscaling methodologies. Figure 7b shows the histogram of the daily T_B differences, with an absolute mean value of 2.92 K. In order to understand to what extent, the 2.92 K cold bias could affect SMAP or SMOS SM retrievals, we analyzed Davenport et al. in [42], who conducted a sensitivity analysis of soil moisture retrieval using the applied tau-omega microwave emission model. A bias of about 3° (K) T_B approximately corresponds to 3–4 (vol.%) error in estimating volumetric soil water content, which would fit to the bias ranges reported in our study.

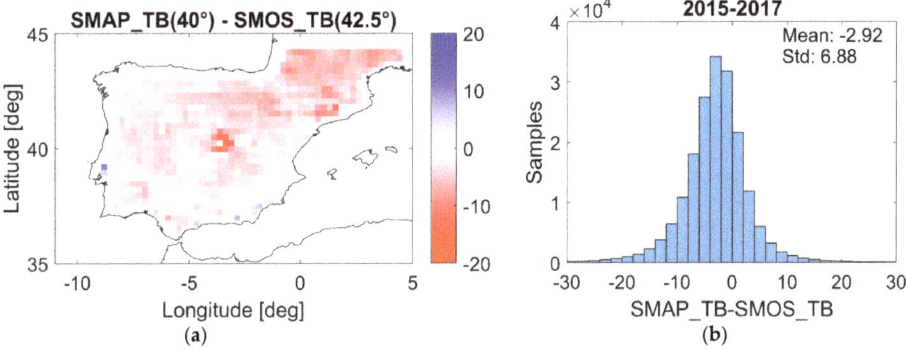

Figure 7. (**a**) Temporally averaged map of daily T_B differences between SMAP (40° incidence angle) and SMOS (42.5° incidence angle) at 25 km and (**b**) histogram of temporally-averaged daily T_B differences, for the period April 2015 to December 2017.

4.2.2. Downscaling Impact on SSM Differences

Figures 8 and 9 show the maps and histograms of daily SSM differences between SMAP_AP1 and SMAPL2 (1 km vs. 36 km), and between SMOSL4 and SMOSL3 (1 km vs. 25 km), respectively. The mean (and std) obtained after calculating the differences along the complete study period are −0.01 m^3/m^3 (0.07 m^3/m^3) for SMAP, and of ~0 m^3/m^3 (0.03 m^3/m^3) for SMOS. Although mean differences are minimal over the whole domain for both sensors, the resulting average map of the SMAP SSM differences reveal some underlying spatial patterns. The highest positive differences obtained for SMAP are concentrated in the forested regions (see Figure 6) and the highest negative differences appear near the coast and in areas of complex topography. This is possibly due to the reduced sensitivity of the Sentinel 1 signal at C-band to soil moisture in presence of significant vegetation backscattering. According to [12], this leads to a decrease in T_B after downscaling and therefore to an increase in estimated soil moisture. Also, the temporally-averaged map of SMOS SSM differences exhibits an underlying boxing effect that can be explained by the use of SMOS SSM at low-resolution as a reference to obtain the downscaling parameters of (2), as previously observed in [13,21]. However, this effect is nonetheless negligible and does not have a significant impact in the enhanced resolution product.

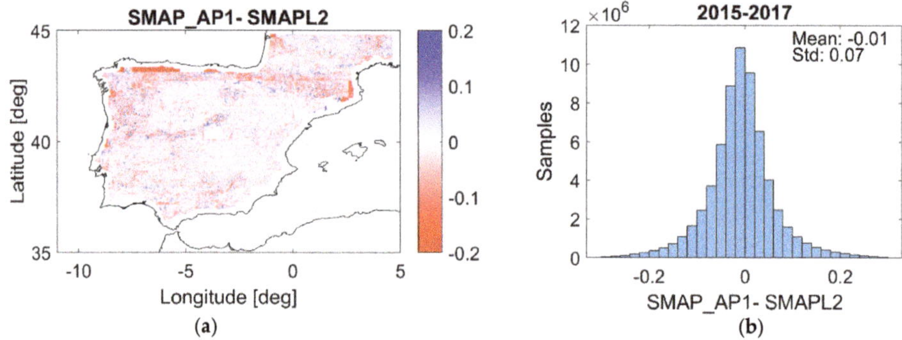

Figure 8. Temporally-averaged map (**a**) and histogram (**b**) of daily SMAP SSM differences (SMAP_AP1 at 1 km minus SMAPL2 at 36 km), for the period April 2015 to December 2017.

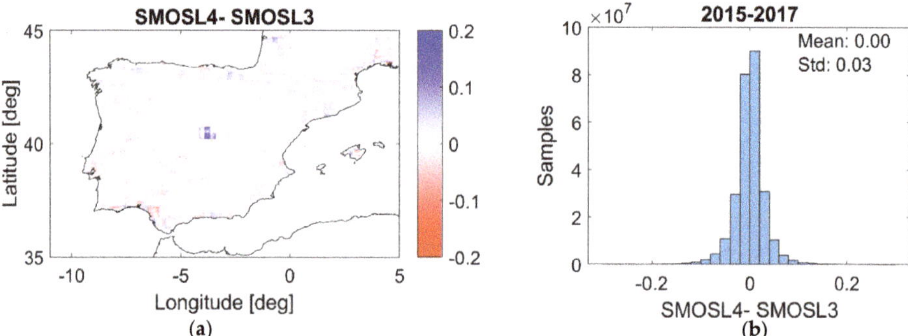

Figure 9. Temporally averaged map (**a**) and histogram (**b**) of daily SMOS SSM differences (SMOSL4 at 1 km minus SMOSL3 at 25 km), for the period April 2015 to December 2017.

5. Discussion

Validation and comparison of SSM satellite products using in situ networks data is a difficult task. Many confounding factors can intervene and must be taken into account when interpreting the obtained results. Importantly for this work, it is essential to understand that the representativeness of each satellite dataset plays a crucial role when validated. In the case of the SMAP and SMOS products presented in this study, the data is stored in maps with spatial grid cells of 36, 25, 9, 3, or 1 km. The information contained in these cells represents an areal-averaged value, while the in situ measurements represent an isolated data point. Measurements of the most representative (in terms of land cover within satellite cells) SSM in situ stations have been averaged (as can be seen in Figure 3 and Table 6) to validate the large scale SSM estimates provided by the satellites [37,43]. It is found that validating satellite-based estimates works best with in situ measurements in locations where the dominant land cover of the satellite footprint prevails.

Besides the representativeness error due to the comparison of point-scale vs. areal-averaged measurements, the mismatch between satellite observations and in situ measurements can also be originated by the penetration depth of microwave frequencies at L-band, which is of 5 cm on average but also depends on the soil moisture content itself (with greater penetration on drier soils) [3]. In contrast, measurements of the REMEDHUS network probes are placed at a depth of 5 cm. This could be one of the reasons explaining the low correlations obtained under water-limited conditions (see Table 6 for JJA), when the soil surface dries out and therefore in situ sensors might measure slightly wetter values than satellites. Moreover, the surface temperature used in SMAP and SMOS retrievals is derived from models. While the SMAP surface temperature is derived from the NASA GEOS-5 model, the SMOS surface temperature is obtained from the ECMWF model. An underestimation of the surface temperature leads to an overestimation of the soil emissivity and, as a result, to an underestimation of the SSM. This could explain the dry bias shown by the SMAP and the SMOS products with respect to the most representative in situ stations in REMEDHUS. Our results are in line with previous studies which have also reported this dry bias when comparing the SSM SMOS products (at high and low-resolution) against the in situ data provided by VAS (Spain), SMOSMANIA (France), and OzNet (Australia) networks [13,21,44]. In Figure 2 SMOS soil moisture estimates with a very low value close to zero can be found, mostly during the summer periods. These values have not been filtered in this study, as the product quality flags do not report they are measurement errors. However, we conducted specific tests and confirmed that they do not affect our overall conclusions (not shown).

Some differences between the SMAP and the SMOS products are intrinsic to the instrument they carry; while SMOS uses an interferometric radiometer with 69 receivers distributed on an Y-shaped antenna array and measurements at different incidence angles are obtained in each snapshot, SMAP uses a large rotating antenna and measurements are performed at a constant incidence angle of 40°. On the other hand, the SSM retrieval algorithms have been tailored to the SMAP and SMOS instrument characteristics, and they involve the use of dedicated techniques to reduce or correct disturbing factors (e.g., surface roughness, soil temperature and vegetation canopy). In a global study conducted by Mariko et al. [45] the SMAP SSM was compared against the one provided by SMOS, Aquarius, Advanced Scatterometer (ASCAT) and Advanced Microwave Scanning Radiometer 2 (AMSR2). Overall, they found that SMAP and SMOS appeared to be the most similar among the five SSM products, in terms of uRMSE and R, excluding forested areas where some discrepancies were found, with SMOS being generally slightly wetter than SMAP. For the particular case of the Iberian Peninsula, which is mostly covered by crops and forested areas, we showed that SMAP is generally wetter than SMOS. Although differences are minor, we showed they are already present at the T_B level (see Figure 7), and are also translated to the SMOS and SMAP derived products at enhanced spatial resolutions (see Figure 5). Mariko et al. [45], indicate that a highly potential cause of the mismatch between the SMAP and the SMOS products is the use of different ancillary data in the retrieval algorithm. Although both algorithms are based on the tau-omega model they use different land cover maps to select the albedo, roughness coefficient and the vegetation opacity; SMAP uses the International Geosphere Biosphere Program (IGBP) [46], and SMOS uses the ECOCLIMAP [47]. This could explain the differences observed between the original SMAP and the SMOS products (~40 km), but also between the satellite observations and the in situ measurements along the whole study.

Focusing on the enhanced resolution SSM products, they allow us to develop applications that will otherwise not be possible using exclusively SMAP and SMOS products in their original resolutions (~40 km). However, we showed that even the higher resolutions (3 and 1 km) of SSM maps may not be completely suitable for local or regional applications if the study area is small and the land cover is not representative of the SMAP or SMOS pixel to which it belongs (Figure 2a,b). In [48], Merlin et al. proposed a performance metric for soil moisture downscaling methods and it was applied to the 1 km disaggregation based on physical and theoretical scale change (DISPATCH) data in central Morocco. They showed that disaggregation applied to irrigated areas surrounded by drylands reduced the negative bias in SMOS observations at 1 km with respect to in situ data, but was not yet fully able to solve the sub-pixels variability in soil moisture. The scientific contribution of the downscaled SMAP and SMOS products is undeniable, adding value in a wide range of applications, such as the prevention management of insect pests [6], the prevention of forest fires [7,8] and the early detection of wild fires [9], but further improvements are needed to reduce the uncertainties when merging information from different sensors.

6. Summary and Conclusions

In this study, several space-borne SSM products (SMAP and SMOS), and their derived products at enhanced spatial resolution (SMAP/Sentinel-1 and SMOS/ERA5/MODIS) have been compared in space, and time.

For the temporal comparison, the in situ information of the REMEDHUS network has been used as a benchmark. In order to study the behavior of the remotely sensed data in different scenarios, we selected a variety of in situ SSM stations located in areas with different land uses (fallow, forest-pasture, vineyard, irrigated, and rainfed). We showed that, independently of the spatial resolution, all the SSM products were able to capture significant rainfall events (e.g., the rainfall event occurring in January 2016, see Figure 2), and seasonality pattern (summers with low and winters with high soil moisture values, see Figures 2 and 3). However, even the highest-resolution product used in this study (1 km) is not fine enough to capture local differences which are not dominant at the pixel scale, like the small irrigated areas where station K13 is located (see Tables 3 and 5). Consequently, when comparing the remotely sensed data with REMEDHUS measurements it is crucial to understand the representativeness of one in situ station within the satellite footprint. In situ SSM measurements are representative at the point scale and are highly sensitive to both the soil characteristics and the effects of precipitation. There are multiple strategies to upscale in situ soil moisture measurements for comparison with satellite-based estimates [49,50]. In this study, we have decided to average the SSM values of the most representative (in terms of prevailing land cover) in situ stations within the satellite footprint. One of the best results when comparing low-resolution as well as high-resolution SMAP/SMOS-based estimates against the in situ measurements, are obtained for the stations H13 and O07, which are located in regions with rainfed or fallow land use (the most common land uses in REMEDHUS). On the contrary, the worst results were obtained over stations J3 (vineyard) and K13 (irrigated), which represent only a minor land cover fraction within the footprints (see Table 4).

Statistically, the differences between the SMAP and SMOS products are considerably low, e.g., at low-resolution, for the station H13 the correlation (and the unbiased error) are 0.83 (0.044 m^3/m^3) for SMAP (SMAPL2_E), 0.8 (0.052 m^3/m^3) for SMOS (SMOSL3); at high-resolution these statistics are 0.81 (0.04 m^3/m^3) for SMAP, and 0.8 (0.045 m^3/m^3) for SMOS. From Figure 3 and Table 6 it can be seen that both SMOS and SMAP have a slightly worst performance in terms of correlation (~0.6) during the summer season. In addition, SMOS shows an important bias (−0.067 m^3/m^3) in this period.

Concerning the spatial analysis, the high-resolution (downscaled) SMAP (passive/active) and SMOS (passive/optical) products have been compared across the Iberian Peninsula. Overall, SMAP is slightly wetter than SMOS, especially in the north, northwest, and west of the Iberian Peninsula. These differences are more pronounced over forested areas, which may be due to the fact that the microwave (radar) signal at C-band used in the SMAP product is not able to penetrate through dense (forested) vegetation [12,51]. Moreover, the differences between the two products can also be seen at the brightness temperature level and therefore are not introduced by the downscaling methodology.

This satellite inter-comparison study has provided and confirmed insights into the SMOS and SMAP multi-scale SSM products that are currently operational. These products are required in a wide spectrum of application and research studies, generally at the best radiometric accuracy and spatial resolution possible. Over the Iberian Peninsula, we showed that all products generally agree in their temporal dynamics, with lowest performances in summer, and SMAP-derived products being wetter than SMOS ones. Yet some differences in spatial patterns are observed in the high-resolution products, linked to the fine-scale information they use and the multi-sensor synergies employed, especially in forested areas. In future studies, the presented analysis can be extended to other regions of the world that have a sufficiently dense soil moisture network to establish reliable estimates at multi-scale resolutions. Also, the proposed spatio-temporal analyses can be widened to global scales with the use of sparse in situ networks.

Remote Sens. **2020**, *12*, 570

Author Contributions: Conceptualization, G.P., T.J., M.P. (Maria Piles), M.V., A.C., and D.E.; Methodology, G.P., T.J., M.P. (Maria Piles), and M.V.; Software, G.P. and M.P. (Maria Piles); Validation, G.P., T.J., and M.P.; Formal analysis, G.P., T.J., M.P., and M.V.; Investigation, G.P., T.J., M.P. (Maria Piles), and M.V.; Resources, T.J. and M.V.; Data curation, G.P.; Writing—original draft preparation, G.P., T.J., M.P. (Maria Piles), and M.V.; Writing—review and editing, G.P., T.J., M.P. (Maria Piles), M.V., A.C., M.P. (Miriam Pablos), and D.E.; Supervision, T.J., M.P. (Maria Piles), M.V., and A.C.; Funding acquisition, M.P. (Maria Piles) and M.V. All authors have read and agreed to the published version of the manuscript.

Funding: This research was funded by the Spanish Ministry of Science, Innovation and Universities, through the coordinated project L-Band (MCIU/AEI/FEDER, UE): Sobre la continuidad de las misiones satelitales de banda L. Nuevos paradigmas en productos y aplicaciones, grant numbers ESP2017-89463-C3-2-R (UPC part) and ESP2017-89463-C3-1-R (ICM part), and the Unidad de Excelencia María de Maeztu MDM-2016-0600. M. Piles is supported by a Ramón y Cajal contract and the project RTI2018-096765-A-100 (MCIU/AEI/FEDER, UE).

Acknowledgments: The authors would like to thank the Water Resources Research group of the University of Salamanca for their support and helpful comments. Thomas J. and Dara E. also want to acknowledge MIT for supporting this research with the MIT-Germany Seed Fund "Global Water Cycle and Environmental Monitoring using Active and Passive Satellite-based Microwave Instruments".

Conflicts of Interest: The authors declare no conflict of interest.

References

1. Lahoz, W.; Blyverket, J.; Hamer, P. Product Validation and Intercomparison Report (PVIR) Revision 3 2018. Available online: https://www.esa-soilmoisture-cci.org/sites/default/files/documents/ESA_CCI_Soil_Moisture_D4.1.2_PVIR_Revision3_v2.6.pdf (accessed on 20 December 2019).

2. Entekhabi, D.; Njoku, E.G.; O'Neill, P.E.; Kellogg, K.H.; Crow, W.T.; Edelstein, W.N.; Entin, J.K.; Goodman, S.D.; Jackson, T.J.; Johnson, J.; et al. The Soil Moisture Active Passive (SMAP) Mission. *Proc. IEEE* **2010**, *98*, 704–716. [CrossRef]

3. Ulaby, F.; Long, D. *Microwave Radar and Radiometric Remote Sensing*; University of Michigan Library: Lansing, MI, USA, 2014.

4. Konings, A.G.; Piles, M.; Das, N.; Entekhabi, D. L-band vegetation optical depth and effective scattering albedo estimation from SMAP. *Remote Sens. Environ.* **2017**, *198*, 460–470. [CrossRef]

5. Jackson, T.; Schmugge, T. Vegetation effects on the microwave emission of soils. *Remote Sens. Environ.* **1991**, *36*, 203–212. [CrossRef]

6. Escorihuela, M.J.; Merlin, O.; Stefan, V.; Moyano, G.; Eweys, O.A.; Zribi, M.; Kamara, S.; Benahi, A.S.; Ebbe, M.A.B.; Chihrane, J.; et al. SMOS based high resolution soil moisture estimates for desert locust preventive management. *Remote Sens. Appl. Soc. Environ.* **2018**, *11*, 140–150.

7. Chaparro, D. Surface moisture and temperature trends anticipate drought conditions linked to wildfire activity in the Iberian Peninsula. *Eur. J. Remote Sens.* **2016**, *49*, 955–971. [CrossRef]

8. Chaparro, D.; Vall-Llossera, M.; Piles, M.; Camps, A.; Rüdiger, C.; Riera-Tatché, R. Predicting the Extent of Wildfires Using Remotely Sensed Soil Moisture and Temperature Trends. *IEEE J. Sel. Top. Appl. Earth Obs. Remote Sens.* **2016**, *9*, 2818–2829. [CrossRef]

9. Chaparro, D.; Vayreda, J.; Vall-Llossera, M.; Banque, M.; Piles, M.; Camps, A.; Martinez-Vilalta, J. The Role of Climatic Anomalies and Soil Moisture in the Decline of Drought-Prone Forests. *IEEE J. Sel. Top. Appl. Earth Obs. Remote Sens.* **2017**, *10*, 503–514. [CrossRef]

10. Sabaghy, S.; Walker, J.P.; Renzullo, L.J.; Jackson, T.J. Spatially enhanced passive microwave derived soil moisture: Capabilities and opportunities. *Remote Sens. Environ.* **2018**, *209*, 551–580. [CrossRef]

11. Peng, J.; Loew, A.; Merlin, O.; Verhoest, N.E.C. A review of spatial downscaling of satellite remotely sensed soil moisture. *Rev. Geophys.* **2017**, *55*, 341–366. [CrossRef]

12. Jagdhuber, T.; Baur, M.; Akbar, R.; Das, N.N.; Link, M.; He, L.; Entekhabi, D. Estimation of active-passive microwave covariation using SMAP and Sentinel-1 data. *Remote Sens. Environ.* **2019**, *225*, 458–468. [CrossRef]

13. Portal, G.; Vall-Llossera, M.; Piles, M.; Camps, A.; Chaparro, D.; Pablos, M.; Rossato, L. A Spatially Consistent Downscaling Approach for SMOS Using an Adaptive Moving Window. *IEEE J. Sel. Top. Appl. Earth Obs. Remote Sens.* **2018**, *11*, 1883–1894. [CrossRef]

14. Portal, G.; Vall-Llosscra, M.; Piles, M.; Camps, A.; Chaparro, D.; Pablos, M.; Rossato, L.; Aabouch, K. Microwave and Optical Data Fusion for Global Mapping of Soil Moisture at High Resolution. *IGARSS 2018 IEEE Int. Geosci. Remote Sens. Symp.* **2018**, *11*, 341–344.

15. Piles, M.; Entekhabi, D.; Camps, A. A Change Detection Algorithm for Retrieving High-Resolution Soil Moisture From SMAP Radar and Radiometer Observations. *IEEE Trans. Geosci. Remote Sens.* **2009**, *47*, 4125–4131. [CrossRef]

16. Jagdhuber, T.; Konings, A.G.; McColl, K.A.; Alemohammad, S.H.; Das, N.N.; Montzka, C.; Link, M.; Akbar, R.; Entekhabi, D. Physics-Based Modeling of Active and Passive Microwave Covariations Over Vegetated Surfaces. *IEEE Trans. Geosci. Remote Sens.* **2019**, *57*, 788–802. [CrossRef]

17. Piles, M.; McColl, K.A.; Entekhabi, D.; Das, N.; Pablos, M. Sensitivity of Aquarius Active and Passive Measurements Temporal Covariability to Land Surface Characteristics. *IEEE Trans. Geosci. Remote Sens.* **2015**, *53*, 1–12. [CrossRef]

18. Das, N.N.; Entekhabi, D.; Dunbar, R.S.; Chaubell, M.J.; Colliander, A.; Yueh, S.; Jagdhuber, T.; Chen, F.; Crow, W.; O'Neill, P.E.; et al. The SMAP and Copernicus Sentinel 1A/B microwave active-passive high resolution surface soil moisture product. *Remote Sens. Environ.* **2019**, *233*, 111380. [CrossRef]

19. Carlson, T.N.; Gillies, R.R.; Perry, E.M. A method to make use of thermal infrared temperature and NDVI measurements to infer surface soil water content and fractional vegetation cover. *Remote Sens. Rev.* **1994**, *9*, 161–173. [CrossRef]

20. Piles, M.; Sánchez, N.; Vall-Llossera, M.; Camps, A.; Martínez-Fernández, J.; Martínez, J.; Gonzalez-Gambau, V. A Downscaling Approach for SMOS Land Observations: Evaluation of High-Resolution Soil Moisture Maps Over the Iberian Peninsula. *IEEE J. Sel. Top. Appl. Earth Obs. Remote Sens.* **2014**, *7*, 3845–3857. [CrossRef]

21. Piles, M.; Petropoulos, G.P.; Sánchez, N.; González-Zamora, Á.; Ireland, G. Towards improved spatio-temporal resolution soil moisture retrievals from the synergy of SMOS and MSG SEVIRI spaceborne observations. *Remote Sens. Environ.* **2016**, *180*, 403–417. [CrossRef]

22. Pablos, M.; Piles, M.; González-Haro, C. *BEC SMOS Land Products Description*. 2019. Available online: http://bec.icm.csic.es/doc/BEC-SMOS-0002-PD-Land.pdf (accessed on 20 December 2019).

23. Entekhabi, D.; Yueh, S.; O'Neill, P.E.; Kellogg, K.H.; Allen, A.; Bindlish, R.; Brown, M.; Chan, S.; Colliander, A.; Crow, W.T.; et al. *SMAP Handbook Soil Moisture Active Passive Mapping Soil Moisture and Freeze/Thaw from Space*; National Aeronautics and Space Administration: Pasadena, CA, USA, 2014.

24. SMAP L2 Radiometer Half-Orbit 36 km EASE-Grid Soil Moisture, Version 6|National Snow and Ice Data Center. Available online: https://nsidc.org/data/SPL2SMP/versions/6 (accessed on 13 October 2019).

25. SMAP Enhanced L2 Radiometer Half-Orbit 9 km EASE-Grid Soil Moisture, Version 3|National Snow and Ice Data Center. Available online: https://nsidc.org/data/SPL2SMP_E/versions/3 (accessed on 13 October 2019).

26. SMAP/Sentinel-1 L2 Radiometer/Radar 30-Second Scene 3 km EASE-Grid Soil Moisture, Version 2|National Snow and Ice Data Center. Available online: https://nsidc.org/data/SPL2SMAP_S/versions/2 (accessed on 13 October 2019).

27. Chan, S. *Level 2 Passive Soil Moisture Product Specification Document.* 2019. Available online: https://nsidc.org/sites/nsidc.org/files/technical-references/D72547%20SMAP%20L2_SM_P%20PSD%20Version%205.1.pdf (accessed on 20 December 2019).

28. Chan, S.; Njoku, E.; Colliander, A. *Algorithm Theoretical Basis Document Level 1C Radiometer Data Product*; 2014. Available online: https://smap.jpl.nasa.gov/system/internal_resources/details/original/279_L1C_TB_ATBD_RevA_web.pdf (accessed on 20 December 2019).

29. El Hajj, M.; Baghdadi, N.; Zribi, M.; Rodriguez-Fernandez, N.; Wigneron, J.P.; Al-Yaari, A.; Al Bitar, A.; Albergel, C.; Calvet, J.-C. Evaluation of SMOS, SMAP, ASCAT and Sentinel-1 Soil Moisture Products at Sites in Southwestern France. *Remote Sens.* **2018**, *10*, 569. [CrossRef]

30. Chan, S. *Enhanced Level 2 Passive Soil Moisture Product Specification Document.* 2019. Available online: https://nsidc.org/sites/nsidc.org/files/technical-references/D56291%20SMAP%20L2_SM_P_E%20PSD%20Version%202.1.pdf (accessed on 20 December 2019).

31. Jagdhuber, T.; Entekhabi, D.; Das, N.; Link, M.; Montzka, C.; Kim, S.; Yueh, S. Microwave covariation modeling and retrieval for the dual-frequency active-passive combination of sentinel-1 and SMAP. In Proceedings of the 2017 IEEE International Geoscience and Remote Sensing Symposium (IGARSS), Fort Worth, TX, USA, 23–28 July 2017; pp. 3996–3999.

32. Kerr, Y.H.; Waldteufel, P.; Wigneron, J.-P.; Delwart, S.; Cabot, F.; Boutin, J.; Escorihuela, M.-J.; Font, J.; Reul, N.; Gruhier, C.; et al. The SMOS Mission: New Tool for Monitoring Key Elements of the Global Water Cycle. *Proc. IEEE* **2010**, *98*, 666–687. [CrossRef]

33. Kerr, Y.; Al-Yaari, A.; Rodriguez-Fernandez, N.; Parrens, M.; Molero, B.; Leroux, D.; Bircher, S.; Mahmoodi, A.; Mialon, A.; Richaume, P.; et al. Overview of SMOS performance in terms of global soil moisture monitoring after six years in operation. *Remote Sens. Environ.* **2016**, *180*, 40–63. [CrossRef]
34. SMOS—eoPortal Directory—Satellite Missions. Available online: https://directory.eoportal.org/web/eoportal/satellite-missions/s/smos (accessed on 3 September 2019).
35. Barcelona Expert Center|Remote Sensing Research, Data Distribution and Visualization Services. Available online: http://bec.icm.csic.es/ (accessed on 19 January 2020).
36. International Soil Moisture Network. Available online: https://ismn.geo.tuwien.ac.at/en/data-access/ (accessed on 19 January 2020).
37. Sanchez, N.; Perez-Gutierrez, C.; Martinez-Fernandez, J.; Scaini, A. Validation of the SMOS L2 Soil Moisture Data in the REMEDHUS Network (Spain). *IEEE Trans. Geosci. Remote Sens.* **2012**, *50*, 1602–1611. [CrossRef]
38. Pablos, M.; Martínez-Fernández, J.; Piles, M.; Sánchez, N.; Vall-Llossera, M.; Camps, A. Multi-Temporal Evaluation of Soil Moisture and Land Surface Temperature Dynamics Using in Situ and Satellite Observations. *Remote Sens.* **2016**, *8*, 587. [CrossRef]
39. ESA/CCI Viewer. Available online: https://maps.elie.ucl.ac.be/CCI/viewer/ (accessed on 20 December 2019).
40. Objective|ESA Climate Change Initiative. Available online: http://cci.esa.int/objective (accessed on 20 December 2019).
41. Entekhabi, D.; Reichle, R.H.; Koster, R.D.; Crow, W.T. Performance Metrics for Soil Moisture Retrievals and Application Requirements. *J. Hydrometeorol.* **2010**, *11*, 832–840. [CrossRef]
42. Davenport, I.; Fernandez-Galvez, J.; Gurney, R. A sensitivity analysis of soil moisture retrieval from the tau-omega microwave emission model. *IEEE Trans. Geosci. Remote Sens.* **2005**, *43*, 1304–1316. [CrossRef]
43. Cosh, M.H.; Jackson, T.J.; Bindlish, R.; Prueger, J.H. Watershed scale temporal and spatial stability of soil moisture and its role in validating satellite estimates. *Remote Sens. Environ.* **2004**, *92*, 427–435. [CrossRef]
44. Chen, Y.; Yang, K.; Qin, J.; Cui, Q.; Lu, H.; La, Z.; Han, M.; Tang, W. Evaluation of SMAP, SMOS, and AMSR2 soil moisture retrievals against observations from two networks on the Tibetan Plateau. *J. Geophys. Res. Atmos.* **2017**, *122*, 5780–5792. [CrossRef]
45. Burgin, M.S.; Colliander, A.; Njoku, E.G.; Chan, S.K.; Cabot, F.; Kerr, Y.H.; Bindlish, R.; Jackson, T.J.; Entekhabi, D.; Yueh, S.H. A Comparative Study of the SMAP Passive Soil Moisture Product With Existing Satellite-Based Soil Moisture Products. *IEEE Trans. Geosci. Remote Sens.* **2017**, *55*, 2959–2971. [CrossRef]
46. Kim, S. *Soil Moisture Active Passive (SMAP)—Ancillary Data Report (Landcover Classification)*; 2013. Available online: https://smap.jpl.nasa.gov/system/internal_resources/details/original/284_042_landcover.pdf (accessed on 20 December 2019).
47. Faroux, S.; Tchuenté, A.T.K.; Roujean, J.-L.; Masson, V.; Martin, E.; Le Moigne, P. ECOCLIMAP-II/Europe: A twofold database of ecosystems and surface parameters at 1 km resolution based on satellite information for use in land surface, meteorological and climate models. *Geosci. Model. Dev.* **2013**, *6*, 563–582. [CrossRef]
48. Merlin, O.; Malbéteau, Y.; Notfi, Y.; Bacon, S.; Khabba, S.E.-R.S.; Jarlan, L.; Er-Raki, S.; Khabba, S. Performance Metrics for Soil Moisture Downscaling Methods: Application to DISPATCH Data in Central Morocco. *Remote Sens.* **2015**, *7*, 3783–3807. [CrossRef]
49. Qin, J.; Yang, K.; Lu, N.; Chen, Y.; Zhao, L.; Han, M. Spatial upscaling of in-situ soil moisture measurements based on MODIS-derived apparent thermal inertia. *Remote Sens. Environ.* **2013**, *138*, 1–9. [CrossRef]
50. Crow, W.T.; Berg, A.A.; Cosh, M.H.; Loew, A.; Mohanty, B.P.; Panciera, R.; De Rosnay, P.; Ryu, D.; Walker, J.P. Upscaling sparse ground-based soil moisture observations for the validation of coarse-resolution satellite soil moisture products. *Rev. Geophys.* **2012**, *50*, 50. [CrossRef]
51. Baur, M.; Jagdhuber, T.; Link, M.; Piles, M.; Akbar, R.; Entekhabi, D. Multi-Frequency Estimation of Canopy Penetration Depths from SMAP/AMSR2 Radiometer and Icesat Lidar Data. *IGARSS 2018 IEEE Int. Geosci. Remote Sens. Symp.* **2018**, 365–368.

Article

Assessment of Root Zone Soil Moisture Estimations from SMAP, SMOS and MODIS Observations

Miriam Pablos *, Ángel González-Zamora, Nilda Sánchez and José Martínez-Fernández

Instituto Hispanoluso de Investigaciones Agrarias (CIALE), University of Salamanca, Duero 12,
37185 Villamayor, Spain; aglezzamora@usal.es (A.G.-Z.); nilda@usal.es (N.S.); jmf@usal.es (J.M.-F.)
* Correspondence: mpablos@usal.es; Tel.: +34-923-294-500

Received: 29 May 2018; Accepted: 18 June 2018; Published: 21 June 2018

Abstract: In this study, six satellite-based root zone soil moisture (RZSM) estimates from March 2015 to December 2016 were evaluated both temporally and spatially. The first two were the Soil Moisture Active Passive (SMAP) and the Soil Moisture and Ocean Salinity (SMOS) L4 RZSM products. The other four were obtained through the Soil Water Index (SWI) approach, which embedded surface soil moisture (SSM). The SMOS-Barcelona Expert Center (BEC) L4 SSM product and the apparent thermal inertia (ATI)-derived SSM from the Moderate Resolution Imaging Spectroradiometer (MODIS) data were used as SSM datasets. In the temporal analysis, the RZSM estimates were compared to in situ RZSM from 14 stations of the Soil Moisture Measurements Station Network of the University of Salamanca (REMEDHUS). Regarding the spatial assessment, the resulting RZSM maps of the Iberian Peninsula were compared between them. All RZSM values followed the temporal evolution of the ground-based measurements well, although SMOS and MODIS showed underestimation while SMAP displayed overestimation. The good results obtained from MODIS ATI are notable, notwithstanding they were not estimated through microwave radiometry. A very high agreement was found in terms of spatial patterns for the whole Iberian Peninsula except for the extreme north area, which is dominated by high mountains and dense forests.

Keywords: soil moisture; root zone; SMAP; SMOS; MODIS

---

1. Introduction

L-band radiometry is the most established technique for remotely measuring soil moisture. Currently, there are two L-band missions in orbit specifically designed to globally monitor soil moisture: the Soil Moisture and Ocean Salinity (SMOS) and the Soil Moisture Active Passive (SMAP). SMOS, launched in 2009 by the European Space Agency (ESA), uses a synthetic aperture radiometer with a spatial resolution of ~35–50 km to provide global soil moisture maps every three days [1]. The Centre Aval de Traitement des Données SMOS (CATDS) provides several soil moisture products (L2, L3, and L4), which are processed with the algorithms developed by the Centre d'Etudes Spatiales de la Biosphere (CESBIO). In 2015, the National Aeronautics and Space Administration (NASA) launched SMAP. It currently employs a real aperture radiometer with ~40 km resolution to retrieve soil moisture maps with a three-day revisit time [2].

While these remote sensing sensors provide soil moisture data at coarse spatial resolutions, a growing number of applications require knowledge of soil moisture at regional or local scales (from a few kilometers down to several meters). To overcome this challenge, some soil moisture disaggregation approaches have been developed, such as those based on the synergy of passive microwaves with ancillary optical visible/infrared (VIS/IR) [3–6] or active data [7]. Optical data have also been used to indirectly estimate soil moisture [8]. For instance, methods based on apparent thermal inertia (ATI) to estimate soil moisture rely upon the fact that wet soils have a higher thermal inertia and a lower temperature fluctuation than dry soils. When soil moisture increases, ATI proportionally increases as well, and there is a short-term reduction in the diurnal land surface temperature (LST) range [9–13].

A main drawback of the L-band observations is that although they have larger penetration than those at higher microwaves (S-, C-, X-, or K-bands), they explore approximately 0–5 cm of the topsoil layer, sensing only surface soil moisture (SSM). However, an increasing number of hydrological and agricultural applications require root zone soil moisture (RZSM) information from the soil profile (0–1 m depth), where plant roots develop [14,15]. Additionally, the SSM may not have significant influence in the soil water availability for plants and crops. For this reason, the RZSM is expected to better reflect the actual soil water content storage of the unsaturated zone than the SSM.

There are several methods for obtaining RZSM. One method comprises in situ soil moisture measurements made at the root zone using probes installed either at a specific depth required or along the whole soil profile. In this regard, although neutron attenuation probes were extensively used in the past, sensors based on the soil dielectric constant—capacitance, Frequency Domain Reflectivity (FDR) and Time Domain Reflectivity (TDR) probes—are now being employed in many networks worldwide [16]. Another method involves using cosmic-ray soil moisture probes. These innovative and noninvasive sensors have been implemented in the Cosmic-ray Soil Moisture Observing System (COSMOS) network [17,18]. The cosmic-ray sensors are placed above the soil surface, but its effective soil measurement depth varies from 12 cm for wet soils to 76 cm for dry soils [19]. Nevertheless, only a limited number of current networks of the International Soil Moisture Network (ISMN) provide RZSM compared to the great amount that provide SSM. Another method relies on the use of Ground Penetration Radar (GPR) measurements. In this case, the GPR sensor can be mounted on a vehicle close the soil surface or an airplane to measure soil moisture during experimental field campaigns. The GPR does not require direct contact with the soil, but its signal can penetrate from one meter to several tens of meters [20]. The last method consists of estimating RZSM through more or less complex models that have this variable as the output.

The models used to estimate RZSM can be classified into two main groups: the so-called land-surface models and the hydrological models. They differ in the detail of description of processes that are taken into account, the parameter estimation approaches, and the spatiotemporal resolutions. Land-surface models describe the vertical exchanges of heat, water, and carbon considering the land-atmosphere couplings and are globally applied. Hydrological models are instead more focused on water resources, are traditionally applied at the basin level, and usually have many parameters that need to be calibrated or estimated regionally [21]. Different data assimilation techniques are used in the two cases to incorporate SSM to them [22–24]. Originally, the assimilated SSM data to estimate RZSM had been measured in situ [25,26], but a variety of satellite SSM data have been assimilated in the last decade [27–29]. In the case of SMAP and SMOS missions, two operational satellite-based RZSM products have recently been developed from the assimilation of SSM measurements into their respective land-surface models [30,31].

The Soil Water Index (SWI) is one of the most common models used to estimate RZSM through SSM remote sensing [32]. This simple model has been able to successfully obtain RZSM over regions with different climatic and soil conditions. Apart from SSM measurements, the SWI requires only an input exponential parameter (T), which is related to the transfer time of water along the soil profile. The T parameter had been calculated in different ways depending on the application, study area, and sensor used [33]. The SWI has been applied to in situ SSM databases in several studies to obtain field scale RZSM [34–36] as well as to active and passive SSM observations to generate several satellite-based RZSM estimates, such as those derived from the European Remote Sensing (ERS) scatterometer [37], the Advanced Scatterometer (ASCAT) [38–40], the Advanced Microwave Scanning Radiometer-Earth Observing System (AMSR-E) [39,41], the SMOS [42,43], and the Climate Change Initiative (CCI) soil moisture database [41,44].

The aim of this work was to evaluate six RZSM estimates obtained from SMAP, SMOS, and Moderate Resolution Imaging Spectroradiometer (MODIS) from 31 March 2015 to 31 December

2016. The study period (one year and nine months) is limited at the beginning by the SMAP launch and at the end by the availability of two SMOS soil moisture products—the SMOS-CESBIO L4 RZSM and the SMOS-Barcelona Expert Centre (BEC) L4 SSM. The first two RZSM estimates came from the SMAP and the SMOS-CESBIO L4 RZSM products. The other four RZSM estimates were customized products generated after applying the SWI model to two SSM datasets—the SMOS-BEC L4 SSM and the MODIS ATI-derived SSM—together with two alternatives for calculating the exponential T parameter of the SWI. Presently, some studies devoted to validate SMAP and SMOS RZSM have been published, but none have been made validating both RZSM products at the same network or showing an intercomparison between them over the same study area, including a RZSM estimation based on ATI. Thus, the present work will constitute a novelty within this research line. In this study, all satellite-based RZSM estimates were analyzed both temporarily and spatially. For the temporal analysis, the RZSM estimates were compared against in situ RZSM measurements from 14 stations of the Soil Moisture Measurements Station Network of the University of Salamanca (REMEDHUS). The spatial analysis was based on comparisons of RZSM maps from all the analyzed estimates over the entire Iberian Peninsula (~582,000 km^2), an area of contrasting environments in both wet and dry periods.

2. Data and Methodology

2.1. REMEDHUS Soil Moisture

REMEDHUS is a soil moisture network located in the central part of the Duero basin in Spain and covering an area of approximately 1300 km^2 (41.1–41.5°N, 5.1–5.7°W). This region has a continental semiarid Mediterranean climate characterized by a clear water deficit, especially during summer. The location of the network was chosen precisely for these criteria, i.e., to monitor crop behavior under the water-limited conditions and over land covers representative of the region. In REMEDHUS, the main land use is agricultural, with rainfed crops such as cereals, legumes, and vineyards. These crops are adapted to the scarcity of water because most of them are fed only by precipitation, which is approximately 381 mm/year on average [45]. The soils are mainly sandy, leading to a limited water holding capacity.

Two sensor types are installed in REMEDHUS. First, there are 22 stations measuring SSM (0–5 cm) using Hydra Probes (Stevens Water Monitoring, Inc., Portland, OR, USA). Out of the 22 stations, 14 also measure soil moisture along the soil profile using EnviroSMART (Sentek Pty. Ltd., Stepney, SA, Australia) sensors deployed at 25, 50, and 100 cm depths. These depth levels ensure a complete measurement of the soil moisture at the root zone in the area [43].

For the assessment of the ATI-derived SSM, the in situ SSM data provided by all 22 stations in REMEDHUS were used. For the analysis of the RZSM estimates, the soil moisture observations at surface and root zone level from the 14 stations were used. Since the observations are on an hourly basis, soil moisture measurements at each station were first daily averaged. A representative estimate of in situ RZSM was obtained for each station by averaging the daily soil moisture measurements at the different depths (5, 25, 50, and 100 cm).

2.2. SMAP L4 Soil Moisture

Among the variety of SMAP products available, the global SMAP L4 soil moisture geophysical data over a 9-km Equal-Area Scalable Earth (EASE)-2 grid [46] were selected. The SMAP brightness temperature—originally at coarser resolution—was downscaled to 9 km using the Backus-Gilbert optimal interpolation. The SMAP L4 soil moisture was derived by assimilating this brightness temperature into the NASA catchment land-surface model, which interpolates and extrapolates the SMAP observations in time and in space [47]. The NASA model describes the vertical transfer of soil moisture between the surface and the root zone, built on a set of Richard's equation calculations under unsaturated conditions. This model is driven by observation-based surface meteorological forcing, including precipitation [30].

The SMAP L4 product provides SSM (0–5 cm) and RZSM (0–100 cm) estimations at a 3 h temporal resolution. In this study, the area corresponding to the Iberian Peninsula (34°N–45°N, 11°W–5°E) was clipped from the global SMAP L4 soil moisture maps. The resulting maps of SSM and RZSM were daily averaged.

2.3. SMOS Soil Moisture

2.3.1. SMOS-CESBIO L3 Surface Soil Moisture

The global SMOS L3 SSM product over a 25-km EASE-2 grid was used. The SMOS L3 processor of CESBIO [48] uses the same physically based forward model as the L2 processor [49], with only minor differences that are mainly related to the gridding system. The L3 SSM retrievals are obtained from the SMOS L1C v.620 brightness temperature. On a given day, the L3 algorithm processes several orbits in the same loop. Later, filtering is applied to select the best estimate when several retrievals are available for the same pixel. Therefore, this product is a daily composite of filtered and binned SMOS SSM data [50].

The SMOS-CESBIO L3 SSM is disseminated in separated daily ascending and descending orbits. Similar to the process used for SMAP L4, the area corresponding to the Iberian Peninsula was clipped. Then, a daily average of the ascending and descending orbit maps was applied.

2.3.2. SMOS-CESBIO L4 Root Zone Soil Moisture

Regarding the current SMOS-derived RZSM products, the global SMOS L4 RZSM at 0–1 m depth over a 25-km EASE-2 grid was used. These data were obtained from the SMOS-CESBIO L3 SSM (using a 3-day average SSM) and other ancillary datasets, such as MODIS observations and climate data from the National Centers for Environmental Prediction (NCEP), among others. The RZSM is computed using a double bucket hydrological model, which is applied daily. The CESBIO model has two soil layers (0–40 cm and 40–100 cm). For the first layer, the selected model is the previously mentioned SWI [32,33]. For the second layer, a water balance model based on a linearized Richard's equation is employed [31].

Since the SMOS-CESBIO L4 RZSM is also distributed separately for daily ascending and descending orbits, the Iberian Peninsula area was clipped from these RZSM maps and a daily average of the ascending and descending orbits was then performed.

2.3.3. SMOS-BEC L4 Surface Soil Moisture

The cloud-free SMOS L4 SSM product—a disaggregated soil moisture at 1 km over the Iberian Peninsula—was used. This product is based on a semi-empirical downscaling approach that combines SMOS and MODIS observations together with data from the European Centre for Medium-Range Weather Forecasts (ECMWF). The product uses a shape-adaptive moving window to integrate the SMOS brightness temperature (L1C v.620) and SSM (L2 v.620) at 25 km, the 16-day Terra MODIS Normalized Difference Vegetation Index (NDVI, MOD13A3 v.5) at 1 km, and the modeled ECMWF Era Retrospective Analysis (ERA)-Interim LST in a linear linking model [51]. The downscaling algorithm is applied separately for daily ascending and descending orbits. As with the previous products, the daily average of ascending and descending data was used.

2.4. MODIS Surface Reflectance and Land Surface Temperature

The daily Aqua MODIS surface reflectance at 500 m (MYD09GA v.6) acquired in bands 1 (620–670 nm), 2 (841–876 nm), 3 (459–479 nm), 4 (545–565 nm), 5 (1230–1250 nm), and 7 (2105–2155 nm) were used. The MODIS reflectance is measured at ground level in the absence of atmospheric scattering or absorption and is already corrected for atmospheric gases, aerosols, and thin cirrus clouds [52].

Regarding the LST, daily Aqua MODIS LST (MYD11A1 v.6) and Terra MODIS LST (MOD11A1 v.6)—both at 1 km spatial resolution—were used. They have an accuracy specification of 1 °C under clear-sky conditions [53].

The three MODIS products are provided into a tile-based sinusoidal projection. The four tiles corresponding to the Iberian Peninsula (h17v04, h17v05, h18v04 and h18v05) were selected. These four tiles of the reflectance maps were first mosaicked. The resulting maps were resampled from the sinusoidal grid to a regular 500 m grid using the nearest neighbor approach. The same processing was applied to the LST but for a regular 1 km grid. The reflectance in all bands was filtered using their corresponding flags (land and highest quality). Finally, the reflectance maps at 500 m were aggregated to the same 1 km grid for the LST using a simple average.

2.5. Estimation of MODIS ATI

The MODIS ATI was computed as the ratio of the daily surface albedo and the diurnal temperature range [10]:

$$ATI = C\frac{1-\alpha}{\Delta LST},\tag{1}$$

where C is the solar correction factor, α is the broadband albedo and ΔLST corresponds to the diurnal temperature range.

The correction factor C is related to the solar flux and therefore compensates for seasonal variation in solar insolation. This factor was obtained following the expression [10]:

$$C = \sin\varphi\sin\delta(1 - \tan^2\varphi\tan^2\delta)^{1/2} + \cos\varphi\cos\delta\arccos(-\tan\varphi\tan\delta),\tag{2}$$

where φ corresponds to the latitude and δ to the solar declination of each pixel for each Julian day of the year.

The broadband albedo can be computed in two different spectral ranges—shortwave and visible in Equations (3) and (4), respectively—from daily Aqua MODIS surface reflectance at 1 km [54]:

$$\alpha_{\text{shortwave}} = 0.160\rho_1 + 0.291\rho_2 + 0.243\rho_3 + 0.116\rho_4 + 0.112\rho_5 + 0.081\rho_7 - 0.0015,\tag{3}$$

$$\alpha_{\text{visible}} = 0.331\rho_1 + 0.424\rho_3 + 0.246\rho_4,\tag{4}$$

where $\rho_1, \rho_2, \rho_3, \rho_4, \rho_5$ and ρ_7 are reflectance in bands 1, 2, 3, 4, 5 and 7, respectively. These two albedo approaches were analyzed.

The diurnal temperature range can be approximated in four different ways. The first method consists of estimating the LST through time as a sinusoid defined by its amplitude, its average temperature, the angular velocity of Earth, and the phase angle (Ψ) corresponding to the time of daily maximum LST [11,13], which coincides with the time of highest correlation between the LST and soil moisture behavior [55]. MODIS provides up to four observations for each day: Aqua nighttime LST at 1:30 h, Terra daytime LST at 10:30 h, Aqua daytime LST at 13:30 h, and Terra nighttime LST at 22:30 h, local time. These four LST at 1 km were used to compute ΔLST as the solution of the amplitude of the sinusoid by means of the least squares method ($\Delta LST_{4\text{values}}$). In this case, the four LST values are required to derive Ψ. Nonetheless, if Ψ was known, the amplitude could be calculated using only 2 LST values, preferable a day-night LST pair, because the difference between 2 daytime or 2 nighttime LSTs is usually low and would result in unrealistic ΔLST estimates. The second method is based on the difference between the Aqua daytime and nighttime LSTs (ΔLST_{Aqua}). The third way employs the same methodology as the second but uses Terra daytime and nighttime LSTs ($\Delta LST_{\text{Terra}}$). The fourth way estimates the ΔLST as the difference between the daily maximum LST—computed from Terra or Aqua daytime LST—and the daily minimum LST—computed from Aqua or Terra nighttime LST ($\Delta LST_{\text{Aqua/Terra}}$). Therefore, four different approaches of the MODIS-based ATI are calculated, and their performances are assessed.

2.6. Estimation of ATI-Derived Surface Soil Moisture

The rationale of the ATI-derived SSM is that high ATI values correspond to maximum soil water content while low ATI values are related to minimum soil water content. The Soil Moisture Saturation Index (SMSI) was used to normalize the MODIS ATI time series at 1 km [12]:

$$SMSI(t) = \frac{ATI(t) - ATI_{min}}{ATI_{max} - ATI_{min}}, \tag{5}$$

where $ATI(t)$ is the ATI at time t, and ATI_{max} and ATI_{min} represent the maximum and minimum values, respectively, of the ATI time series during the study period.

Because the SMSI varies from 0 to 1, the ATI-derived SSM was finally estimated after applying a change of dynamic range [12]:

$$SSM(t) = SMSI(t)(SSM_{max} - SSM_{min}) + SSM_{min}, \tag{6}$$

where SSM_{max} and SSM_{min} are the maximum and minimum SSM values of a reference soil moisture dynamic range, respectively.

To obtain an adequate soil moisture dynamic range and to account for a SSM dataset independent of SMAP and SMOS, two different data sources were used. The first data were the saturation (SAT), field capacity (FC), and wilting point (WP) water content maps at 1 km from the European 3D Soil Hydraulic Database (SHD) at 5 cm depth [56]. For each pixel over the Iberian Peninsula, the SSM_{max} was calculated as the mean of SAT and FC, whereas the SSM_{min} was calculated as the half value of the WP. This was similar to previous studies [12,57]. The second dataset was the combined CCI SSM v3.2 product at 25 km from 1978 to 2015 [58]. In this case, the SSM_{max} and SSM_{min} were computed as the maximum and minimum values, respectively, of the long-term time series of each pixel over the Iberian Peninsula. These SSM_{max} and SSM_{min} maps from CCI were resampled from 25 to 1 km resolution using the nearest neighbor method. Thus, the two different reference dynamic ranges were assessed.

2.7. Assessment of MODIS ATI Surface Soil Moisture

Prior to using the resulting ATI-derived SSM to estimate RZSM, all alternative SSM estimations based on MODIS data were tested, i.e., varying the albedo ($\alpha_{shortwave}$ or $\alpha_{visible}$), the diurnal temperature range ($\Delta LST_{4values}$, ΔLST_{Aqua}, ΔLST_{Terra}, or $\Delta LST_{Aqua/Terra}$), and the reference dynamic range (SHD or CCI). The assessment of these SSM estimations was performed for the 22 stations in the REMEDHUS network. The in situ SSM time series were compared with the MODIS ATI-derived SSM time series of the 1 km pixel that overlapped the corresponding station. To evaluate the level of agreement of both time series, a set of statistical metrics—namely the Pearson correlation coefficient (R), the root mean square difference (RMSD), the unbiased or centered $RMSD$ (cRMSD), and the bias—was used. They were computed following these equations:

$$R = \frac{\sum_{i=1}^{n} (y_i - \bar{y})(x_i - \bar{x})}{\sqrt{\sum_{i=1}^{n} (y_i - \bar{y})^2}\sqrt{\sum_{i=1}^{n} (x_i - \bar{x})^2}}, \tag{7}$$

$$RMSD = \sqrt{\frac{\sum_{i=1}^{n} (y_i - x_i)^2}{n}}, \tag{8}$$

$$cRMSD = \sqrt{\frac{\sum_{i=1}^{n} [(y_i - \bar{y}) - (x_i - \bar{x})]^2}{n}}, \tag{9}$$

$$\text{bias} = \frac{\sum_{i=1}^{n} (y_i - x_i)}{n}, \tag{10}$$

where y is the soil moisture to be analyzed, x is the soil moisture used as benchmark, i corresponds to each day of the study period with coincident data; the average value of both soil moisture datasets are indicated by a bar.

Since the in situ soil moisture did not have data gaps along the study period, the number of coincident days—in which there are both in situ and satellite data (N, expressed in percentage referred to a total number of 642 days)—was also computed to give some insight about the coverage of each satellite dataset.

2.8. Estimation of Root Zone Soil Moisture from SMOS-BEC and MODIS ATI Surface Soil Moisture

The SWI model was used to estimate the RZSM from the SMOS-BEC and MODIS ATI SSM. This model consists of two soil layers. The first corresponds to the surface topsoil, while the second extends from the bottom of the first layer downward [32]. These two layers are related by an exponential formulation, thus simulating the dynamics of water within the soil profile. The SWI is computed recursively, where each RZSM value depends on the previous one [33]. The advantage of the SWI compared to other models is its simplicity. Moreover, the SWI uses only the SSM as input, together with the exponential T parameter. This T is interpreted as the characteristic time length that defines the rate of water transfer of each type of soil, increasing with the thickness of the soil layer and decreasing with the soil diffusivity constant [32,33,40,43].

For this study, the method to obtain the optimal T values relies on the comparison of SSM and RZSM products from SMAP and SMOS. Two alternative T maps over the Iberian Peninsula were computed: T_{SMAP} at 9 km from the SMAP L4 SSM and RZSM, and T_{SMOS} at 25 km from the SMOS-CESBIO L3 SSM and L4 RZSM. First, different T values ranging from 1 to 100 days were introduced into the SWI model with the SSM (SMAP L4 SSM or SMOS-CESBIO L3 SSM), obtaining 100 SWI time series for each pixel. Later, all 100 SWI time series were compared with their corresponding RZSM (SMAP L4 RZSM or SMOS-CESBIO L4 RZSM). The optimal T of each pixel was selected from these comparisons based on the criterion of highest correlation [43]. The two resulting T maps were resampled into the regular 1 km grid using the nearest neighbor technique. Then, the T_{SMAP} and T_{SMOS} were combined with the SMOS-BEC L4 and MODIS ATI SSM, both at 1 km, into the SWI to obtain four possible RZSM estimates at 1 km.

For a comprehensive understanding, a flowchart (Figure 1) summarizes all the data and the methodology applied to obtain the six different RZSM estimates to be analyzed in this study: (1) SMAP L4 RZSM; (2) SMOS-CESBIO L4 RZSM; (3) SMOS-BEC SWI (TSMAP); (4) SMOS-BEC SWI (TSMOS); (5) MODIS ATI SWI (TSMAP); and (6) MODIS ATI SWI (TSMOS).

2.9. Comparison of Root Zone Soil Moisture Estimates

The RZSM measured in REMEDHUS was used as the benchmark dataset to be compared with the six different RZSM estimates. Owing to the different scales of in situ and remotely sensed datasets, two strategies were used. On the one hand, a comparison between each station-pixel pairwise was performed. Thus, the in situ RZSM time series of each station was compared with the satellite-based RZSM time series of the overlapping pixel at its spatial resolution (9 km for SMAP L4 RZSM, 25 km for SMOS-CESBIO L4 RZSM, and 1 km for SMOS-BEC and MODIS ATI SWI). On the other hand, a validation based on the area averages was done. Then, the average of the 14 RZSM stations was compared with the average of 10 pixels for SMAP L4 RZSM, 4 pixels for SMOS-CESBIO L4 RZSM, and 14 pixels of SMOS-BEC and MODIS ATI SWI. Similar to the assessment suggested for the MODIS ATI SSM, the R, RMSD, cRMSD, and bias were used to evaluate the agreement of the RZSM estimates with the in situ measurements as well as the number of coinciding data days.

In addition, with the aim of analyzing the impact of the exponential T parameter (T_{SMAP} and T_{SMOS}) in the SWI-based estimations, temporal correlation maps between the SMOS-BEC SWI (T_{SMAP}), the SMOS-BEC SWI (T_{SMOS}), the MODIS ATI SWI (T_{SMAP}), and the MODIS ATI SWI (T_{SMOS}) were obtained at 9 km and at 25 km. To do this, the four different SWI maps were aggregated from 1 to 9 km, using the average value of the 1 km pixels that span the 9 km SMAP pixel. Then, these SWI maps at 9 km were compared with the SMAP L4 RZSM product and between them at this spatial resolution. The same aggregation technique was also applied for the comparison with the SMOS-CESBIO L4 RZSM product at 25 km.

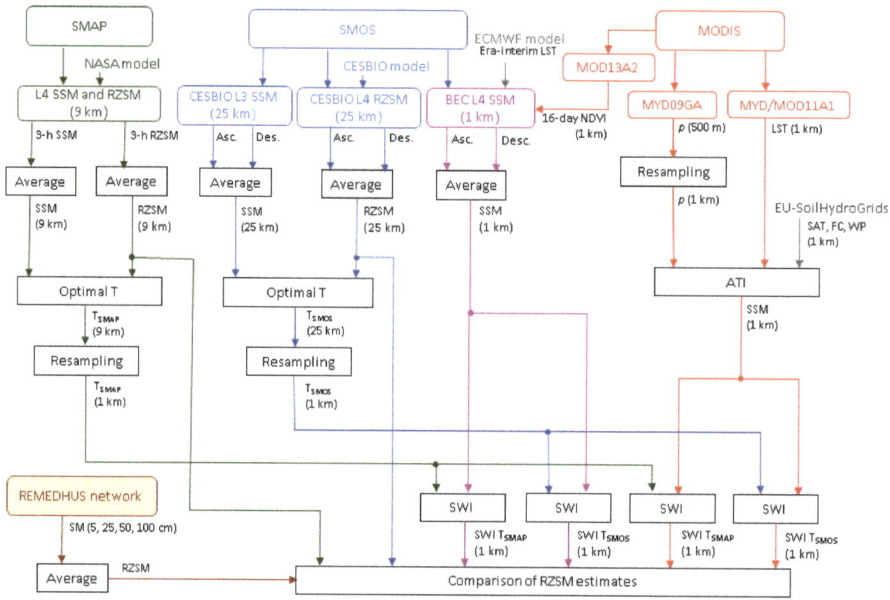

Figure 1. Flowchart showing all data and methodology applied to obtain the RZSM estimates.

3. Results and Discussion

3.1. Preliminary Assessment of MODIS ATI Surface Soil Moisture

To decide and select the best approach for the MODIS ATI SSM before applying the SWI model, different MODIS ATI SSM estimations were tested against the in situ SSM across REMEDHUS (Table 1). In general, the correlations obtained from the MODIS ATI SSM (R ~0.32 to 0.70) were lower than those obtained from other SSM products, such as SMOS, SMAP, and CCI SSM, confirmed in previous validation studies [1,45,59,60]. However, the errors were very similar (cRSMD ~0.04 m^3/m^3).

Analyzing the different albedo approaches, no significant differences were found between using the $\alpha_{shortwave}$ or $\alpha_{visible}$ in terms of correlation (maximum R ~0.66), errors (minimum RMSD ~0.040 m^3/m^3 and minimum cRMSD ~0.030 m^3/m^3), bias, and coverage. The capabilities of both albedo approaches to characterize the full range of vegetation were comparable, as shown in other studies [61]. Therefore, because the $\alpha_{shortwave}$ computation required reflectance data from a higher number of MODIS bands, $\alpha_{visible}$ was preferred.

Considering the different methods to compute the diurnal temperature range, it was clearly seen that the $\Delta LST_{4values}$ displayed correlation and bias similar to those of the remaining approximations although with lower errors (RMSD ~0.035 to 0.133 m^3/m^3 and cRMSD ~0.019 to 0.085 m^3/m^3). However, higher coverages were obtained when using ΔLST_{Aqua} (N ~35 %), ΔLST_{Terra} (N ~40 %), and $\Delta LST_{Aqua/Terra}$ (N ~50 %). These results mean that the availability of the MODIS ATI SSM was mainly limited by the availability of the LST observations. Previous research used similar approximations for ΔLST [57] or complex methods to interpolate the LST data [12,13,62,63]. In this study, the $\Delta LST_{\text{Aqua/Terra}}$ was chosen for the estimation of the diurnal temperature range.

Regarding the different dynamic ranges used as reference, the Soil Hydrologic Database (SHD) afforded statistics similar to those from the CCI approach. This fact, together with the higher spatial resolution of SHD compared to CCI (1 vs. 25 km), supported the selection of the SHD for the dynamic range.

Table 1. Statistics obtained from the comparison of the different MODIS ATI SSM estimations with the in situ SSM over all the REMEDHUS stations. N indicates the number of coincident data days.

MODIS ATI SSM Estimation	R	RMSD (m^3/m^3)	cRMSD (m^3/m^3)	Bias (m^3/m^3)	N (%)
$\alpha_{\text{shortwave}}$ $\Delta LST_{\text{Aqua/Terra}}$ SHD	0.33 to 0.66	0.039 to 0.153	0.029 to 0.097	-0.133 to 0.099	45.0 to 53.9
α_{visible} $\Delta LST_{\text{Aqua/Terra}}$ SHD	0.34 to 0.66	0.042 to 0.147	0.032 to 0.096	-0.127 to 0.108	45.2 to 54.2
α_{visible} ΔLST_{Aqua} SHD	0.32 to 0.66	0.038 to 0.138	0.027 to 0.107	-0.125 to 0.107	35.8 to 42.2
α_{visible} $\Delta LST_{\text{Terra}}$ SHD	0.39 to 0.69	0.042 to 0.132	0.025 to 0.090	-0.110 to 0.128	31.6 to 38.5
α_{visible} $\Delta LST_{\text{4values}}$ SHD	0.38 to 0.70	0.035 to 0.133	0.019 to 0.085	-0.122 to 0.101	25.1 to 30.7
α_{visible} $\Delta LST_{\text{4values}}$ CCI	0.34 to 0.66	0.049 to 0.134	0.030 to 0.097	-0.111 to 0.125	45.2 to 54.2

All values are significant (p-value < 0.05).

3.2. Temporal Analysis of Root Zone Soil Moisture Estimates

The pixel-average time series of the different RZSM estimates across REMEDHUS were compared against the station-average time series (Figure 2). At first glance, it is remarkable that all satellite datasets followed a temporal pattern similar to that of the in situ RZSM evolution, in agreement with the seasonality and climate of this region. It is highlighted that SMAP L4 RZSM and SMOS-BEC SWI had the best agreement with the dry-down and wetting-up events indicated by the in situ RZSM. To a lesser extent, this agreement also occurred for the MODIS ATI SWI, which fluctuated less. The SMOS-CESBIO L4 RZSM exhibited a certain time lag, especially during the rising periods, compared to the rest of the estimates. The reason could be the three-day window average of this product, which also smoothed the curve. While the SMOS SSM has proven more variable than the ground-based SSM itself [45,64], the SMOS RZSM showed more stability, which is consistent with the expected behavior of the deep soil moisture [43] that is typically more steady than the SSM.

In general, the SMAP L4 RZSM slightly overestimated the in situ RZSM but captured its overall variability well, with the bias almost constant. By contrast, the SMOS-CESBIO L4 RZSM underestimated the in situ RZSM, and the differences are higher during the dry season compared to the other estimates except for the SMOS-BEC SWI. This fact could be explained from the SMOS SSM underestimation previously detected in this area [45,64,65] and in other regions [66–68]. Both MODIS ATI SWIs also underestimated the in situ RZSM, although to a minor degree, while their agreement was better than with the other estimates, particularly during the dry season.

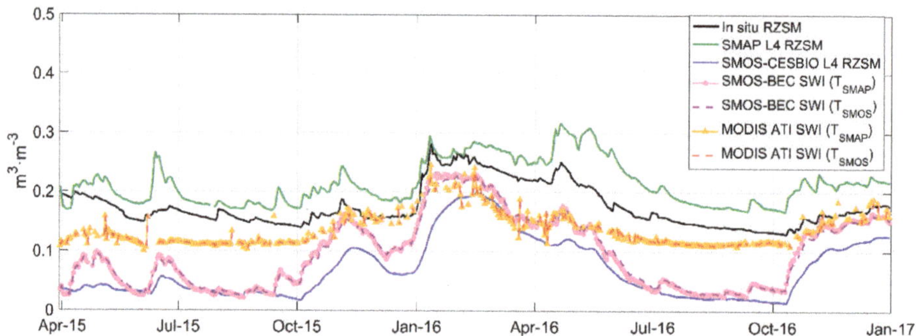

Figure 2. Average time series of the different RZSM estimates across REMEDHUS.

Regarding the impact of the T parameter, there were no differences in the use of T_{SMAP} and T_{SMOS} to derive the SWI, both with SMOS-BEC and MODIS ATI. This result could have occurred because they were very similar in the REMEDHUS area. Indeed, when computing the statistics of the T maps, the mean values of T_{SMAP} and T_{SMOS} were 14.5 and 17 days, the median values were 13 and 16.5 days, and the mode values were 13 and 16 days, respectively. It has been demonstrated that the T parameter is the response time of the autocorrelation function of soil moisture [37]. The T was affected by soil depth, soil hydraulic properties, and different physical processes such as evaporation and runoff. However, other processes such as transpiration or the soil hydraulic conductivity variation depending on soil moisture conditions were not considered in the SWI model [32,33]. Additionally, the potential dependence of T on soil texture was previously suggested [69]. Unfortunately, no significant relations were observed between T and soil properties (sand and clay fractions, bulk density, and organic matter content) [33]. The variability of the T parameter has previously been shown in several studies [32,33,37–40,42,43,69]. However, the SWI performed with similar good results in all cases. The robustness of SWI on the T parameter may rely on the recursive expression of the exponential filter, where the SSM has the highest influence and T only represents a timescale value.

The statistics obtained from the validation of the different RZSM estimates across REMEDHUS showed very similar correlation coefficients for all the estimates (by stations; Table 2 and Figure 3), although the correlations for MODIS ATI SWI (area-average) were slightly lower (R ~0.75 to 0.77). All correlations were significant at a 95 % confidence level. Similar correlation coefficient values were found in previous research when the SWI model was used. For instance, R ~0.45 to 0.88 was obtained when the SWI was applied to in situ SSM measurements of the Soil Climatic Analysis Network (SCAN) and compared to soil moisture measurements at 10–90 cm depth [34]. Likewise, R ~0.49 to 0.75 was obtained when the SWI was applied to the SMOS SSM and compared to the soil moisture of the 25–60 cm profile layer at the Oklahoma Mesonet and the Nebraska Automated Weather Data Network (AWDN) [42] and also compared to the soil moisture of the 0–50 cm depth of the same REMEDHUS network [43].

Regarding the differences from the in situ RZSM (Table 2), the values obtained for MODIS ATI SWI (RMSD ~0.031 to 0.110 m^3/m^3 and cRMSD ~0.017 to 0.054 m^3/m^3, by stations) were slightly lower than those for the other estimates (RMSD ~0.027 to 0.180 m^3/m^3 and cRMSD ~0.019 to 0.058 m^3/m^3) except for the SMAP L4 RZSM product, as expected from Figure 2. Both SMAP L4 RZSM and MODIS ATI SWI had the lowest differences (RMSD ~0.044 m^3/m^3 and cRMSD ~0.020–0.021 m^3/m^3, area-average), while SMOS-CESBIO L4 RZSM and SMOS-BEC SWI had the highest. These results suggested that MODIS ATI SWI could be a good estimator of RZSM at a spatial resolution of 1 km.

The bias (Table 2) of SMAP L4 RZSM exhibited positive values (by stations and area-average), indicating an overestimation of the in situ RZSM, whereas SMOS-CESBIO L4 RZSM, SMOS-BEC SWI and MODIS ATI SWI had dry biases in both cases, as observed in Figure 2. Different studies have shown that SMAP overestimates the SSM [2,59]. In this line, the SMAP L4 RZSM showed a slightly higher positive bias (0.063 m^3/m^3) during the validation of this product over the Little River. The irrigated crops and wetlands in the surroundings of this site, which are not considered in the SMAP model system, were first assumed as possible reasons for this wet bias, but REMEDHUS has no wetlands in its vicinity. Moreover, the wet bias over the Little River also appeared in the SMAP model-only simulations, suggesting that errors in the NASA catchment model parameters may be the main reason [30]. By contrast, the SMOS-derived RZSM estimates exhibited a dry bias (−0.15 to +0.05 m^3/m^3), with values similar to those previously observed [43].

Table 2. Statistics obtained from comparison of the six RZSM estimates (by stations and area-average) with the in situ RZSM across REMEDHUS. N indicates the number of coincident data days.

	RZSM Estimation		R	RMSD (m^3/m^3)	cRMSD (m^3/m^3)	bias (m^3/m^3)	N (%)
	SMAP L4 RZSM		0.39 to 0.89	0.027 to 0.180	0.019 to 0.058	−0.036 to 0.177	86.3 to 99.7
	SMOS-CESBIO L4 RZSM		0.33 to 0.89	0.036 to 0.179	0.023 to 0.061	−0.174 to 0.006	86.5 to 99.8
By stations	SMOS-BEC	SWI (T$_{SMAP}$)	0.30 to 0.93	0.030 to 0.148	0.028 to 0.063	−0.143 to 0.033	78.3 to 91.1
		SWI (T$_{SMOS}$)	0.31 to 0.93	0.027 to 0.148	0.024 to 0.061	−0.144 to 0.032	78.3 to 91.1
	MODIS ATI	SWI (T$_{SMAP}$)	0.19 to 0.88	0.032 to 0.110	0.017 to 0.054	−0.098 to 0.057	44.9 to 53.4
		SWI (T$_{SMOS}$)	0.18 to 0.87	0.031 to 0.110	0.017 to 0.053	−0.098 to 0.056	44.9 to 53.4
	SMAP L4 RZSM		0.86	0.044	0.020	0.040	99.7
	SMOS-CESBIO L4 RZSM		0.84	0.109	0.028	−0.105	99.8
Area-average	SMOS-BEC	SWI (T$_{SMAP}$)	0.81	0.086	0.039	−0.077	92.7
		SWI (T$_{SMOS}$)	0.82	0.086	0.037	−0.077	92.7
	MODIS ATI	SWI (T$_{SMAP}$)	0.75	0.045	0.022	−0.038	68.1
		SWI (T$_{SMOS}$)	0.77	0.044	0.021	−0.039	68.1

All values are significant (p-value < 0.05).

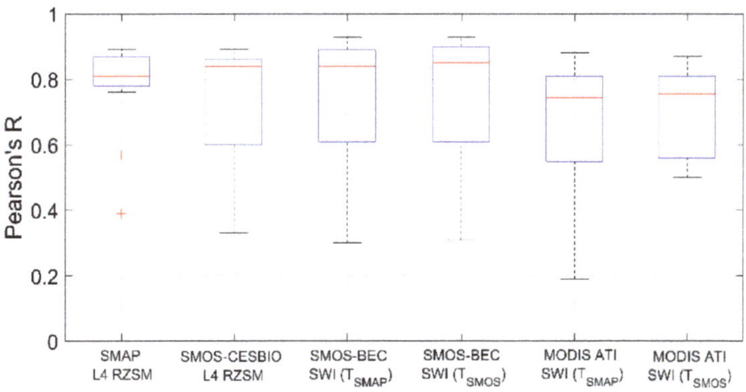

Figure 3. Pearson's correlation coefficients (R) between the in situ RZSM and the six RZSM estimates across REMEDHUS (by stations).

In summary, it is remarkable that the results for the SMAP RZSM and SMOS SWI (R > 0.8 and cRMSD ~0.02 m^3/m^3) were better than those found in some works about SSM over the REMEDHUS network. For the SMOS L2 SSM product, the correlation was R ~0.78, and the errors were beyond cRMSD ~0.04 m^3/m^3 [45]. In the case of the SMAP L2 SSM product, the correlation was R ~0.58, and the errors were also close to cRMSD ~0.04 m^3/m^3 [2]. Moreover, the results obtained for SMAP RZSM in this research were near those found for the SSM product in REMEDHUS (R ~0.9) but for a longer period of validation [59].

A previous study suggested that the inherent differences between SMOS and in situ were larger than the errors produced by the SWI model [42]. This situation was also true here, where the RMSD and bias were smaller for the SWI-derived estimations than for the SMOS RZSM product itself. The impact of using T_{SMOS} or T_{SMAP} in the SWI model was negligible for both SMOS-BEC and MODIS ATI, and non-significant differences were found during the validation across REMEDHUS. It is highlighted that the coverage of MODIS ATI SWI was clearly lower than that of the other estimations by both stations and area-average, which evidenced the most important drawback of the MODIS ATI-derived RZSM owing the use of VIS/IR observations.

3.3. Spatial Analysis of Root Zone Soil Moisture Estimates

Six maps of RZSM over the Iberian Peninsula—one for each different RZSM estimate—are displayed during a summer and a winter day (9 July and 31 December 2016) in Figures 4 and 5, respectively. No significant differences were observed between the SMOS-BEC and MODIS ATI SWI calculated with T_{SMAP} and T_{SMOS}, as it was seen in the REMEDHUS area (Table 2 and Figure 3). This result suggests that the T parameter could have a very small influence in the SWI model results.

On both dates, there was a high similarity between the SMOS-CESBIO L4 RZSM and the SMOS-BEC SWI maps because both estimations used data from the same radiometer and similar approaches were employed to obtain the RZSM. Note that the higher spatial resolution of the SMOS-BEC SWI maps allowed observing more details than that of the SMOS-CESBIO L4 RZSM maps. In general, although the SMOS-BEC SWI used both ascending and descending orbits to increase its coverage, the SMOS orbital path of a unique day usually does not cover the entire Iberian Peninsula and some data gaps are still present. Instead, the SMOS-CESBIO L4 RZSM covered the entire spatial domain since it was produced by a three day composite.

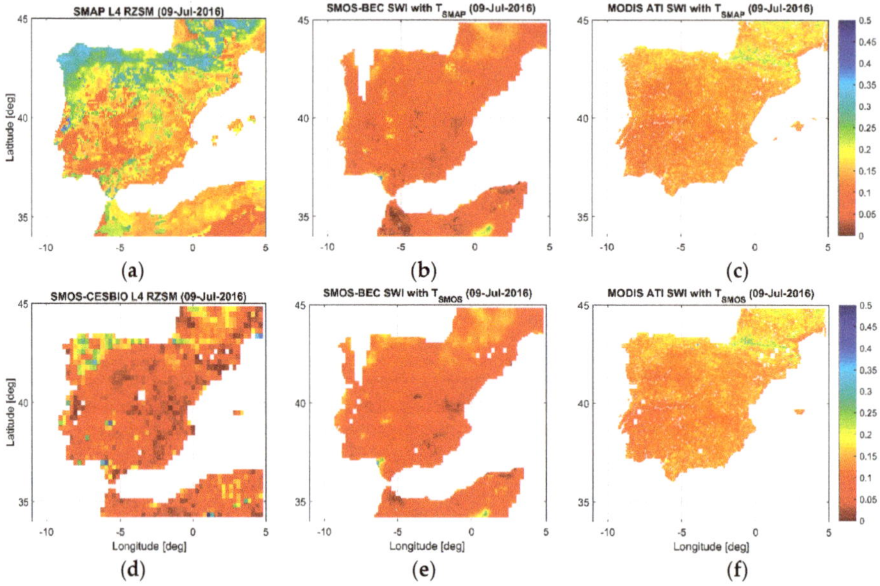

Figure 4. RZSM maps corresponding to a summer day (9 July 2016): (**a**) SMAP L4 RZSM; (**b**) SMOS-BEC SWI (T_{SMAP}); (**c**) MODIS ATI SWI (T_{SMAP}); (**d**) SMOS-CESBIO L4 RZSM; (**e**) SMOS-BEC SWI (T_{SMOS}); and (**f**) MODIS ATI SWI (T_{SMOS}).

In general, the SMAP L4 RZSM product displayed wetter values than all the other estimates on both days, confirming the overestimation found in REMEDHUS. Additionally, the SMAP L4 RZSM maps clearly exhibited wetter values in the north than in the south, which was in accordance with the climatic patterns over the Iberian Peninsula. This effect was especially noticeable for the Pyrenees and the Cantabrian ranges in the north and the Galicia region and the northern part of Portugal in the northwestern Iberian Peninsula. The SMOS-CESBIO L4 RZSM maps also displayed this synoptic situation in the summer. In summer as well as winter, both SMAP and SMOS-CESBIO L4 RZSM were able to detect the Doñana National Park—the most important wetland in Spain—located at the Guadalquivir River mouth (southern Spain). This area was also captured by the SMOS-BEC SWI, whereas the MODIS ATI SWI did not show these distinctive features. There was no marked spatial contrast in RZSM for the MODIS ATI SWI maps throughout the Iberian Peninsula on both days except the high values in the north during the wet period. The MODIS ATI SWI can be masked not only by clouds but also by fog banks due to the use of optical observations. This effect was captured in the MODIS ATI SWI maps for the winter day where some river valleys in the center of the Iberian Peninsula did not have data.

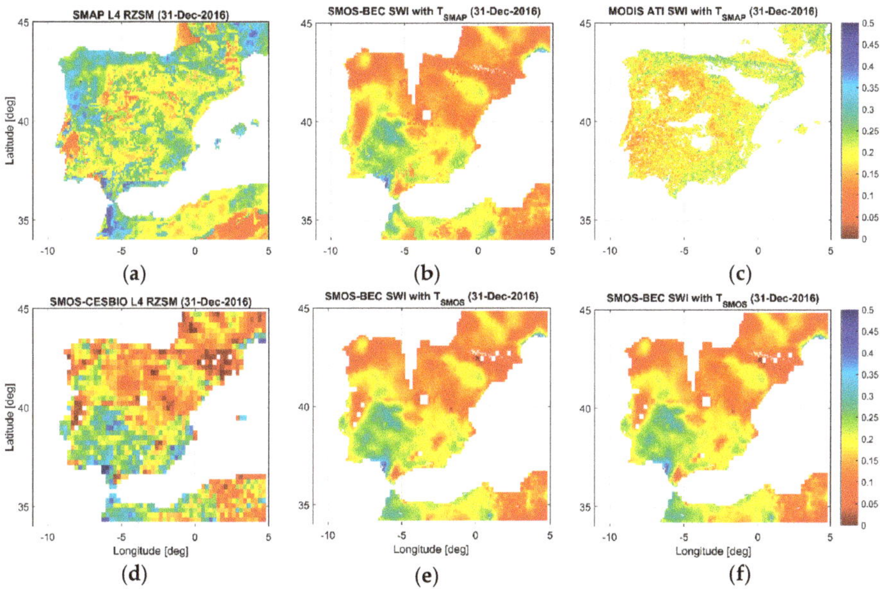

Figure 5. RZSM maps corresponding to a winter day (31 December 2016): (**a**) SMAP L4 RZSM; (**b**) SMOS-BEC SWI (T_{SMAP}); (**c**) MODIS ATI SWI (T_{SMAP}); (**d**) SMOS-CESBIO L4 RZSM; (**e**) SMOS-BEC SWI (T_{SMOS}); and (**f**) MODIS ATI SWI (T_{SMOS}).

The temporal correlation maps of the SMAP and SMOS-CESBIO L4 RZSM with both SMOS-BEC and MODIS ATI SWI aggregated to 9 km and 25 km are shown in Figures 6 and 7, respectively. Only pixels with significant correlation (p-value < 0.05) are shown. It can be seen that microwave-based RZSM estimates (the SMAP L4 RZSM product at 9 km against SMOS-BEC SWI estimates) were highly correlated over most parts of the Iberian Peninsula ($R \geq 0.7$), regardless of the T parameter used. Only some areas displayed low values of correlation ($R \sim 0.2$ to 0.3). These areas, such as the Pyrenees, had a low number of days with data because the abrupt topography decreased the number of soil moisture retrievals for these pixels. However, when comparing the SMAP L4 RZSM against the MODIS ATI SWI, the correlations over all the extreme north of the peninsula were low and even reached negative values, especially in regions with abundant and dense vegetation such as the Pyrenees and Cantrabrian ranges. The same patterns were obtained when comparing the MODIS ATI SWI against the SMOS-BEC SWI with both T parameter values. This result may be related to the distinctive vegetation of these regions because there are other mountainous areas in the Iberian Peninsula, but the north ranges are also the more densely forested ones.

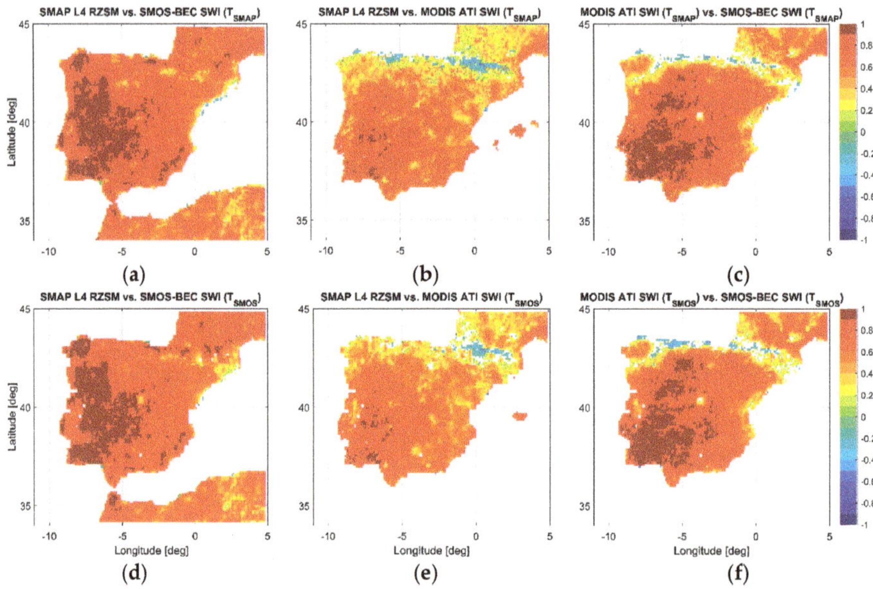

Figure 6. Temporal correlation maps at 9 km between: (**a**) SMAP L4 RZSM and SMOS-BEC SWI (T_{SMAP}); (**b**) SMAP L4 RZSM and MODIS ATI SWI (T_{SMAP}); (**c**) MODIS ATI SWI (T_{SMAP}) and SMOS-BEC (T_{SMAP}); (**d**) SMAP L4 RZSM and SMOS-BEC SWI (T_{SMOS}); (**e**) SMAP L4 RZSM and MODIS ATI SWI (T_{SMOS}); and (**f**) MODIS ATI SWI (T_{SMOS}) and SMOS-BEC (T_{SMOS}).

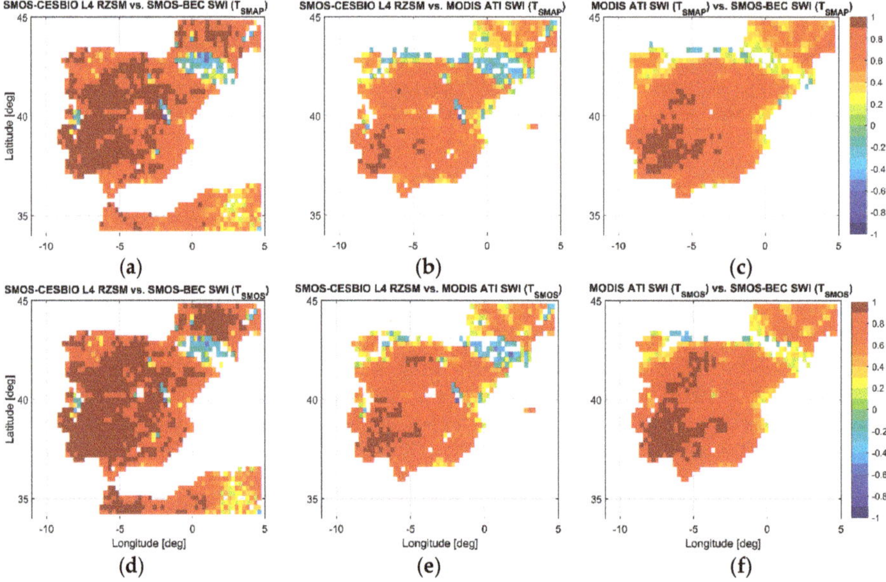

Figure 7. Temporal correlation maps at 25 km between: (**a**) SMOS-CESBIO L4 RZSM and SMOS-BEC SWI (T_{SMAP}); (**b**) SMOS-CESBIO L4 RZSM and MODIS ATI SWI (T_{SMAP}); (**c**) MODIS ATI SWI (T_{SMAP}) and SMOS-BEC (T_{SMAP}); (**d**) SMOS-CESBIO L4 RZSM and SMOS-BEC SWI (T_{SMOS}); (**e**) SMOS-CESBIO L4 RZSM and MODIS ATI SWI (T_{SMOS}); and (**f**) MODIS ATI SWI (T_{SMOS}) and SMOS-BEC (T_{SMOS}).

In addition, a similar comparison was made with the SMOS-CESBIO L4 RZSM at 25 km, and the same patterns were distinguished irrespective of its lower spatial resolution compared to SMAP. The same patterns were found in the evapotranspiration and Fractional Vegetation Cover products from the Spinning Enhanced Visible and Infrared Imager (SEVIRI) sensor on board the Meteosat Second Generation (MSG) [70]. Then, the VIS/IR-based observations produced negative correlations in northern Spain. This result could be due to the MODIS variables involved in the ATI-derived soil moisture estimation (reflectance or LST). The reflectance was directly linked to the MODIS NDVI that was used to obtain the SMOS-BEC L4 SSM from its downscaling algorithm but not the MODIS LST. However, the observed correlation patterns were not seen in the correlations maps of the SMOS-BEC SWI with other passive RZSM estimates. Due to this fact, the reflectance was discarded as the cause of the negative correlations. Therefore, the reason for these correlation patterns may be the different characterization of the MODIS LST reflecting the north-south temperature gradient compared with the LSTs from the NASA catchment model and the ECMWF ERA-Interim model used in SMAP L4 RZSM and SMOS-BEC L4 SSM, respectively.

4. Conclusions

In recent years, the use of RZSM has become increasingly important in several hydrological and agricultural applications. In this research, six different RZSM estimates provided by three satellites—SMAP and SMOS operating at the microwave L-band and MODIS operating at the optical frequency range—were evaluated across the Iberian Peninsula from 31 March 2015 to 31 December 2016.

A preliminary analysis of the MODIS ATI-derived SSM estimation was carried out. It correctly reproduced the in situ SSM measurements with slightly lower correlations but a similar performance in terms of errors and bias than estimates obtained from the L-band. Because there were no remarkable differences between all possible alternatives to calculate the ATI-derived SSM, the approach that was easiest to implement, had the lowest computational cost, and with the best available spatial and temporal resolution was selected. This approach used the visible albedo reflectance, the combination of the Aqua/Terra LST, and the Soil Hydraulic Database to set the reference thresholds for soil moisture.

In the temporal analysis of the six RZSM estimates performed across the REMEDHUS area, all of them followed a similar evolution. The SMOS-CESBIO L4 RZSM product, the SMOS-BEC SWI, and MODIS ATI SWI showed underestimations with respect to the in situ measurements, while the SMAP L4 RZSM product showed overestimation. The different alternatives for T parameters used in the SWI model did not seem to impact the results.

In the spatial analysis performed across the Iberian Peninsula, a high level of similarity was observed between the SMOS-based estimations, with clearly higher spatial detail for the SMOS-BEC SWI than for the other estimates. The most distinctive spatial variability between wet and dry regions was displayed by the SMAP L4 RZSM, whereas the MODIS ATI SWI had the lowest spatial variability.

Regarding the correlation spatial patterns, the RZSM estimates from the microwave sensors were highly correlated across the Iberian Peninsula except in areas with low amounts of data. The comparison between microwave and optical observations showed low or negative correlation values all over the extreme north of the peninsula, coinciding with the location of higher mountains and more densely forested areas.

Although two different spectral regions were used (microwave or optical frequencies), all six different RZSM estimates reproduced the temporal patterns of the in situ RZSM well and could be considered as good RZSM estimators. The SWI model, which required only SSM and the T parameter as inputs, worked as well as more complex land surface or hydrological models for estimating RZSM, as in the cases of the SMOS and SMAP L4 RZSM products. The good results based on the MODIS ATI must be highlighted, despite its lesser temporal coverage, given MODIS is not a satellite mission specifically devoted to soil moisture retrieval.

The use of remote sensing as an alternative to ground-based measurements for RZSM monitoring offers a new opportunity for hydrological studies and agricultural applications at the global scale, improving their spatial coverage and temporal resolution.

Author Contributions: The initial idea for this research was conceived by J.M.-F. and N.S. The in situ data were prepared by Á.G.-Z. The satellite data were downloaded by M.P. The exponential T parameters were computed by Á.G.-Z. All other data processing was performed by M.P., who also collected all the results. The four authors have equally contributed to the interpretation of the results. The first manuscript was prepared by M.P. in collaboration with the other authors. All the authors revised the final manuscript and approved it.

Funding: This research was supported by the Junta de Castilla y León (Project SA007U16), the Spanish Ministry of Economy and Competitiveness (Projects ESP2015-67549-C3-3-R and ESP2017-89463-C3-3-R), and the European Regional Development Fund (ERDF).

Acknowledgments: The in situ SSM data are available at the International Soil Moisture Network (ISMN) database (https://ismn.geo.tuwien.ac.at) and the rest of in situ soil moisture data are available upon request. The SMAP L4 soil moisture (both SSM and RZSM variables within the SPL4SMGP v.3 product) are accessible at the National Snow and Ice Data Center Distributed Active Archive Center (NSIDC DAAC, https://nsidc.org/data/SPL4SMGP/versions/3). The SMOS-CESBIO L3 SSM (SM_RE04/OPER_MIR_ CLF31A/D v.300 product) was distributed by the Centre Aval de Traitement des Données SMOS (CATDS, http://www.catds.fr/Products/Available-products-from-CPDC). The SMOS-CESBIO L4 RZSM (SM_SCIE_MIR_ CLF4RA/D v.300 product) was also disseminated by CATDS (http://www.catds.fr/Products/Available-products-from-CEC-SM/L4-Land-research-products). The authors especially thank Gerard Portal and Mercè Vall-llossera from the Technical University of Catalonia (UPC) and the Barcelona Expert Centre (BEC, http://bec.icm.csic.es) for providing the new cloud-free SMOS-BEC L4 SSM v.3 product. The Aqua MODIS surface reflectance (MYD09GA v.6 product) and both Aqua and Terra LST (MYD11A1 and MOD11A1 v.6 products) were provided by the NASA Land Processes Distributed Active Archive Center (LP DAAC, https://lpdaac.usgs.gov).

Conflicts of Interest: The authors declare no conflict of interest.

References

1. Kerr, Y.H.; Al-Yaari, A.; Rodríguez-Fernández, N.; Parrens, M.; Molero, B.; Leroux, D.; Bircher, S.; Mahmoodi, A.; Mialon, A.; Richaume, P.; et al. Overview of SMOS performance in terms of global soil moisture monitoring after six years in operation. *Remote Sens. Environ.* **2016**, *180*, 40–63. [CrossRef]

2. Chan, S.K.; Bindlish, R.; O'Neill, P.E.; Njoku, E.; Jackson, T.J.; Colliander, A.; Chen, F.; Burgin, M.; Dunbar, S.; Piepmeier, J.; et al. Assessment of the SMAP passive sol moisture product. *IEEE Trans. Geosci. Remote Sens.* **2016**, *54*, 4994–5007. [CrossRef]

3. Piles, M.; Sánchez, N.; Vall-llossera, M.; Camps, A.; Martínez-Fernández, J.; Martínez, J.; González-Gambau, V. A downscaling approach for SMOS land observations: Evaluation of high-resolution soil moisture maps over the Iberian Peninsula. *IEEE J. Sel. Top. Appl. Earth Observ. Remote Sens.* **2014**, *7*, 3845–3857. [CrossRef]

4. Merlin, O.; Malbéteau, Y.; Notfi, Y.; Bacon, S.; Khabba, S.E.R.; Jarlan, L. Performance metrics for soil moisture downscaling methods: Application to DISPATCH data in central Morocco. *Remote Sens.* **2015**, *7*, 3783–3807. [CrossRef]

5. Pablos, M.; Piles, M.; Sánchez, N.; Vall-llossera, M.; Martínez-Fernández, J.; Camps, A. Impact of day/night time land surface temperature in soil moisture disaggregation algorithms. *Eur. J. Remote Sens.* **2016**, *49*, 899–916. [CrossRef]

6. Piles, M.; Petropoulos, G.P.; Sánchez, N.; González-Zamora, A.; Ireland, G. Towards improved spatio-temporal resolution soil moisture retrievals from the synergy of SMOS and MSG SEVIRI spaceborne observations. *Remote Sens. Environ.* **2016**, *180*, 403–417. [CrossRef]

7. Das, N.N.; Entekhabi, D.; Dunbar, R.S.; Njoku, E.G.; Yueh, S.H. Uncertainty estimates in the SMAP combined active & passive downscaled brightness temperature. *IEEE Trans. Geosci. Remote Sens.* **2016**, *54*, 640–650. [CrossRef]

8. Petropoulos, G.P.; Ireland, G.; Barrett, B. Surface soil moisture retrievals from remote sensing: Current status, products & future trends. *Phy. Chem. Earth Parts A/B/C* **2015**, *83–84*, 36–56. [CrossRef]

9. Price, J.C. Thermal inertia mapping: A new view of the Earth. *J. Geophys. Res.* **1977**, *82*, 2582–2590. [CrossRef]

10. Short, N.M.; Stuart, L.M. *The Heat Capacity Mapping Mission (HCMM) Anthology*; Scientific and technical information banch; National Aeronautics and Space Administration (NASA): Washington, DC, USA, 1983; p. 264.

11. Sobrino, J.A.; El Kharraz, M.H. Combining afternoon and morning NOAA satellites for thermal inertia estimation: 1. Algorithm and its testing with Hydrologic Atmospheric Pilot Experiment-Sahel data. *J. Geophys. Res. Atmos.* **1999**, *104*, 9445–9453. [CrossRef]

12. Verstraeten, W.W.; Veroustraete, F.; van der Sande, C.J.; Grootaers, I.; Feyen, J. Soil moisture retrieval using thermal inertia, determined with visible and thermal spaceborne data, validated for European forests. *Remote Sens. Environ.* **2006**, *101*, 299–314. [CrossRef]

13. Van doninck, J.; Peters, J.; De Baets, B.; De Clercq, E.M.; Ducheyne, E.; Verhoest, N.E.C. The potential of multitemporal Aqua and Terra MODIS apparent thermal inertia as a soil moisture indicator. *Int. J. Appl. Earth Observ. Geoinf.* **2011**, *13*, 934–941. [CrossRef]

14. Vereecken, H.; Huisman, J.A.; Bogena, H.; Vanderborght, J.; Vrugt, J.A.; Hopmans, J.W. On the value of soil moisture measurements in vadose zone hydrology: A review. *Water Resour. Res.* **2008**, *44*, 1–21. [CrossRef]

15. Mohanty, B.P.; Cosh, M.H.; Lakshmi, V.; Montzka, C. Soil moisture remote sensing: State-of-the-science. *Vadose Zone J.* **2017**, *16*, 1–9. [CrossRef]

16. Dorigo, W.A.; Wagner, W.; Hohensinn, R.; Hahn, S.; Paulik, C.; Xaver, A.; Gruber, A.; Drusch, M.; Mecklenburg, S.; van Oevelen, P.; et al. The International Soil Moisture Network: A data hosting facility for global in situ soil moisture measurements. *Hydrol. Earth Syst. Sci.* **2011**, *15*, 1675–1698. [CrossRef]

17. Zreda, M.; Shuttleworth, W.J.; Zeng, X.; Zweck, C.; Desilets, D.; Franz, T.; Rosolem, R. COSMOS: The COsmic-ray Soil Moisture Observing System. *Hydrol. Earth Syst. Sci.* **2012**, *16*, 4079–4099. [CrossRef]

18. Kędzior, M.; Zawadzki, J. Comparative study of soil moisture estimations from SMOS satellite mission, GLDAS database, and cosmic-ray neutrons measurements at COSMOS station in Eastern Poland. *Geoderma* **2016**, *283*, 21–31. [CrossRef]

19. Zreda, M.; Desilets, D.; Ferré, T.P.A.; Scott, R.L. Measuring soil moisture content non-invasively at intermediate spatial scale using cosmic-ray neutrons. *Geophys. Res. Lett.* **2008**, *35*, 1–5. [CrossRef]

20. Liu, X.; Chen, J.; Cui, X.; Liu, Q.; Cao, X.; Chen, X. Measurement of soil water content using ground-penetrating radar: A review of current methods. *Int. J. Dig. Earth* **2017**, 1–24. [CrossRef]

21. Haddeland, I.; Clark, D.B.; Franssen, W.; Ludwig, F.; Voß, F.; Arnell, N.W.; Bertrand, N.; Best, M.; Folwell, S.; Gerten, D.; et al. Multimodel estimate of the global terrestrial water balance: Setup and first results. *J. Hydrometeorol.* **2011**, *12*, 869–884. [CrossRef]

22. Muñoz-Sabater, J.; Jarlan, L.; Calvet, J.C.; Bouyssel, F.; De Rosnay, P. From near-surface to root-zone soil moisture using different assimilation techniques. *J. Hydrometeorol.* **2007**, *8*, 194–206. [CrossRef]

23. Das, N.N.; Mohanty, B.P.; Njoku, E.G. Profile soil moisture across spatial scales under different hydroclimatic conditions. *Soil Sci.* **2010**, *175*, 315–319. [CrossRef]

24. Dumedah, G.; Walker, J.P. Evaluation of model parameter convergence when using data assimilation for soil moisture estimation. *J. Hydrometeorol.* **2014**, *15*, 359–375. [CrossRef]

25. Calvet, J.-C.; Noilhan, J. From near-surface to root-zone soil moisture using year-round data. *J. Hydrometeorol.* **2000**, *1*, 393–411. [CrossRef]

26. Montaldo, N.; Albertson, J.D.; Mancini, M.; Kiely, G. Robust simulation of root zone soil moisture with assimilation of surface soil moisture data. *Water Resour. Res.* **2001**, *37*, 2889–2900. [CrossRef]

27. Reichle, R.H.; Koster, R.D.; Liu, P.; Mahanama, S.P.P.; Njoku, E.G.; Owe, M. Comparison and assimilation of global soil moisture retrievals from the Advanced Microwave Scanning Radiometer for the Earth Observing System (AMSR-E) and the Scanning Multichannel Microwave Radiometer (SMMR). *J. Geophys. Res. Atmos.* **2007**, *112*, 1–14. [CrossRef]

28. Draper, C.; Mahfouf, J.-F.; Calvet, J.-C.; Martin, E.; Wagner, W. Assimilation of ASCAT near-surface soil moisture into the SIM hydrological model over France. *Hydrol. Earth Syst. Sci.* **2011**, *15*, 3829–3841. [CrossRef]

29. Dumedah, G.; Walker, J.P.; Merlin, O. Root-zone soil moisture estimation from assimilation of downscaled Soil Moisture and Ocean Salinity data. *Adv. Water Resour.* **2015**, *84*, 14–22. [CrossRef]

30. Reichle, R.H.; De Lannoy, G.J.M.; Liu, Q.; Ardizzone, J.V.; Colliander, A.; Conaty, A.; Crow, W.; Jackson, T.J.; Jones, L.A.; Kimball, J.S.; et al. Assessment of the SMAP level-4 surface and root-zone soil moisture product using in situ measurements. *J. Hydrometeorol.* **2017**, *18*, 2621–2645. [CrossRef]

31. Al Bitar, A.; Kerr, Y.H.; Merlin, O.; Cabot, F.; Wigneron, J.P. Global Drought Index from SMOS Soil Moisture. In Proceedings of the IEEE International Geoscience and Remote Sensing Symposium (IGARSS), Melbourne, Australia, 21–26 July 2013.

32. Wagner, W.; Lemoine, G.; Rott, H. A method for estimating soil moisture from ERS scatterometer and soil data. *Remote Sens. Environ.* **1999**, *70*, 191–207. [CrossRef]

33. Albergel, C.; Rüdiger, C.; Pellarin, T.; Calvet, J.-C.; Fritz, N.; Froissard, F.; Suquia, D.; Petitpa, A.; Piguet, B.; Martin, E. From near-surface to root-zone soil moisture using an exponential filter: An assessment of the method based on in-situ observations and model simulations. *Hydrol. Earth Syst. Sci. Discuss.* **2008**, *12*, 1323–1337. [CrossRef]

34. Manfreda, S.; Brocca, L.; Moramarco, T.; Melone, F.; Sheffield, J. A physically based approach for the estimation of root-zone soil moisture from surface measurements. *Hydrol. Earth Syst. Sci.* **2014**, *18*, 1199–1212. [CrossRef]

35. Qiu, J.; Crow, W.T.; Nearing, G.S.; Mo, X.; Liu, S. The impact of vertical measurement depth on the information content of soil moisture times series data. *Geophys. Res. Lett.* **2014**, *41*, 4997–5004. [CrossRef]

36. Peterson, A.M.; Helgason, W.D.; Ireson, A.M. Estimating field-scale root zone soil moisture using the cosmic-ray neutron probe. *Hydrol. Earth Syst. Sci.* **2016**, *20*, 1373–1385. [CrossRef]

37. Ceballos, A.; Scipal, K.; Wagner, W.; Martínez-Fernández, J. Validation of ERS scatterometer-derived soil moisture data in the central part of the Duero Basin, Spain. *Hydrol. Process.* **2005**, *19*, 1549–1566. [CrossRef]

38. Brocca, L.; Melone, F.; Moramarco, T.; Wagner, W.; Hasenauer, S. ASCAT soil wetness index validation through in situ and modeled soil moisture data in central Italy. *Remote Sens. Environ.* **2010**, *114*, 2745–2755. [CrossRef]

39. Brocca, L.; Hasenauer, S.; Lacava, T.; Melone, F.; Moramarco, T.; Wagner, W.; Dorigo, W.; Matgen, P.; Martínez-Fernández, J.; Llorens, P.; et al. Soil moisture estimation through ASCAT and AMSR-E sensors: An intercomparison and validation study across Europe. *Remote Sens. Environ.* **2011**, *115*, 3390–3408. [CrossRef]

40. Paulik, C.; Dorigo, W.; Wagner, W.; Kidd, R. Validation of the ASCAT Soil Water Index using in situ data from the International Soil Moisture Network. *Int. J. Appl. Earth Observ. Geoinf.* **2014**, *30*, 1–8. [CrossRef]

41. Tobin, K.J.; Torres, R.; Crow, W.T.; Bennett, M.E. Multi-decadal analysis of root-zone soil moisture applying the exponential filter across CONUS. *Hydrol. Earth Syst. Sci.* **2017**, *21*, 4403–4417. [CrossRef]

42. Ford, T.W.; Harris, E.; Quiring, S.M. Estimating root zone soil moisture using near-surface observations from SMOS. *Hydrol. Earth Syst. Sci.* **2014**, *18*, 139–154. [CrossRef]

43. González-Zamora, A.; Sánchez, N.; Martínez-Fernández, J.; Wagner, W. Root-zone plant available water estimation using the SMOS-derived soil water index. *Adv. Water Resour.* **2016**, *96*, 339–353. [CrossRef]

44. González-Zamora, A.; Martínez-Fernández, J.; Sánchez, N.; Pablos, M. Estimación de la humedad en la zona radicular a partir de observaciones remotes de humedad superficial de larga duración. In *Las XIII de Jornadas de Investigación de la Zona No Saturada*; Moret-Fernández, D., López, M.V., Eds.; Consejo Superior de Investigaciones Científicas (CSIC): Zaragoza, Spain, 2017; Volume XIII, pp. 493–504.

45. González-Zamora, A.; Sánchez, N.; Martínez-Fernández, J.; Gumuzzio, A.; Piles, M.; Olmedo, E. Long-term SMOS soil moisture products: A comprehensive evaluation across scales and methods in the Duero Basin (Spain). *Phys. Chem. Earth Parts A/B/C* **2015**, *83–84*, 123–136. [CrossRef]

46. Reichle, R.H.; De Lannoy, G.; Koster, R.D.; Crow, W.T.; Kimball, J.S. *SMAP L4 9 km EASE-Grid Surface and Root Zone Soil Moisture Geophysical Data, Version 3*; National Snow and Ice Data Center Distributed Active Archive Center (NSIDC DAAC): Boulder, CO, USA, 2017.

47. Reichle, R.H.; Koster, R.D.; De Lannoy, G.J.M.; Crow, W.T.; Kimball, J. *Algorithm Theoretical Basis Document Level 4 Surface and Root Zone Soil Moisture (L4_SM) Data Product*; National Aerounautics and Space Administration (NASA): Greenbelt, MD, USA, 2014; pp. 1–65.

48. Kerr, Y.H.; Jacquette, E.; Al Bitar, A.; Cabot, F.; Mialon, A.; Richaume, P.; Quesney, A.; Berthon, L. *CATDS SMOS L3 Soil Moisture Retrieval Processor Algorithm Theoretical Baseline Document (ATBD)*; Centre Aval de Traitement des Données SMOS (CATDS): Toulouse, France, 2013.

49. Kerr, Y.H.; Waldteufel, P.; Richaume, P.; Wigneron, J.P.; Ferrazzoli, P.; Mahmoodi, A.; Bitar, A.A.; Cabot, F.; Gruhier, C.; Juglea, S.E.; et al. The SMOS soil moisture retrieval algorithm. *IEEE Trans. Geosci. Remote Sens.* **2012**, *50*, 1384–1403. [CrossRef]

50. Al Bitar, A.; Mialon, A.; Kerr, Y.H.; Cabot, F.; Richaume, P.; Jacquette, E.; Quesney, A.; Mahmoodi, A.; Tarot, S.; Parrens, M.; et al. The global SMOS Level 3 daily soil moisture and brightness temperature maps. *Earth Syst. Sci. Data* **2017**, *9*, 293–315. [CrossRef]

51. Portal, G.; Vall-llossera, M.; Piles, M.; Camps, A.; Chaparro, D.; Pablos, M.; Rossato, L. A spatially consistent downscaling approach for SMOS using an adaptive moving window. *IEEE J. Sel. Top. Appl. Earth Observ. Remote Sens.* **2018**, in press. [CrossRef]

52. Vermote, E.F.; Vermeulen, A. *Atmospheric Correction Algorithm: Spectral Reflectances (MOD09) Version 4.0*; National Aeronautics and Space Administration (NASA): Washington, DC, USA; Department of Geography, University of Maryland: College Park, MD, USA, 1999.

53. Wan, Z. *MODIS Land-Surface Temperature Algorithm Theoretical Basis Document (LST ATBD) Version 3.3*; National Aeronautics and Space Administration (NASA): Washington, DC, USA; Institute for Computational Earth System Science, University of California: Santa Barbara, CA, USA, 1999.

54. Liang, S. Narrowband to broadband conversions of land surface albedo I: Algorithms. *Remote Sens. Environ.* **2001**, *76*, 213–238. [CrossRef]

55. Pablos, M.; Martínez-Fernández, J.; Piles, M.; Sánchez, N.; Vall-llossera, M.; Camps, A. Multi-temporal evaluation of soil moisture and land surface temperature dynamics using in situ and satellite observations. *Remote Sens.* **2016**, *8*, 587. [CrossRef]

56. Toté, C.; Patricio, D.; Boogaard, H.; van der Wijngaart, R.; Tarnavsky, E.; Funk, C. Evaluation of satellite rainfall estimates for frought and food monitoring in Mozambique. *Remote Sens.* **2015**, *7*, 1758–1776. [CrossRef]

57. Chang, T.Y.; Wang, Y.C.; Feng, C.C.; Ziegler, A.D.; Giambelluca, T.W.; Liou, Y.A. Estimation of root zone soil moisture using apparent thermal inertia with MODIS imagery over a tropical catchment in northern Thailand. *IEEE J. Sel. Top. Appl. Earth Observ. Remote Sens.* **2012**, *5*, 752–761. [CrossRef]

58. Dorigo, W.; Wagner, W.; Albergel, C.; Albrecht, F.; Balsamo, G.; Brocca, L.; Chung, D.; Ertl, M.; Forkel, M.; Gruber, A.; et al. ESA CCI soil moisture for improved Earth system understanding: State-of-the art and future directions. *Remote Sens. Environ.* **2017**, *203*, 185–215. [CrossRef]

59. Colliander, A.; Jackson, T.J.; Bindlish, R.; Chan, S.; Das, N.; Kim, S.B.; Cosh, M.H.; Dunbar, R.S.; Dang, L.; Pashaian, L.; et al. Validation of SMAP surface soil moisture products with core validation sites. *Remote Sens. Environ.* **2017**, *191*, 215–231. [CrossRef]

60. González-Zamora, Á.; Sánchez, N.; Pablos, M.; Martínez-Fernández, J. CCI soil moisture assessment with SMOS soil moisture and in situ data under different environmental conditions and spatial scales in Spain. *Remote Sens. Environ.* **2018**. [CrossRef]

61. Wang, K.; Liang, S.; Schaaf, C.L.; Strahler, A.H. Evaluation of Moderate Resolution Imaging Spectroradiometer land surface visible and shortwave albedo products at FLUXNET sites. *J. Geophys. Res. Atmos.* **2010**, *115*, 1–8. [CrossRef]

62. Qin, J.; Yang, K.; Lu, N.; Chen, Y.; Zhao, L.; Han, M. Spatial upscaling of in-situ soil moisture measurements based on MODIS-derived apparent thermal inertia. *Remote Sens. Environ.* **2013**, *138*, 1–9. [CrossRef]

63. Gao, S.; Zhu, Z.; Weng, H.; Zhang, J. Upscaling of sparse in situ soil moisture observations by integrating auxiliary information from remote sensing. *Int. J. Remote Sens.* **2017**, *38*, 4782–4803. [CrossRef]

64. Sánchez, N.; Martinez-Fernández, J.; Scaini, A.; Pérez-Gutiérrez, C. Validation of the SMOS L2 soil moisture data in the REMEDHUS network (Spain). *IEEE Trans. Geosci. Remote Sens.* **2012**, *50*, 1602–1611. [CrossRef]

65. Gumuzzio, A.; Brocca, L.; Sánchez, N.; González-Zamora, A.; Martínez-Fernández, J. Comparison of SMOS, modelled and in situ long-term soil moisture series in the northwest of Spain. *Hydrol. Sci. J.* **2016**, *61*, 2610–2625. [CrossRef]

66. Dall'Amico, J.T.; Schlenz, F.; Loew, A.; Mauser, W. First results of SMOS soil moisture validation in the Upper Danube Catchment. *IEEE Trans. Geosci. Remote Sens.* **2012**, *50*, 1507–1516. [CrossRef]

67. Dente, L.; Su, Z.; Wen, J. Validation of SMOS soil moisture products over the Maqu and Twente regions. *Sensors* **2012**, *12*, 9965–9986. [CrossRef] [PubMed]

68. Djamai, N.; Magagi, R.; Goïta, K.; Hosseini, M.; Cosh, M.H.; Berg, A.; Toth, B. Evaluation of SMOS soil moisture products over the CanEx-SM10 area. *J. Hydrol.* **2015**, *520*, 254–267. [CrossRef]

69. de Lange, R.; Beck, R.; van de Giesen, N.; Friesen, J.; de Wit, A.; Wagner, W. Scatterometer-Derived Soil Moisture Calibrated for Soil Texture with a One-Dimensional Water-Flow Model. *IEEE Trans. Geosci. Remote Sens.* **2008**, *46*, 4041–4049. [CrossRef]
70. Petropoulos, G.P.; Ireland, G.; Lamine, S.; Griffiths, H.M.; Ghilain, N.; Anagnostopoulos, V.; North, M.R.; Srivastava, P.K.; Georgopoulou, H. Operational evapotranspiration estimates from SEVIRI in support of sustainable water management. *Int. J. Appl. Earth Observ. Geoinf.* **2016**, *49*, 175–187. [CrossRef]

Article

Dominant Features of Global Surface Soil Moisture Variability Observed by the SMOS Satellite

Maria Piles [1,*], **Joaquim Ballabrera-Poy** [2] **and Joaquín Muñoz-Sabater** [3]

[1] Image Processing Laboratory, Universitat de València, C/Catedrático José Beltrán 2, 46980 València, Spain
[2] Institut de Ciències del Mar, CSIC, Pg. Maritim de la Barceloneta 37-49, 08003 Barcelona, Spain; joaquim@icm.csic.es
[3] European Centre for Medium-Range Weather Forecasts (ECMWF), Copernicus department, Shinfield Park, Reading RG2 9AX, UK; Joaquin.Munoz@ecmwf.int
* Correspondence: maria.piles@uv.es; Tel.: +34-96-354-4161

Received: 23 November 2018; Accepted: 29 December 2018; Published: 8 January 2019

Abstract: Soil moisture observations are expected to play an important role in monitoring global climate trends. However, measuring soil moisture is challenging because of its high spatial and temporal variability. Point-scale in-situ measurements are scarce and, excluding model-based estimates, remote sensing remains the only practical way to observe soil moisture at a global scale. The ESA-led Soil Moisture and Ocean Salinity (SMOS) mission, launched in 2009, measures the Earth's surface natural emissivity at L-band and provides highly accurate soil moisture information with a 3-day revisiting time. Using the first six full annual cycles of SMOS measurements (June 2010–June 2016), this study investigates the temporal variability of global surface soil moisture. The soil moisture time series are decomposed into a linear trend, interannual, seasonal, and high-frequency residual (i.e., subseasonal) components. The relative distribution of soil moisture variance among its temporal components is first illustrated at selected target sites representative of terrestrial biomes with distinct vegetation type and seasonality. A comparison with GLDAS-Noah and ERA5 modeled soil moisture at these sites shows general agreement in terms of temporal phase except in areas with limited temporal coverage in winter season due to snow. A comparison with ground-based estimates at one of the sites shows good agreement of both temporal phase and absolute magnitude. A global assessment of the dominant features and spatial distribution of soil moisture variability is then provided. Results show that, despite still being a relatively short data set, SMOS data provides coherent and reliable variability patterns at both seasonal and interannual scales. Subseasonal components are characterized as white noise. The observed linear trends, based upon one strong El Niño event in 2016, are consistent with the known El Niño Southern Oscillation (ENSO) teleconnections. This work provides new insight into recent changes in surface soil moisture and can help further our understanding of the terrestrial branch of the water cycle and of global patterns of climate anomalies. Also, it is an important support to multi-decadal soil moisture observational data records, hydrological studies and land data assimilation projects using remotely sensed observations.

Keywords: SMOS; soil moisture; climatology; trends; signal decomposition

---

1. Introduction

During the last decade, the interest in low-frequency microwave remote sensing and technological advances in instrumentation and space technology have resulted in a series of new mission concepts to measure key components of the water cycle. Soil moisture is one of these key components, controlling the partition of energy at the surface and the interactions between the land surface and the atmosphere at varying temporal and spatial scales [1–4]. Theoretical and experimental evidence supports the idea that L-band (1 to 2 GHz) microwave radiometry is the optimal technology for measuring global surface soil moisture (SM) on an operational basis. The emission of thermal microwave radiation from soils is strongly dependent on soil moisture content. L-band measurements are insensitive to cloud liquid water and, compared to higher microwave frequencies, they are more sensitive to deeper soil moisture layers (up to 5 cm) and penetrate through denser layers of vegetation canopy [5,6]. The latter is especially important for improvements in climate, numerical weather prediction, water and energy cycle science. As a result, the first two satellite missions specifically designed to measuring SM have an L-band radiometer on-board: ESA launched the Soil Moisture and Ocean Salinity (SMOS) mission in November 2009 [7] and NASA launched the Soil Moisture Active-Passive (SMAP) satellite in early 2015 [8]. SMOS unique payload is an L-band synthetic aperture radiometer with multi-angular and full-polarimetric capabilities that provides synoptic views of the Earth's global SM with a spatial resolution of ~40 km and a 3-day revisit. SMAP has a real aperture L-band radiometer and an L-band radar to enhance the spatial resolution of the estimates from 36 km (radiometer only) to 9 km (radar-radiometer). However, operations of active-passive products ceased abruptly with the failure of the SMAP radar after about ten weeks of operations. Nonetheless, the radiometer is continuing to make measurements and four annual cycles of measurements are about to be completed.

Also during the last decade, satellite sensors with long technological heritage operating in the low-frequency microwave spectrum, that were initially devoted to atmospheric and/or oceanic sensing, have proved suitable for SM retrieval. As a result, several SM datasets from active and passive microwave sensors at C-band (6 GHz) and X-band (10 GHz) partially covering nearly the last 4 decades have been published and shared openly with the international community. Although these sensors are only sensitive to the top 1 cm of soil and have a larger attenuation in presence of vegetation, they can complement recent L-band missions and allow for a multi-decadal soil moisture observational data record. Spaceborne SM data sets from low-frequency microwaves have been widely validated under different biomes and climate conditions by comparison with ground-based observations (e.g., [9–16] and outputs of land surface models ([17–20]). Relevant for this work, Polcher et al. [20] showed that the rainfall driven structures of SM captured by the ORCHIDEE land surface model and SMOS are compatible, and comprehensive validations of SMOS retrievals have been undertaken showing good agreement with other sensors and consistent results over all surfaces, from very dry (Arizona, Sahel) to wet (tropical rain forests) [15].

Soil moisture was recognized by the Global Climate Observing System (GCOS) to be an Essential Climate Variable (ECV) in 2010. This underscores the potential of SM data sets to support the work of the United Nations Framework Convention on Climate Change (UNFCCC) and the Intergovernmental Panel on Climate Change (IPCC). In this context, the ESA's soil moisture Climate Change Initiative (CCI) is one of the first initiatives to merge the different microwave products available into a single soil moisture climate data record (http://www.esa-soilmoisture-cci.org/). One of the key steps in building a multi-decadal soil moisture data record is that, since different products display different ranges of soil moisture values, data have to be harmonized first using a common climatology. In its current version, the ESA CCI SM product uses the climatology provided by the Global Land Data Assimilation System (GLDAS-Noah) as a reference to scale the individual products. Despite some limitations, the ESA CCI SM product is nowadays the most complete and consistent long-term soil moisture data record available, covering (almost) the 40-year period from 1978 to June 2018 [21]. Recent research has been focused into incorporating SMOS to the climate data record [22]. However, the impact of using a model's climatology in absolute retrievals remains unclear, and the climate community has repeatedly argued for the need for a satellite-based SM record to serve as a reference for verifying land surface model performance and trends.

This study presents an SMOS-based L-band climatology that could potentially serve as a reference for the readily available microwave-based soil moisture data sets (from X-, C- and L-bands) into a long-term climate data record exclusively based on observational data sets. Ideally, an L-band climatology should be built from both SMOS and SMAP data. However, since a combined product is not yet available, and SMAP only covers a short observation period, this paper is focused on an L-band climatology solely based on SMOS observations. We show that this observation-driven climatology provides a close representation of the dominant features of temporal variability in the Earth's SM for the period 2010–2016, and allows identifying areas subjected to seasonal, subseasonal and long-term variability. Previous research showed that, despite being a short data set, SMOS provides coherent and reliable SM variability patterns at both seasonal and interannual scales [23,24]. In this work, the Seasonal Trend decomposition using Loess (STL) procedure [25] is implemented and tailored to the first six annual cycles of SMOS data to decompose the temporal variability of the signal. Our analysis quantifies how much SM variation is due to long-term influences, how much is due to seasonal cycle and how much is dominated by subseasonal short-term influences. This knowledge is critical for understanding how well climate data and land surface models compare with the remotely sensed variable. It also allows distinguishing between linear trends and interannual variability, which could be later related to main phenomena of global weather alterations over land (e.g., El Niño Southern Oscillation (ENSO)).

This paper is organized as follows. The SM dataset and the methodology followed to build the SM climatology and to provide a global assessment of SM variability are described in Section 2. Main results are shown in Section 3. The distribution of SM variance among temporal components is first analyzed at selected target sites representative of terrestrial biomes with distinct vegetation type and seasonality. The SMOS temporal series at these sites are compared to GLDAS-Noah and ERA5 modeled SM and to ground-based estimates available at one of the sites (REMEDHUS network, in Spain) to identify consistencies and potential shortcomings of building a climatology with observational data alone. Subsequently, the main features of SM temporal variability are analyzed at the global scale. Conclusions and perspectives from this work are given in Section 4.

2. Materials and Methods

2.1. SMOS Soil Moisture

This study uses six years of global SMOS-BEC L3 v.2 daily SM retrievals, starting on 1 June 2010. This product is provided in the 25 km EASE2 equal-area grid and is obtained from the ESA L2 v.620 data, after discarding SM retrievals being potentially affected by radio frequency interferences (RFI flag) or with a Data Quality Index (DQX) greater than 0.07 $m^3 \cdot m^{-3}$. The DQX is an estimate of the error in the SM retrieval and the brightness temperature measurement accuracy. Products are provided separately from ascending and descending orbits. Data as well as further processing details are available at http://bec.icm.csic.es/. In this work, daily products from both ascending and descending orbits are combined to maximize coverage (further pre-processing is detailed in Section 2.5).

It is relevant to note the impact that RFI has on L-band satellite measurements. Although L-band is an internationally protected band for radio astronomy, it was soon clear after the SMOS launch that anthropogenic RFI exceeded expected levels in many regions worldwide [26]. The situation has now improved, but the presence of RFI still masks SMOS observations, severely limiting its coverage in some regions [27]. The impact of RFI is considerably reduced in SMAP, since a number of hardware and software measures were implemented to detect and where possible mitigate its effects [28]. In this regard, recent studies combining the brightness temperatures of the two missions are particularly promising [29].

2.2. Selection of Target Sites

A set of representative target sites were selected to characterize and illustrate the SMOS SM climatology under contrasting vegetation types and climatic conditions. The analysis of the SMOS STL decomposition at the target sites allowed us to analyze major features of the STL decomposition procedure and adapt its parameterization to SMOS measurements. This is described in detail in

Section 2.6. In addition, the target sites were used as a basis for the inter-comparison of SMOS with GLDAS-Noah and ERA5 SM time series.

The 11 terrestrial Transcom regions were used as an initial segmentation of global continental land from where to select representative sites. Transcom regions are based on a 1-degree land cover map and are delimited by climate zones [30]. The advantage of using this classification is that it encloses climates with similar vegetation seasonality into a limited number of classes. After careful analysis of the SMOS SM time series in each region, eight target sites were chosen covering Transcom regions North America Boreal, North America Temperate, South America Tropical, South America Temperate, Europe, Northern Africa, Southern Africa, and Australia. No target sites were chosen from Transcom regions Eurasia Boreal, Eurasia Temperate and Asia Tropical, since availability of SMOS SM data in these regions is very limited due to combined effects of RFI, the presence of snow and high topography. A global map showing the selected target sites and the terrestrial Transcom regions is shown in Figure 1. The specific location of each target site is provided in Table 1. One of the SMOS and SMAP Core Cal/Val sites (REMEDHUS, site E) was chosen as a target site for further analysis *vs* ground-based measurements.

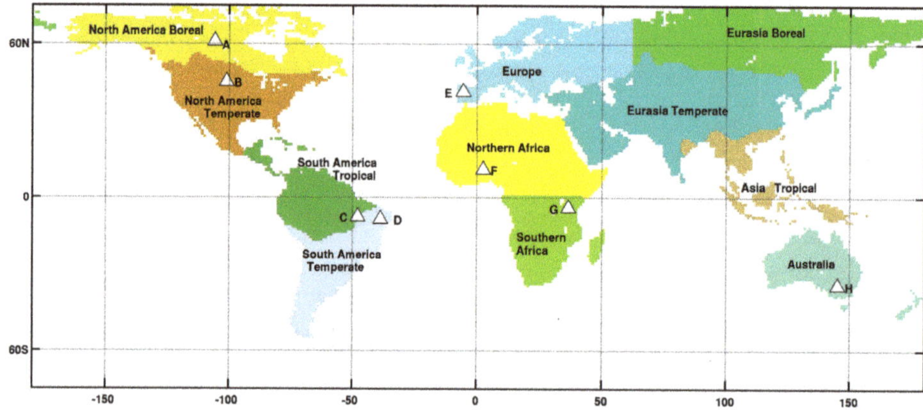

Figure 1. Geographical position of the eight locations (triangles, letters A to H) selected to illustrate the main features of SMOS climatology for a variety of vegetation and climatic conditions. Colors represent the 11 terrestrial Transcom regions [30].

Table 1. Target sites: name and location.

Site	Latitude	Longitude
A – North America Boreal	60.93°N	105.68°W
B – North America Temperate	45.27°N	101.00°W
C – South America Tropical	7.54°S	47.90°W
D – South America Temperate	8.52°S	38.52°W
E – Europe - REMEDHUS	41.3°N	5.4°W
F – Northern Africa	10.90°N	2.60°E
G – Southern Africa	4.20°S	36.70°E
H – Australia	34.56°S	145.25°E

To establish the actual spatial representativeness of the target sites, a temporal correlation analysis was performed at each selected location. Figure 2 shows, for each target site, the correlation map of its SMOS SM time series with the SMOS SM time series of all the pixels on the globe. It can be seen that the highest correlation is located in the neighborhoods of the chosen pixel (within the Transcom

Region), which indicates the target site is representative of their surrounding area, but, as expected, not of the whole region.

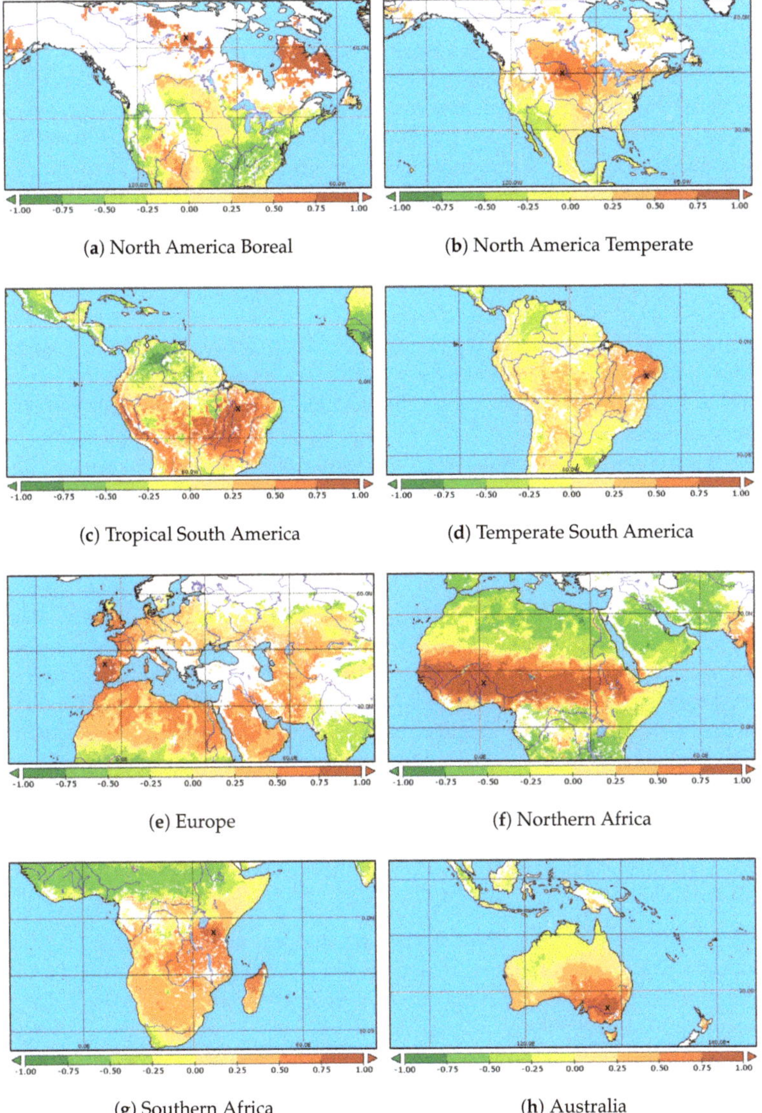

(**a**) North America Boreal

(**b**) North America Temperate

(**c**) Tropical South America

(**d**) Temperate South America

(**e**) Europe

(**f**) Northern Africa

(**g**) Southern Africa

(**h**) Australia

Figure 2. Correlation maps showing the representativeness of the SMOS time series for the study period at the selected target sites. The exact location of each target site (see Table 1) is marked with a black cross.

2.3. Modeled Soil Moisture

2.3.1. GLDAS-Noah

Global Land Data Assimilation System (GLDAS) Noah model v.1 top 10 cm soil moisture estimates were used in this study [31]. GLDAS-Noah land surface state and flux products are provided by the Hydrology Data and Information Services Center (HDISC), a component of the NASA Goddard Earth Sciences Data and Information Services Center (GES DISC). GLDAS-Noah is available with a 3-h temporal interval and 0.25° spatial resolution, covering dates from 2000 to the present. Data was extracted for the eight selected target sites (see Figure 1) and the 6-year study period. The original unit of GLDAS-Noah SM is $kg \cdot m^{-2}$, which was converted to volumetric SM ($m^3 \cdot m^{-3}$) for this study by considering a top soil layer depth (10 cm) and assuming the density of water in the soil is $1000\ kg \cdot m^{-3}$. After averaging for daily values, 1-day time series were used to construct 18-day temporal average fields every 5 days (the rationale for this filtering is explained in Section 2.5).

2.3.2. ERA5

ERA5 is the most modern reanalysis produced by ECMWF, using a recent version of the ECMWF Integrated Forecasting System. The data cover the Earth on a 30 km grid and resolve the atmosphere using 137 levels from the surface up to a height of 80 km. It uses a vast number of observations, including several reprocessed datasets [32]. Although ERA5 is currently in production, the data for the period of this study is already available; top 7 cm soil moisture fields have been extracted for the selected target sites and the 6-year study period. After averaging 00 UTC and 12 UTC to obtain the daily values, 1-day time series were used to construct 18-day temporal average fields every 5 days.

2.4. Ground-Based Soil Moisture

In-situ soil moisture from REMEDHUS (target site E, see Figure 1, Table 1) was obtained through the International Soil Moisture Network [33]. The REMEDHUS soil moisture monitoring network is composed of 23 automated stations deployed within an area of $1300\ km^2$ in a semi-arid sector of the Duero basin in Spain. Each station is equipped with capacitance probes providing hourly measurements over the top soil 5 cm with a reported accuracy of $0.003\ m^3 \cdot m^{-3}$. This network has been continuously operating and quality-controlled since 2005 and is therefore ideally suited for validation of multi-year satellite time series. Further details on the network can be found in [12]. For the period of study, data from 17 REMEDHUS stations were available. Data from these stations was first daily averaged and then used to construct 18-day temporal average fields every 5 days (see Section 2.5).

2.5. Temporal Averaging and Filtering of SMOS Data

SMOS L3 daily SM maps need to be pre-processed to ensure smooth spatio-temporal transitions and representative soil moisture states in the climatology. To this aim, outliers were first detected and screened out from the 1-day maps. This was done by comparing the SM retrieved at each pixel with the retrievals embedded within the one-degree box centered at the considered pixel. Any pixel value failing to pass the Tukey outlier test [34] were removed from the dataset. In the next step, the filtered 1-day SM maps were used to construct 18-day temporal average fields every five days. Although the daily SMOS SM maps are useful for the monitoring and evaluation of episodic events, we opted for an average period of 18 days to increase the spatial coverage of the resulting maps, reduce random retrieval errors and filter out high-frequency modes that can be considered as noise when calculating climatological averages. The temporal window of 18-day is the closest to SMOS repeat cycle and the one generally chosen to avoid orbital artifacts (e.g., [35]).

As shown in detail by Robock et al. [36], there are two distinct scales that determine the variations of SM in time and space. The small scale, referred to as hydrological or land surface related scale, is on the order of days and tens of meters. Soil moisture can vary on this scale due to variations of soil properties, vegetation, and topography or drainage patterns. This small-scale variability is intertwined with a much larger scale on the order of weeks-to-months and tens-to-hundreds of kilometers that is mainly due to atmospheric forcing. Microwave satellite measurements integrate over relative large-scale areas on the order of ∼25 km with a typical revisit of 3-days. At these spatio-temporal scales, the short-term (up to 3 days) and small-scale (tens of meters) land surface component of soil moisture variability appears as random (white) noise in comparison with the long-term (about 1–4 months) and large-scale (about 400–800 km) signal related to atmospheric forcing.

Examples of the time series of the original 1-day SM, together with the 18-day average, are shown in Figure 3 for the 8 target sites, which cover a variety of vegetation seasonality and climatic conditions (Figure 1, Table 1). Dots represent the daily values and the solid line corresponds to the 18-day average every five days. It can be seen that the variability of the daily signal is captured in the filtered time series, where the six SM annual cycles can be clearly identified. Difference in magnitude and extent of rainy seasons among years are more clearly distinguished in the filtered series (e.g., sites E and H). Also, notice the opposite timing of wet and dry seasons in Northern and Southern Africa (sites F and G), which reflect the displacement of the Inter-Tropical Convergence Zone (ITZC). Site D exhibits limited temporal variability in both the original and the filtered series.

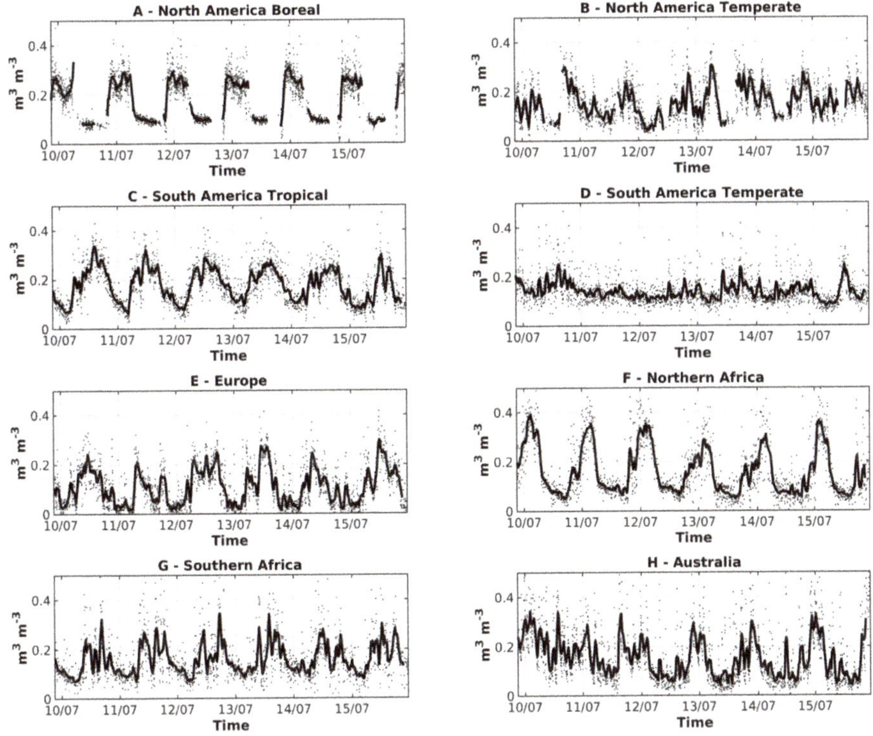

Figure 3. Time series of the 1-day SMOS-based soil moisture retrievals (dots) and overlapped the 18-day average every five days (solid line) at the target locations.

Notice that not all the 25-km pixels in this 18-day averaged SM fields are continuously observed. SMOS's wide swath (1000 km) and polar orbit allow for a 3-day global revisit period. However, there is an important amount of missing data due to the presence of radio frequency interferences masking L-band measurements, particularly in South-East Asia [27]. In addition, no retrievals are attempted in areas of high topography (e.g., Himalaya) and in soils covered by snow. The latter strongly reduces the data availability in high latitudes and especially during the fall and winter seasons. The global map of Figure 4 shows the SMOS temporal coverage for the study period, with 1 representing a 100% coverage of the filtered and temporally-averaged SMOS signal. In this study, only the pixels with a minimum of 80% temporal coverage were used in order to ensure representativeness and robustness of the signal decomposition and of the results presented. As we will show later in the study, this threshold does not exclude areas with limited data in winter due to snow. The impact of these "intermittent" data gaps in the STL decomposition is discussed in Sections 3 and 4.

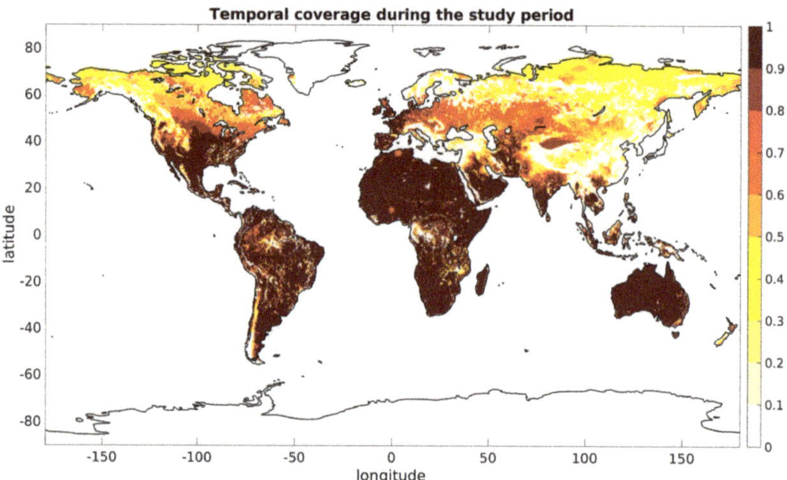

Figure 4. Global map of SMOS temporal coverage during the study period.

2.6. Signal Decomposition

The STL geostatistical procedure is used in this work to decompose the SMOS signal into its temporal components and build an observation-based SM climatology. The STL technique was originally introduced by Cleveland et al. [25], and adapted by Humphrey et al. [37] to evaluate the seasonal cycle of unevenly spaced time series using locally weighted regression, or Loess. This technique has already been used in several research studies to extract the seasonal and interannual components of GRACE time series [37–39]. In this work, the STL procedure has been used to decompose the filtered and temporally-averaged SMOS SM signal (SM_{tot}) as the sum of a seasonal component (SM_{seas}), a low-frequency component ($SM_{long-term}$) and remaining high-frequency residuals (SM_{res}):

$$SM_{tot} = SM_{long-term} + SM_{seas} + SM_{res} \tag{1}$$

The low-frequency component $SM_{long-term}$ contains only periodicities larger than a season and is further decomposed into linear trends (SM_{trend}) and the anomalies with respect to this linear trend or interannual variability ($SM_{interannual}$). The high-frequency residual is expected to be both a real signal representing subseasonal variability and noise present in SMOS data. For a detailed description of the method, refer to [25,37]. In short, STL is a double recursive approach: an inner iteration cycle is used to recover the seasonal cycle from the low-frequency component using Loess; the outer iteration cycle is used to recalculate the Loess weights and to separate the signal between low- and high-frequency components again using Loess. At the end of the process, the low-frequency signal remaining after removal of the seasonal cycle is decomposed as a trend and an interannual signal. The residual component is equal to the resulting high-frequency component.

The application of the STL algorithm requires the specification of six smoothing filter parameters that need to be optimized to minimize spectral leakage between high- and low-frequency components and control the possible influence of outliers in the time series. These are: (1) the length of the seasonal cycle; (2) the degree of the weighted polynomial regression; (3) the number of cycles of the inner loop (used to estimate the trend, the seasonal and the interannual components); (4) the number of cycles of the outer loop (used to estimate the residual signal, i.e., the subseasonal component); (5) the maximum time lag for the seasonal component, and (6) the maximum time lag for the long-term component. Here, the seasonal cycle is taken as exactly 365 days, a multiple of the 5-day map interval (which was constructed disregarding the occurrence of the leap day, 29 February 2012). Following Humphrey et al. [37], the number of inner and outer loops is set to 2 and 3, respectively, a quadratic fit is used for the seasonal cycle and a linear fit for the long-term components. The optimal values for the maximum time lag of the seasonal and long-term components have been selected after a comprehensive analysis carried out at the 8 target sites that is reported hereafter. The role of the maximum time lag for the seasonal component λ_{per} is illustrated in Figure 5 for site A. It is shown that the smoothness of the seasonal cycle increases with the value of the maximum seasonal time-lag period. Although large values are recommended when using noisy data, they may hide key details of the seasonal cycle. A maximum seasonal time lag of 90 days fails to reconstruct the double-maximum SM in June and September. Instead, it results in an erroneous maximum during July. On the other hand, too short maximum seasonal time lags tend to over fit the noisy data. A similar effect was observed for sites C, E and G, whereas there was not a strong impact in B, D and F (not shown). These results indicate that a maximum seasonal time lag of 45 days is a reasonable compromise for SMOS SM data and it is the value used for the rest of this study.

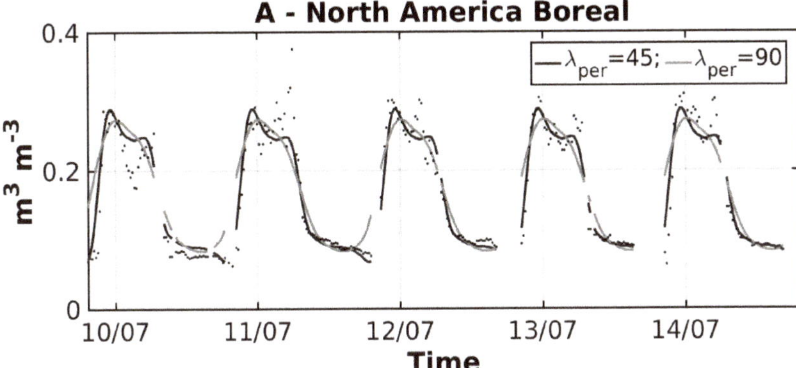

Figure 5. Analysis of the extracted soil moisture seasonal component at location A. Two seasonal components are shown depending on the value of the maximum seasonal time lag λ_{per}. Black and gray lines corresponds the values of 45 days and 90 days, respectively.

The role of the maximum time lag for the long-term or low-frequency component is to separate the seasonal anomalies between a low-frequency (that will be further decomposed as a trend and an interannual component) and a residual high-frequency component. Figure 6 shows the spectra of the interannual and subseasonal SM components for a long-term maximum time lag of 0.20×365 (top) and of 0.10×365 (bottom). It is observed that the larger lag leads to a subseasonal component having maximums for cycles beyond 90 days, whereas the shorter lag leads to a subseasonal variability closer to white noise. Here, the lag of 0.10×365 has been selected since it allows a more appropriate decomposition of long-term variability and residual components (which integrates both subseasonal variability and instrumental noise). Results shown in Figure 6 are for site A, results obtained for the other sites are consistent (not shown).

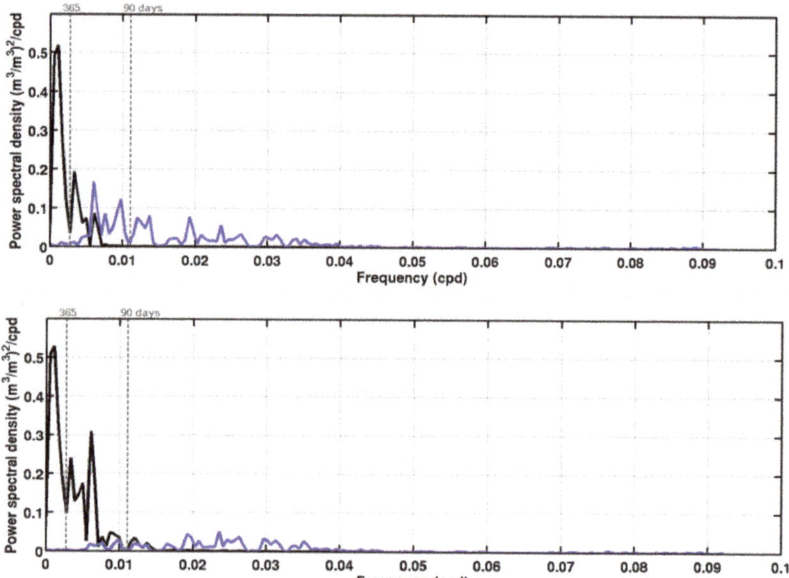

Figure 6. Spectra $(m^3 \cdot m^{-3})^2 \cdot cpd^{-1}$ (*cpd* = cycles per day) versus frequency (cycles per day) for soil moisture interannual (black line) and residual (blue line) temporal components at location A. The seasonal maximum time-lag parameter is 45 days. Top and bottom plots are obtained using a long-term maximum time lag equal to 0.20×365 and 0.10×365 days, respectively. The time lag of 0.10×365 is selected to ensure the residual component contains only subseasonal variability and instrumental noise i.e., it has no maximums for cycles beyond 90 days.

2.7. Analyses at Target Sites

The STL procedure was applied to the SMOS time series at the target sites. The obtained distribution of SMOS SM variance among its temporal components was analyzed in detail. Subsequently, SMOS, GLDAS-Noah and ERA5 SM time series at these locations were inter-compared to identify consistencies and potential shortcomings of building a climatology with satellite data alone. A comparison of the three data sets to ground-based SM was performed for one of the target sites (REMEDHUS, in Europe). Statistical scores of the comparisons are provided.

2.8. Analyses at the Global Scale

Global maps of the long-term average and standard deviation of SM were computed on a pixel basis over the filtered and temporally-averaged SMOS signal. The STL decomposition was subsequently applied globally to each individual pixel with a temporal coverage greater than 80% for the study period (see Figure 4). The relative magnitude of the extracted long-term, seasonal and residual components with respect to the total variance was computed to assess the dominant modes of temporal variability in global SM during the study period. The magnitude of the linear trends within the long-term component were also evaluated per pixel at the global scale.

3. Results

3.1. Soil Moisture Temporal Decomposition at Target Sites

Figure 7 illustrates the STL decomposition of the SMOS signal into the different subcomponents at the 8 target regions. In general, the temporal series show an almost negligible linear trend, with Australia (H) and South America Temperate (D) presenting a slight trend towards drier conditions and North America Temperate (B) and Southern Africa (G) revealing a slight trend towards wetter conditions. This will be examined in detail later in this section. A clear seasonal signal is extracted for most of the sites, except for South America Temperate (D), which exhibits limited temporal variability in the original series (see also Figure 3). The residual component shows a temporal behaviour similar to that of white noise in all the sites. It is especially high in Australia (H) and, to a lesser extent, in Southern Africa (G). The analysis for Europe (E) reflects a regular seasonal cycle, with exceptionally dry conditions in winter of years 2011–2012 and 2014–2015, and exceptionally wet conditions during short periods of spring 2012, winter 2013–2014 and end of 2015, as suggested by the analysis of the interannual component. The dry anomaly in winter 2011–2012 has also been reported in previous studies devoted to detection of agricultural drought [40,41]. The interannual component presents also prominent fluctuations in North America Temperate (B), Southern Africa (G) and Australia (H). In particular, the dry anomaly observed in North America Temperate (B) in 2012 reflects the strong summer drought suffered in the contiguous US due to unusual high temperatures in spring and summer 2012 combined with record low rainfall [42]. The analysis of North America Boreal (A) exemplifies the non-negligible impact of temporal gaps in the decomposition procedure. The wet peak measured by SMOS in October 2010 and followed by a short no-data is assigned mainly to the subseasonal component. The same applies to the wet peak measured in October 2013, which is not affected by the absence of data. In contrast, there is a no-data corresponding to the winter months of the six annual cycles that are assigned to the interannual component. This reveals a limitation of the method to correctly differentiate short-term from long-term variability in the extremes of temporal gaps. This same effect was observed in initial tests conducted at target sites within Eurasia Boreal, Eurasia Temperate and Asia Tropical (see Figure 1), which were discarded from the analysis. Although we imposed an 80% minimum temporal coverage, the impact of no data in the time series needs to be taken with some caution when interpreting overall results.

3.2. Comparison of SMOS, GLDAS-Noah, ERA5 and In-Situ at Target Sites

Time series of SMOS-based SM together with top 10 cm soil moisture GLDAS-Noah and top 7 cm soil moisture analysis from ERA5 are shown in Figure 8. It can be observed that in regions C to H, SMOS, GLDAS and ERA5 soil moisture time series are comparable in terms of temporal phase (Pearson correlation of 0.7–0.9). In regions A and B, however, the SMOS observations are uncorrelated with the uppermost soil moisture modeled estimates. This is mostly due to the presence of snow, which masks satellite measurements and can also affect the uncertainty of model predictions. Further research is needed in these areas of mutual disagreement to identify potential deficiencies in the satellite and/or in the model estimates. Similar results were reported by a previous study comparing SMOS with MERRA reanalysis [43]. Note that the correlation metric benefits from the seasonal cycle, which for the period of this study is included. Correlation of time series at daily time scales lead to similar results (not shown).

There are notable differences in the absolute values of satellite and modeled data (e.g., sites E and H), which remarks the need to use bias correction procedures when assimilating soil moisture satellite observations in land surface simulations [44]. The assimilation of satellite data in land surface models has proven to be a powerful technique to leverage from the two sources of information. A recent study has shown that assimilating SMOS SM data into the Noah Land Information System after bias correction significantly increased the anomaly correlation of modeled top soil moisture estimates with station measurements [45]. Also, Pinnington et al. [46] showed that assimilation of (bias-corrected) satellite rainfall and SM data had the greatest impact on model estimates during the seasonal wetting-up and drying-down of the soil, respectively. While L-band satellites have been designed for measuring soil moisture, land surface models have been designed for a much wider purpose, including ecological, hydrological or climate applications. Modeled soil moisture is generally highly sensitive to the meteorological forcing data used and the land surface model encoded physics [47,48]. This makes comparison of absolute values of observed and modeled SM very challenging, and therefore studies generally focus on the comparison of SM temporal anomalies (e.g., [19,20]), or even in the comparison of observed and modeled brightness temperatures directly (e.g., [49]). The differences found between modeled and observed SM and also among different models support the idea proposed here of leveraging from the natural soil moisture variability captured by satellite observations as a reference for harmonizing SM climate data records so that they become model-independent.

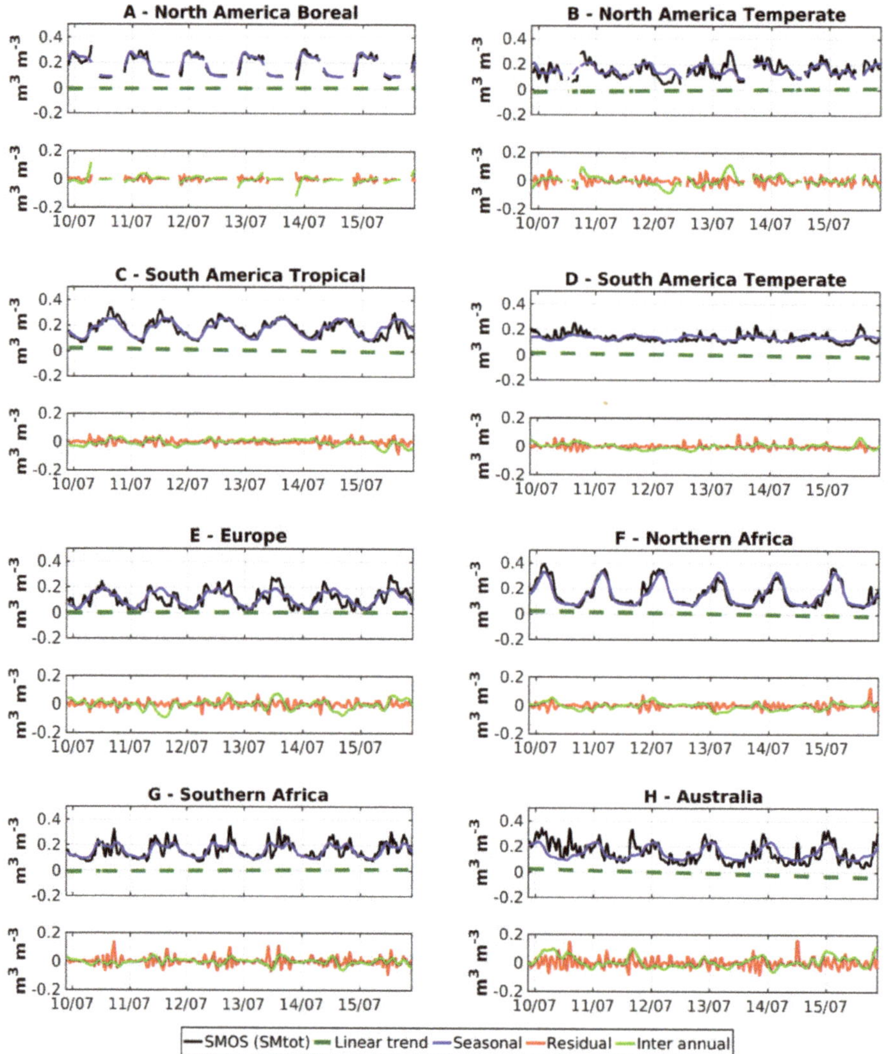

Figure 7. STL decomposition of the time series at the eight target locations (from Figure 1 and Table 1). The smoothed SMOS signal (in black) is decomposed in its seasonal cycle (in blue), linear trend (dashed green line), interannual variability (green), and residual component (red).

Ground-based soil moisture data from REMEDHUS (average of 17 in-situ stations) are also included in Europe time series (Figure 8E). It can be seen that both satellite and models generally capture the temporal dynamics measured by the in-situ sensors. A statistical analysis has been undertaken following the recommended performance metrics in [50]. Results are shown in Table 2. The temporal correlation (R) and the unbiased Root-Mean-Squared difference (ubRMSD) between the in-situ, remotely sensed and modeled SM are satisfactory (R > 0.85, ubRMSD < 0.003). As in the previous analyses, it should be noted that the R metric benefits from the seasonal cycle, which for the period of this study is included. In terms of accuracy, SMOS shows a low dry bias of 0.02 while the two models present a large wet bias of about 0.11 m^3·m^{-3}. As expected, the models capture reasonably well the temporal dynamics but not the absolute magnitude [47]. The statistical scores when calculated at daily time scales lead to similar results (not shown).

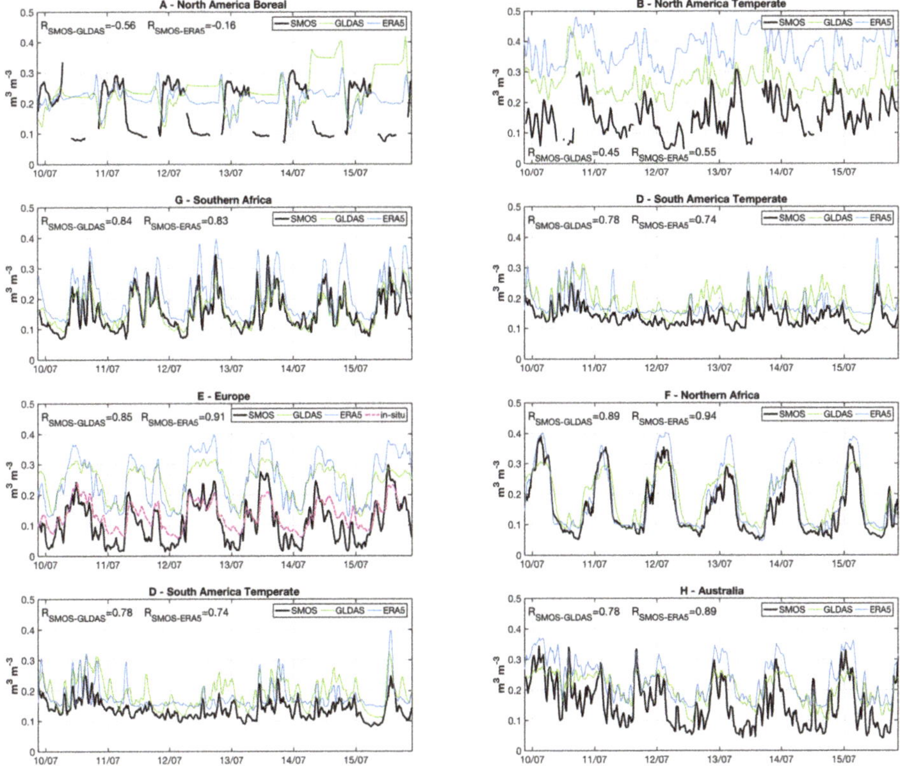

Figure 8. Time series of SMOS (in black), top 10 cm GLDAS-Noah (in green) and top 7 cm ERA5 (in blue) soil moisture at the eight target locations (from Figure 1 and Table 1). Target location E also includes time series of collocated ground-based soil moisture (dashed line in magenta, average of 17 REMEDHUS in-situ stations).

Table 2. Statistical scores from comparison to in-situ at REMEDHUS (time series on Figure 8E): bias, unbiased Root-Mean-Squared difference (ubRMSD) and Pearson correlation (R).

	Bias (m^3·m^{-3})	ubRMSD (m^3·m^{-3})	R
SMOS	−0.022	0.003	0.88
GLDAS-Noah	0.110	0.002	0.85
ERA5	0.116	0.003	0.91

3.3. Temporal Mean and Variance of SMOS Soil Moisture Retrievals at the Global Scale

The temporal average global map of SMOS SM surface volumetric soil water content for the study period (pixel average of SM_{tot} in Equation (1)) is shown in Figure 9. The average conditions observed during the study period exhibit a mode of 0.1 $m^3 \cdot m^{-3}$ (inset figure). It shows the expected spatial patterns of SM, from the dry arid regions to the wet forested areas. Note that the average conditions shown in areas with limited availability of data may not be representative (e.g., northern latitudes and tropical Africa, see coverage map on Figure 4).

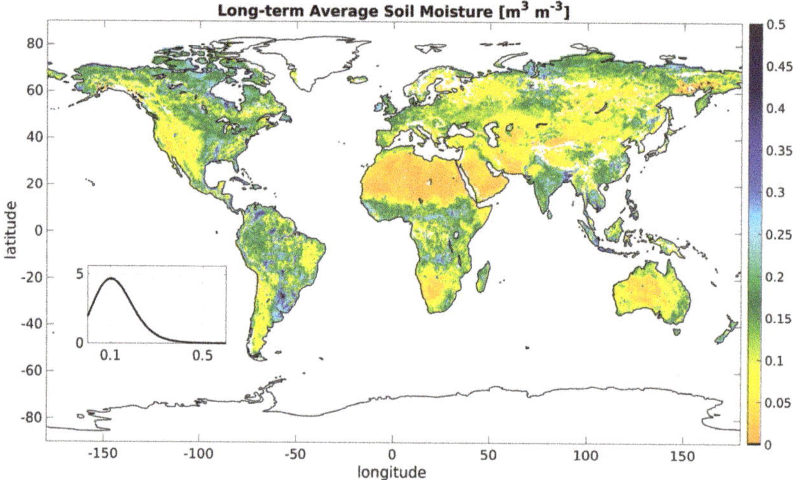

Figure 9. Global distribution of time-average soil moisture based on six years of SMOS observations, starting in 1 June 2010. The inset figure is an estimate of its marginal probability density.

The global map of SMOS SM standard deviation for the study period (standard deviation of SM_{tot} in Equation (1)) is shown in Figure 10. As expected, areas with higher variability are concentrated in the tropics, where there is a strong seasonality dominated by the position of the ITCZ [51]. Other areas with strong variability include India and South-East Asia, strongly affected by Monsoon rainfall, and some regions in the Southern Hemisphere like eastern Australia and South America that could be related to ENSO. In particular, the high SM variability observed in the so-called Southeastern South America (SESA) region can be explained by the intense summer precipitation over this region [52]. This pattern has also been observed with satellite SM products derived from higher microwave frequencies and climate models [53], and responds to strong land-atmosphere interactions in the region. The variability observed in northern latitudes is probably due to imperfect detection of ice/snow and poor temporal coverage (see Figure 4).

3.4. Analysis of the Dominant Features of Global Soil Moisture Variability

In this study we decomposed the total variance of the soil moisture signal (Figure 10) by breaking it down into: (1) a seasonal component, (2) a long-term component, and (3) a high-frequency residual or subseasonal component. The relative magnitude of each of these three components is shown as an RGB triplet in Figure 11. It reveals the seasonal cycle is dominant in many tropical regions such as Brazil, Central Africa, India and Northern Australia. Wet tropical forests such as the Amazon and Congo exhibit high spatial heterogeneity in its dominant components, probably due to the combined effects of human activities and climate variability. In contrast, the soil moisture variability in dry tropical forests, including the savannahs south of Central Africa and several regions in Southwestern Brazil, is clearly dominated by seasons. Though dry tropical forests may receive several hundred centimeters of rain per year, they have long dry seasons which last several months and vary with geographic location. Forests in dry tropics show great diversity of phenological patterns and large interannual variation, with predominance of deciduous tree species [54]. Our analysis indicates that L-band is capturing the emissivity from the soil and the seasonal drought through the forested canopy. Two regimes can be identified in Europe: whereas western countries are governed by seasonal variability, long-term variability predominates in the eastern countries. In Australia, seasonality dominates the north and the south-east, the eastern region is dominated by long-term variability and the western by subseasonal. The Indo-Australian archipelago is also dominated by long-term variability. It is interesting to note that subseasonal variability is predominant in regions where the SMOS signal has already a relatively low variance (Figure 10) and is most likely influenced by noise such as the Sahara desert, the Arabian Peninsula and Western Australia, i.e., where the noise is at least of the same magnitude that the annual variance. Indeed, despite having carefully filtered the data, some of the SMOS retrievals in Asia and Europe may still be affected by undetected RFI contamination [27]. Also, results may be affected by the intrinsic uncertainty of microwave satellite SM retrievals, which is higher in presence of dense vegetation canopies, heterogeneous landscapes, and high topography [6,11,55].

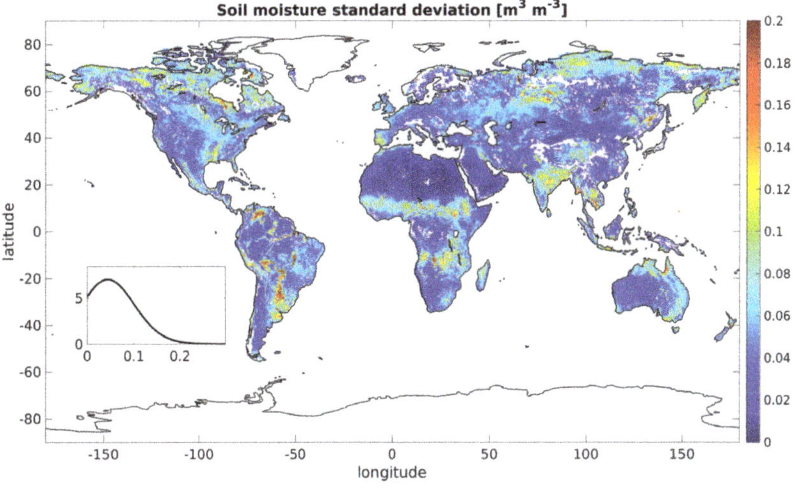

Figure 10. Global map of SMOS soil moisture standard deviation for the six-year period of this study. The inset figure is an estimate of its marginal probability density.

The linear trends within the long-term variability component are further examined in Figure 12. The linear trends observed in the SMOS signal illustrate the ENSO conditions during the six-year study period. During the first five annual cycles -from 2010 to 2015- the equatorial Pacific Ocean was mostly in a cold phase (La Niña); however, a warm event (El Niño) occurred during the last cycle (i.e., in 2016). This explains why SMOS-derived linear trends reproduce known ENSO teleconnection spatial patterns. During El Niño, limitations in terrestrial moisture supply result in vegetation water stress and reduced evaporation in eastern and central Australia, southern Africa and Eastern South America (areas in red). The contrary situation is experienced in Argentina, Tanzania and southeastern US (areas in blue), where there is a regime of above-average rainfall. This result is in line with a previous study that showed that multi-decadal (1980–2011) variability in SM and terrestrial evaporation was dominated by ENSO dynamics [56]. The linear trends in Figure 12 also provide evidence of strong dry/wet patterns in regions which have not been previously related to ENSO precipitation patterns (e.g., western Europe, Northern Africa, California). Still, results should be taken with caution, and further analysis of the long-term variability is needed to identify new climate patterns at short and median temporal scales. Investigating the relation of ENSO to global soil moisture variability and the existence of potential new teleconnection patterns is recommended for future research.

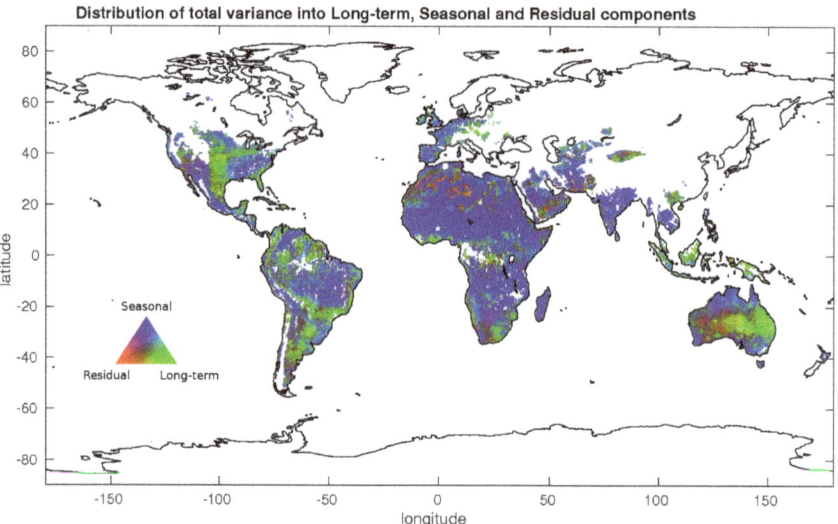

Figure 11. Distribution of the total SMOS variance among the long-term (green), seasonal (blue) and subseasonal (red) components, expressed in per cent of the total variance, indicating the dominant models of temporal variability in soil moisture for different regions. Each vertex in the triangle corresponds to 100%. Empty areas correspond to locations with less than 80% SMOS temporal coverage that have been masked out.

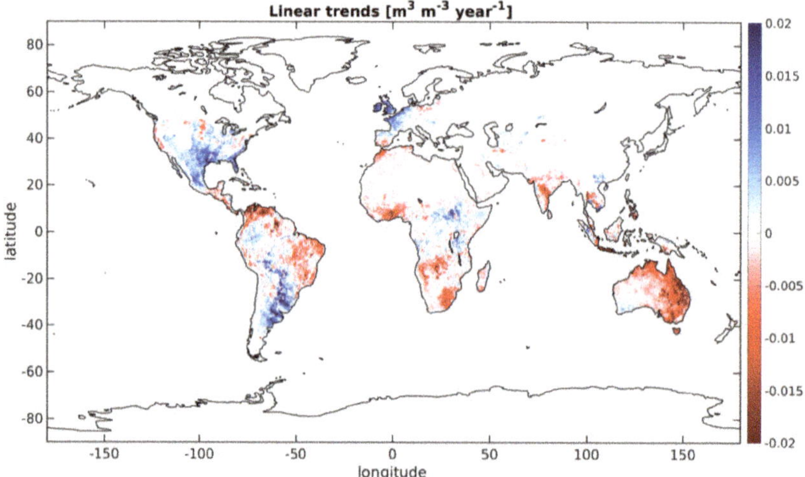

Figure 12. Magnitude of linear trends in the SMOS signal (expressed in $m^3 \cdot m^{-3} \cdot year^{-1}$). The trends reflect that during the SMOS period (2010–2016) the equatorial Pacific Ocean was mostly in a cold phase (La Niña) with a transition to a warm phase (El Niño) in 2015–2016.

4. Discussion and Final Remarks

The two first space missions dedicated to measuring the Earth's surface soil moisture have been launched in the last decade: SMOS in 2009 and SMAP in 2015. They are providing L-band measurements that, combined with C- and X-band measurements available since the 1980s, allow generating, for the first time, observational soil moisture climate data records combining the three microwave frequency bands. However, the synergies of microwave measurements across different frequencies and the potential of the L-band data record to serve as a common reference to harmonize the long-term data record have not yet been fully exploited. This study presents first evidence that the relatively short data record of available SMOS L-band observations allows characterizing the main modes and spatial distribution of the Earth's surface soil moisture variability. This information could serve as a reference to harmonize and construct a model-independent SM climate data record. In its current version, the ESA CCI soil moisture product uses a climatology obtained from GLDAS-Noah to harmonize the individual products for the 40-year period of record. However, given the differences found between modeled and observed SM, and among SM from different models, there is a clear need for a SM climate data record based solely on observational data sets. Such a record will be instrumental to verify land surface model performance and trends.

This study is a first attempt to derive a global L-band climatology from observational data alone. The STL geostatistical procedure was implemented and tailored to the first six annual cycles of SMOS data to decompose the temporal variability of the signal. For the period of study (2010–2016) this analysis allowed identifying regions where soil moisture variability was dominated by seasonal cycles, regions that did not exhibit a clear seasonal pattern and with likely subseasonal variability, and regions where the long-term variability dominated. Results show that the seasonal cycle was dominant in the tropics (Brazil, Central Africa, India and Northern Australia), and in dry tropical forests, including the savannahs south of Central Africa and several regions in Southwestern Brazil. Wet tropical forests, in turn, exhibited high spatial heterogeneity in its dominant components, probably due to the combined effects of human activities and climate variability. In Europe, western countries were governed by seasonal variability whereas the long-term variability predominated in the eastern countries. In Australia, seasonality dominated the north and the south-east and the eastern region was dominated by long-term variability. Interestingly, in regions where SMOS has very limited variance (e.g., Sahara desert, Arabian Peninsula, Western Australia), the subseasonal or residual component was dominant. This is probably due to the fact that the noise of the observations is at least of the same magnitude that the annual variance and the dominant variability is identified as noise. During the study period (2010–2016), the equatorial Pacific Ocean was mostly in a cold phase until 2015–2016 when there was a transition to a warm phase. The observed global linear trends, based upon the strong El Niño event in 2016, are shown to be consistent with the ENSO teleconnections calculated over multiple events.

This study has demonstrated that L-band observations provide a reliable source of data to monitor the distribution of shallow water content in continental surfaces from space platforms. However, this study cannot conclude on the presence of a clear climate trend in the water content of the soil and results should be taken with caution, since they are limited by the length of the L-band data record available ($\sim$8 years vs. the 30 years which are generally considered for climate studies). Still, our results are consistent with a previous study that showed that multi-decadal (1980–2011) variability in SM and terrestrial evaporation was dominated by ENSO dynamics [56]. Also, there are relevant studies considering the trends of SM and terrestrial evapotranspiration at scales smaller than 30 years (e.g., 10 years in [57]).

Our results showed that the STL procedure, after adequate parameterization, was a solid means to build a soil moisture climatology based solely on the temporal dynamics of the data. Yet, it would be recommendable to assess in future research the additional benefits of using techniques exploiting the temporal and spatial components of the data (e.g., [24,58,59]). The global STL parameterization was thoughtfully chosen after an analysis at 8 selected sites with distinct vegetation seasonality and climatic conditions. However, they do not likely cover each climate on the Earth continents and may not be optimal under specific conditions (e.g., areas where small SMOS variance is dominated by noise). A comparison with modeled GLDAS-Noah and ERA5 soil moisture reanalysis in the selected sites, and with in-situ data at the REMEDHUS site (Europe), provides confidence in the obtained results. The exception is in areas where the presence of satellite data gaps -although limited to 20% of the study period- leads to mutual disagreement of model and satellite estimates. It should be noted that the presence of observational data gaps in the satellite time series can potentially bias the obtained climatology and severely limit the applicability of the method to Northern latitudes, where seasonal snow masks soil emissivity and soil moisture retrievals cannot be performed. In this regard, recent studies on the use of Gaussian process regression techniques to mitigate the effect of missing information in Earth observation data are very promising (e.g., [60]).

This work has shown that the relatively short SMOS data record allows providing insight into the dominant modes of temporal variability in the Earth's surface soil moisture. Also, although previous studies have shown some weaknesses of SMOS retrievals (e.g., [11,27,55]), the good correlation obtained with ERA5 modern reanalysis and GLDAS-Noah indicates that an L-band climatology, such as the one proposed in this study, is a reasonable reference to be used for a climate data record exclusively based on remote sensing data. The presented SMOS-based climatology offers a unique view of recent processes governing freshwater fluxes in the water cycle, and allows observing specific phenomena, as the different variability regimes present within the Amazon and Congo basins and the dominance of the interannual variability in wide regions within Europe, United States, Eastern Australia and South America. It also opens the path to forthcoming studies focused on the analysis of the interannual variability and the impact of ENSO in areas that have not been documented so far.

Author Contributions: All authors contributed significantly to this research. The initial idea was conceived by M.P. and J.B.-P.; M.P. carried out the data analysis and led the manuscript writing; J.B.-P. implemented the STL decomposition technique and adapted it to SMOS data; J.M.-S. fully contributed to the design and development of the study and provided the ERA5 data. All the authors contributed to the interpretation of the results, participated in the manuscript and approved it.

Funding: This research received no external funding.

Acknowledgments: M.P. is supported by a Ramón y Cajal Contract (Spanish Ministry of Science, Innovation and Universities). The authors would like to thank the support received from the ESA CCI Soil Moisture project team.

Conflicts of Interest: The authors declare no conflict of interest.

References

1. Entin, J.; Robock, A.; Vinnikov, K.; Hollinger, S.; Liu, S.; Namkhai, A. Temporal and spatial scales of observed soil moisture variations in extratropics. *J. Geophys. Res.* **2000**, *105*, 11865–11877. [CrossRef]
2. Seneviratne, S.; Corti, T.; Davin, E.; Hirschi, M.; Jaeger, E.; Lehner, I.; Orlowsky, B.; Teuling, A. Investigating soil moisture-climate interactions in changing climate: A review. *Earth Sci. Rev.* **2010**, *99*, 125–161. [CrossRef]
3. Taylor, C.M.; Gounou, A.; Guichard, F.; Harris, P.P.; Ellis, R.J.; Couvreux, F.; De Kauwe, M. Frequency of Sahelian storm initiation enhanced over mesoscale soil-moisture patterns. *Nat. Geosci.* **2011**, *4*, 430. [CrossRef]
4. Koster, R.D.; Mahanama, S.P.P.; Yamada, T.J.; Balsamo, G.; Berg, A.A.; Boisserie, M.; Dirmeyer, P.A.; Doblas-Reyes, F.J.; Drewitt, G.; Gordon, C.T.; et al. The Second Phase of the Global Land–Atmosphere Coupling Experiment: Soil Moisture Contributions to Subseasonal Forecast Skill. *J. Hydrometeorol.* **2011**, *12*, 805–822. [CrossRef]
5. Njoku, E.; Entekhabi, D. Passive microwave remote sensing of soil moisture. *J. Hydrol.* **1996**, *184*, 101–129. [CrossRef]
6. Ulaby, F.; Long, D.; Blackwell, W.; Elachi, C.; Fung, A.; Ruf, C.; Sarabandi, K.; van Zyl, J.; Zebker, H. *Microwave Radar Radiometric Remote Sensing*; University of Michigan Press: Ann Arbor, MI, USA, 2014.
7. Kerr, Y.; Waldteufel, P.; Wigneron, J.P.; Delwart, S.; Cabot, F.; Boutin, J.; Escorihuela, M.J.; Font, J.; Reul, N.; Gruhier, C.; et al. The SMOS mission: New Tool for Monitoring Key Elements of the Global Water Cycle. *Proc. IEEE* **2010**, *98*, 666–687. [CrossRef]

8. Entekhabi, D.; Njoku, E.; O'Neill, P.; Kellogg, K.; Crow, W.; Edelstein, W.; Entin, J.; Goodman, S.; Jackson, T.; Johnson, J.; et al. The Soil Moisture Active Passive (SMAP) mission. *Proc. IEEE* **2010**, *98*, 704–716. [CrossRef]

9. Gruhier, C.; de Rosnay, P.; Hasenauer, S.; Holmes, T.; de Jeu, R.; Kerr, Y.; Mougin, E.; Njoku, E.; Timouk, F.; Wagner, W.; et al. Soil moisture active and passive microwave products: Intercomparison and evaluation over a Sahelian site. *Hydrol. Earth Syst. Sci.* **2010**, *14*, 141–156. [CrossRef]

10. Brocca, L.; Hasenauer, S.; Lacava, T.; Melone, F.; Moramarco, T.; Wagner, W.; Dorigo, W.; Matgen, P.; Martínez-Fernández, J.; Llorens, P.; et al. Soil moisture estimation through ASCAT and AMSR-E sensors: An intercomparison and validation study across Europe. *Remote Sens. Environ.* **2011**, *115*, 3390–3408. [CrossRef]

11. Albergel, C.; de Rosnay, P.; Gruhier, C.; Muñoz-Sabater, J.; Hasenauer, S.; Isaksen, L.; Kerr, Y.; Wagner, W. Evaluation of remotely sensed and modelled soil moisture products using global ground-based in-situ observations. *Remote Sens. Environ.* **2012**, *118*, 215–226. [CrossRef]

12. Sánchez, N.; Martínez-Fernández, J.; Scaini, A.; Pérez-Gutiérrez, C. Validation of the SMOS L2 soil moisture data in the REMEDHUS Network (Spain). *IEEE Trans. Geosci. Remote Sens.* **2012**, *50*, 1602–1611. [CrossRef]

13. Parrens, M.; Zakharova, E.; Lafont, S.; Calvet, J.C.; Kerr, Y.; Wagner, W.; Wigneron, J.P. Comparing soil moisture retrievals from SMOS and ASCAT over France. *Hydrol. Earth Syst. Sci.* **2012**, *16*, 423–440. [CrossRef]

14. González-Zamora, Á.; Sánchez, N.; Martínez-Fernández, J.; Gumuzzio, A.; Piles, M.; Olmedo, E. Long-term SMOS soil moisture products: A comprehensive evaluation across scales and methods in the Duero Basin (Spain). *Phys. Chem. Earth* **2015**, *83–84*, 123–136. [CrossRef]

15. Kerr, Y.; Al-Yaari, A.; Rodríguez-Fernández, N.; Parrens, M.; Molero, B.; Leroux, D.; Bircher, S.; Mahmoodi, A.; Mialon, A.; Richaume, P.; et al. Overview of SMOS performance in terms of global soil moisture monitoring after six years in operation. *Remote Sens. Environ.* **2016**, *180*, 40–63. [CrossRef]

16. Rodríguez-Fernández, N.J.; Muñoz Sabater, J.; Richaume, P.; de Rosnay, P.; Kerr, Y.H.; Albergel, C.; Drusch, M.; Mecklenburg, S. SMOS near-real-time soil moisture product: Processor overview and first validation results. *Hydrol. Earth Syst. Sci.* **2017**, *21*, 5201–5216. [CrossRef]

17. Albergel, C.; Dorigo, W.; Balsamo, G.; Muñoz-Sabater, J.; de Rosnay, P.; Isaksen, L.; Brocca, L.; de Jeu, R.; Wagner, W. Monitoring multi-decadal satellite earth observation of soil moisture products through land surface reanalyses. *Remote Sens. Environ.* **2013**, *138*, 77–89. [CrossRef]

18. Parinussa, R.M.; Holmes, T.R.H.; Wanders, N.; Dorigo, W.A.; de Jeu, R.A.M. A Preliminary Study toward Consistent Soil Moisture from AMSR2. *J. Hydrometeorol.* **2015**, *16*, 932–947. [CrossRef]

19. Grings, F.; Bruscantini, C.A.; Smucler, E.; Carballo, F.; Dillon, M.E.; Collini, E.A.; Salvia, M.; Karszenbaum, H. Validation Strategies for Satellite-Based Soil Moisture Products Over Argentine Pampas. *IEEE J. Sel. Top. Appl. Earth Obs. Remote Sens.* **2015**, *8*, 4094–4105. [CrossRef]

20. Polcher, J.; Piles, M.; Gelati, E.; Barella-Ortiz, A.; Tello, M. Comparing surface-soil moisture from the SMOS mission and the ORCHIDEE land-surface model over the Iberian Peninsula. *Remote Sens. Environ.* **2016**, *174*, 69–81. [CrossRef]

21. Dorigo, W.; Gruber, A.; Jeu, R.D.; Wagner, W.; Stacke, T.; Loew, A.; Albergel, C.; Brocca, L.; Chung, D.; Parinussa, R.; et al. Evaluation of the ESA CCI soil moisture product using ground-based observations. *Remote Sens. Environ.* **2015**, *162*, 380–395. [CrossRef]

22. van der Schalie, R.; de Jeu, R.; Kerr, Y.; Wigneron, J.P.; Rodríguez-Fernández, N.; Al-Yaari, A.; Parinussa, R.; Susanne, M.; Drusch, M. The merging of radiative transfer based surface soil moisture data from SMOS and AMSR-E. *Remote Sens. Environ.* **2017**, *189*, 180–193. [CrossRef]

23. Piles, M.; Martínez, E.; Ballabrera-Poy, J.; Martínez, J.; Vall-llossera, M.; Font, J. Estimation of global soil moisture seasonal variability using SMOS satellite observations. In Proceedings of the 4th International Symposium on Recent Advances in Quantitative Remote Sensing, Torrent, Spain, 18–22 September 2014; pp. 151–152.

24. Bueso, D.; Piles, M.; Camps-Valls, G. Nonlinear Complex PCA for Spatio-Temporal Analysis of Global Soil Moisture. In Proceedings of the 2018 IEEE International Geoscience and Remote Sensing Symposium, Valencia, Spain, 22–27 July 2018.

25. Cleveland, R.B.; Cleveland, W.S.; McRae, J.E.; Terpenning, I. STL: A Seasonal-Trend Decomposition Procedure Based on Loess. *J. Off. Stat.* **1990**, *6*, 3–73.

26. Oliva, R.; Daganzo, E.; Kerr, Y.H.; Mecklenburg, S.; Nieto, S.; Richaume, P.; Gruhier, C. SMOS Radio Frequency Interference Scenario: Status and Actions Taken to Improve the RFI Environment in the 1400–1427-MHz Passive Band. *IEEE Trans. Geosci. Remote Sens.* **2012**, *50*, 1427–1439. [CrossRef]

27. Oliva, R.; Daganzo, E.; Richaume, P.; Kerr, Y.; Cabot, F.; Soldo, Y.; Anterrieu, E.; Reul, N.; Gutierrez, A.; Barbosa, J.; et al. Status of Radio Frequency Interference (RFI) in the 1400–1427 MHz passive band based on six years of SMOS mission. *Remote Sens. Environ.* **2016**, *180*, 64–75. [CrossRef]

28. Mohammed, P.N.; Aksoy, M.; Piepmeier, J.R.; Johnson, J.T.; Bringer, A. SMAP L-Band Microwave Radiometer: RFI Mitigation Prelaunch Analysis and First Year On-Orbit Observations. *IEEE Trans. Geosci. Remote Sens.* **2016**, *54*, 6035–6047. [CrossRef]

29. Lannoy, G.J.M.D.; Reichle, R.H.; Peng, J.; Kerr, Y.; Castro, R.; Kim, E.J.; Liu, Q. Converting Between SMOS and SMAP Level-1 Brightness Temperature Observations Over Nonfrozen Land. *IEEE Geosci. Remote Sens. Lett.* **2015**, *12*, 1908–1912. [CrossRef]

30. Gurney, K.; Law, R.; Rayner, P.; Denning, A. *TransCom 3 Experimental Protocol*; Technical Report 707; Colorado State University: Fort Collins, CO, USA, 2000.

31. Rodell, M.; Houser, P.R.; Jambor, U.; Gottschalck, J.; Mitchell, K.; Meng, C.J.; Arsenault, K.; Cosgrove, B.; Radakovich, J.; Bosilovich, M.; et al. The Global Land Data Assimilation System. *Bull. Am. Meteorol. Soc.* **2004**, *85*, 381–394. [CrossRef]

32. Hersbach, H.; Dee, D. ERA5 reanalysis is in production. *ECMWF Newslett.* **2016**, *147*, 7.

33. Dorigo, W.A.; Wagner, W.; Hohensinn, R.; Hahn, S.; Paulik, C.; Xaver, A.; Gruber, A.; Drusch, M.; Mecklenburg, S.; van Oevelen, P.; et al. The International Soil Moisture Network: A data hosting facility for global in-situ soil moisture measurements. *Hydrol. Earth Syst. Sci.* **2011**, *15*, 1675–1698. [CrossRef]

34. Tukey, J. *Exploratory Data Analysis*; Addison-Wesley: Reading, PA, USA, 1977.

35. Hernandez, O.; Boutin, J.; Kolodziejczyk, N.; Reverdin, G.; Martin, N.; Gaillard, F.; Reul, N.; Vergely, J.L. SMOS salinity in the subtropical North Atlantic salinity maximum: 1. Comparison with Aquarius and in-situ salinity. *J. Geophys. Res. Oceans* **2014**, *119*, 8878–8896. [CrossRef]

36. Robock, A.; Schlosser, C.; Vinnikov, K.; Speranskaya, N.; Entin, J.; Qiu, S. Evaluation of AMIP soil moisture simulations. *Glob. Planet Chang.* **1998**, *19*, 181–208. [CrossRef]

37. Humphrey, V.; Gudmundsson, L.; Seneviratne, S.I. Assessing Global Water Storage Variability from GRACE: Trends, Seasonal Cycle, Subseasonal Anomalies and Extremes. *Surv. Geophys.* **2016**, *37*, 357–395. [CrossRef]

38. Bergmann, I.; Ramillien, G.; Frappart, F. Climate-driven interannual ice mass evolution in Greenland. *Glob. Planet. Chang.* **2012**, *82–83*, 1–11. [CrossRef]

39. Hassan, A.A.; Jin, S. Lake level change and total water discharge in East Africa Rift Valley from satellite-based observations. *Glob. Planet. Chang.* **2014**, *117*, 79–90. [CrossRef]

40. Martínez-Fernández, J.; González-Zamora, A.; Sánchez, N.; Gumuzzio, A. A soil water based index as a suitable agricultural drought indicator. *J. Hydrol.* **2015**, *522*, 265–273. [CrossRef]

41. Sánchez, N.; González-Zamora, A.; Piles, M.; Martínez-Fernández, J. A New Soil Moisture Agricultural Drought Index (SMADI) Integrating MODIS and SMOS Products: A Case of Study over the Iberian Peninsula. *Remote Sens.* **2016**, *8*, 287. [CrossRef]

42. Karl, T.R.; Gleason, B.E.; Menne, M.J.; McMahon, J.R.; Heim, R.R.; Brewer, M.J.; Kunkel, K.E.; Arndt, D.S.; Privette, J.L.; Bates, J.J.; et al. U.S. temperature and drought: Recent anomalies and trends. *Trans. Am. Geophys.* **2012**, *93*, 473–474. [CrossRef]

43. Al-Yaari, A.; Wigneron, J.P.; Ducharne, A.; Kerr, Y.; Wagner, W.; Lannoy, G.D.; Reichle, R.; Bitar, A.A.; Dorigo, W.; Richaume, P.; et al. Global-scale comparison of passive (SMOS) and active (ASCAT) satellite based microwave soil moisture retrievals with soil moisture simulations (MERRA-Land). *Remote Sens. Environ.* **2014**, *152*, 614–626. [CrossRef]

44. Reichle, R.H.; Koster, R.D. Bias reduction in short records of satellite soil moisture. *Geophys. Res. Lett.* **2004**, *31*. [CrossRef]

45. Blankenship, C.B.; Case, J.L.; Zavodsky, B.T.; Crosson, W.L. Assimilation of SMOS Retrievals in the Land Information System. *IEEE Trans. Geosci. Remote Sens.* **2016**, *54*, 6320–6332. [CrossRef]

46. Pinnington, E.; Quaife, T.; Black, E. Impact of remotely sensed soil moisture and precipitation on soil moisture prediction in a data assimilation system with the JULES land surface model. *Hydrol. Earth Syst. Sci.* **2018**, *22*, 2575–2588. [CrossRef]

47. Koster, R.D.; Guo, Z.; Yang, R.; Dirmeyer, P.A.; Mitchell, K.; Puma, M.J. On the Nature of Soil Moisture in Land Surface Models. *J. Clim.* **2009**, *22*, 4322–4335. [CrossRef]

48. Traore, A.K.; Ciais, P.; Vuichard, N.; Poulter, B.; Viovy, N.; Guimberteau, M.; Jung, M.; Myneni, R.; Fisher, J.B. Evaluation of the ORCHIDEE ecosystem model over Africa against 25 years of satellite-based water and carbon measurements. *J. Geophys. Res. Biogeosci.* **2014**, *119*, 1554–1575. [CrossRef]

49. Barella-Ortiz, A.; Polcher, J.; de Rosnay, P.; Piles, M.; Gelati, E. Comparison of measured brightness temperatures from SMOS with modelled ones from ORCHIDEE and H-TESSEL over the Iberian Peninsula. *Hydrol. Earth Syst. Sci.* **2017**, *21*, 357–375. [CrossRef]

50. Entekhabi, D.; Reichle, R.; Koster, R.; Crow, W. Performance metrics for soil moisture retrievals and applications requirements. *J. Hydrometeorol.* **2010**, *11*, 832–840. [CrossRef]

51. Oliver, J. *Encyclopedia of World Climatology*, 2nd ed.; Springer: Berlin/Heidelberg, Germany, 2006.

52. Ruscica, R.C.; Sörensson, A.A.; Menéndez, C.G. Hydrological links in Southeastern South America: Soil moisture memory and coupling within a hot spot. *Int. J. Climatol.* **2014**, *34*, 3641–3653. [CrossRef]

53. Spennemann, P.; Salvia, M.; Ruscica, R.C.; Sörensson, A.; Grings, F.; Karszenbaum, H. Land-atmosphere interaction patterns in southeastern South America using satellite products and climate models. *Int. J. Appl. Earth Obs. Geoinf.* **2017**, *64*. [CrossRef]

54. Singh, K.; Kushwaha, C. Emerging paradigms of tree phenology in dry tropics. *Curr. Sci.* **2005**, *89*, 964–975.

55. Gherboudj, I.; Magagi, R.; Goita, K.; Berg, A.A.; Toth, B.; Walker, A. Validation of SMOS Data Over Agricultural and Boreal Forest Areas in Canada. *IEEE Trans. Geosci. Remote Sens.* **2012**, *50*, 1623–1635. [CrossRef]

56. Miralles, D.G.; van den Berg, M.J.; Gash, J.H.; Parinussa, R.M.; de Jeu, R.A.M.; Beck, H.E.; Holmes, T.R.H.; Jiménez, C.; Verhoest, N.E.C.; Dorigo, W.A.; et al. El Niño-La Niña cycle and recent trends in continental evaporation. *Nat. Clim. Chang.* **2013**, *4*, 122. [CrossRef]

57. Jung, M.; Reichstein, M.; Ciais, P.; Seneviratne, S.I.; Sheffield, J.; Goulden, M.L.; Bonan, G.; Cescatti, A.; Chen, J.; de Jeu, R.; et al. Recent decline in the global land evapotranspiration trend due to limited moisture supply. *Nature* **2010**, *467*, 951. [CrossRef] [PubMed]

58. Horel, J.D. Complex Principal Component Analysis: Theory and Examples. *J. Clim. Appl. Meteorol.* **1984**, *23*, 1660–1673. [CrossRef]

59. Arenas-Garcia, J.; Petersen, K.B.; Camps-Valls, G.; Hansen, L.K. Kernel Multivariate Analysis Framework for Supervised Subspace Learning: A Tutorial on Linear and Kernel Multivariate Methods. *IEEE Signal Process. Mag.* **2013**, *30*, 16–29. [CrossRef]

60. Camps-Valls, G.; Verrelst, J.; Muñoz-Marí, J.; Laparra, V.; Mateo-Jimenez, F.; Gomez-Dans, J. A Survey on Gaussian Processes for Earth-Observation Data Analysis: A Comprehensive Investigation. *IEEE Geosci. Remote Sens. Mag.* **2016**, *4*, 58–78. [CrossRef]

MDPI
St. Alban-Anlage 66
4052 Basel
Switzerland
Tel. +41 61 683 77 34
Fax +41 61 302 89 18
www.mdpi.com

Remote Sensing Editorial Office
E-mail: remotesensing@mdpi.com
www.mdpi.com/journal/remotesensing